W0080315

CELL and MUSCLE MOTILITY

MOTILITY

Volume 2

Cell and Muscle Motility

Advisory Editors:

B. R. Brinkley, *Baylor College of Medicine, Houston*

Setsuro Ebashi, *University of Tokyo, Tokyo*

Robert D. Goldman, *Northwestern Medical School, Chicago*

Raymond J. Lasek, *Case Western Reserve University, Cleveland*

Frank A. Pepe, *University of Pennsylvania, Philadelphia*

Keith R. Porter, *University of Colorado, Boulder*

Andrew G. Szent-Gyorgyi, *Brandeis University, Waltham*

Edwin W. Taylor, *University of Chicago, Chicago*

CELL and MUSCLE MOTILITY

MOTILITY
Volume 2

Edited by
Robert M. Dowben
Baylor University Medical Research Institute
and
University of Texas Health Science Center
Dallas, Texas

and
Jerry W. Shay
University of Texas Health Science Center
Dallas, Texas

PLENUM PRESS ● *NEW YORK AND LONDON*

ISBN-13: 978-1-4684-4039-3   e-ISBN-13: 978-1-4684-4038-6
DIO:10.1007/978-1-4684-4038-6

© 1982 Plenum Press, New York
Softcover reprint of the haradcover 1st edition 1982

A Division of Plenum Publishing Corporation
233 Spring Street, New York, N.Y. 10013
All rights reserved

No part of this book may be reproduced, stored in a retrieval system, or transmitted,
in any form or by any means, electronic, mechanical, photocopying, microfilming,
recording, or otherwise, without written permission from the publisher

Contributors

Ueli Aebi, Department of Cell Biology and Anatomy, The Johns Hopkins University School of Medicine, Baltimore, Maryland 21218

David Alcorta, Department of Biological Sciences, Columbia University, New York, New York 10027

Stephen H. Blose, Cold Spring Harbor Laboratory, Cold Spring Harbor, New York 11724

Arthur P. Bollon, Department of Molecular Genetics, Wadley Institutes of Molecular Medicine, Dallas, Texas 75235

B. R. Brinkley, Department of Cell Biology, Baylor College of Medicine, Houston, Texas 77030

Joseph Bryan, Department of Cell Biology, Baylor College of Medicine, Houston, Texas 77030

Keith Burridge, Cold Spring Harbor Laboratory, Cold Spring Harbor, New York 11724

Anne Bushnell, Cold Spring Harbor Laboratory, Cold Spring Harbor, New York 11724

J. Cartwright, Jr., Section of Cardiovascular Sciences, Department of Medicine, Baylor College of Medicine, and The Methodist Hospital, Houston, Texas 77030

Herbert C. Cheung, Biophysics Section, Department of Biomathematics, University of Alabama in Birmingham, Birmingham, Alabama 35294

Mike A. Clark, Department of Cell Biology, The University of Texas Health Science Center at Dallas, Dallas, Texas 75235

John A. Cooper, Department of Cell Biology and Anatomy, The Johns Hopkins University School of Medicine, Baltimore, Maryland 21218

Barry S. Eckert, Department of Anatomical and Biophysical Sciences, School of Medicine, State University of New York at Buffalo, Buffalo, New York 14214

Marshall Elzinga, Biology Division, Brookhaven National Laboratories, Upton, New York 11973

Howard Feit, Department of Neurology, The University of Texas Health Science Center at Dallas, Dallas, Texas 75235

James R. Feramisco, Cold Spring Harbor Laboratory, Cold Spring Harbor, New York 11724

Walter E. Fowler, Department of Cell Biology and Anatomy, The Johns Hopkins University School of Medicine, Baltimore, Maryland 21218

Dixie W. Frederiksen, Department of Biochemistry, Vanderbilt University, Nashville, Tennessee 37232

John W. Fuseler, Department of Cell Biology, The University of Texas Health Science Center at Dallas, Dallas, Texas 75235

M. A. Goldstein, Section of Cardiovascular Sciences, Department of Medicine, Baylor College of Medicine, and The Methodist Hospital, Houston, Texas 77030

Linda M. Griffith, Department of Cell Biology and Anatomy, The Johns Hopkins University School of Medicine, Baltimore, Maryland 21218

William F. Harrington, Department of Biology and McCollum–Pratt Institute, The Johns Hopkins University, Baltimore, Maryland 21218

David J. Hartshorne, Muscle Biology Group, Department of Nutrition and Food Science, University of Arizona, Tucson, Arizona 85721

Stephen C. Harvey, Biophysics Section, Department of Biomathematics, University of Alabama in Birmingham, Birmingham, Alabama 35294

Ira M. Herman, Department of Cell Biology and Anatomy, The Johns Hopkins University School of Medicine, Baltimore, Maryland 21218

John Heuser, Department of Physiology and Biophysics, Washington University School of Medicine, St. Louis, Missouri 63110

Michael J. Holroyde, Department of Physiology, University of Cincinnati College of Medicine, Cincinnati, Ohio 45267

Gerhard Isenberg, Department of Cell Biology and Anatomy, The Johns Hopkins University School of Medicine, Baltimore, Maryland 21218

J. David Johnson, Section of Contractile Proteins, Department of Pharmacology and Cell Biophysics, University of Cincinnati College of Medicine, Cincinnati, Ohio 45267

Daniel P. Kiehart, Department of Cell Biology and Anatomy, The Johns Hopkins University School of Medicine, Baltimore, Maryland 21218

Stephen J. Koons, Department of Anatomical and Biophysical Sciences, School of Medicine, State University of New York, Buffalo, New York 14214

Evangelia G. Kranias, Section of Contractile Proteins, Department of Pharmacology and Cell Biophysics, University of Cincinnati College of Medicine, Cincinnati, Ohio 45267

I. Lacko, The Department of Biological Sciences and The Genetics and Aging Centers, North Texas State University, Denton, Texas 76203

Janelle Levy, Department of Cell Biology and Anatomy, The Johns Hopkins University School of Medicine, Baltimore, Maryland 21218

Jim Jung-Ching Lin, Cold Spring Harbor Laboratory, Cold Spring Harbor, New York 11724

Gay Lorkowski, Department of Cell Biology, The University of Texas Health Science Center at Dallas, Dallas, Texas 75235

R. Lynch, The Department of Biological Sciences and The Genetics and Aging Centers, North Texas State University, Denton, Texas 76203

Susan MacLean-Fletcher, Department of Cell Biology and Anatomy, The Johns Hopkins University School of Medicine, Baltimore, Maryland 21218

M. Mattson, The Department of Biological Sciences and The Genetics and Aging Centers, North Texas State University, Denton, Texas 76203

Pamela Maupin, Department of Cell Biology and Anatomy, The Johns Hopkins University School of Medicine, Baltimore, Maryland 21218

Robert A. Mendelson, Department of Biochemistry and Biophysics and The Cardiovascular Research Institute, University of California, San Francisco, California 94143

Mark S. Mooseker, Department of Biology, Yale University, New Haven, Connecticut 06510

J. Mrotek, The Department of Biological Sciences and The Genetics and Aging Centers, North Texas State University, Denton, Texas 76203

Ursula Neudeck, Department of Neurology, The University of Texas Health Science Center at Dallas, Dallas, Texas 75235

Nancy Nicholson, Department of Biological Sciences, Columbia University, New York, New York 10027

Lee D. Peachey, Department of Biology G7, University of Pennsylvania, Philadelphia, Pennsylvania 19104

Suzanne M. Pemrick, Department of Biochemistry, Downstate Medical Center, State University of New York, Brooklyn, New York 11203

Frank A. Pepe, Department of Anatomy, School of Medicine, University of Pennsylvania, Philadelphia, Pennsylvania 19104

George Perry, Department of Cell Biology, Baylor College of Medicine, Houston, Texas 77030

Robert Pollack, Department of Biological Sciences, Columbia University, New York, New York 10027

Thomas D. Pollard, Department of Cell Biology and Anatomy, The Johns Hopkins University School of Medicine, Baltimore, Maryland 21218

Rhonda R. Porterfield, Department of Cell Biology, The University of Texas Health Science Center at Dallas, Dallas, Texas 75235

James D. Potter, Section of Contractile Proteins, Department of Pharmacology and Cell Biophysics, University of Cincinnati College of Medicine, Cincinnati, Ohio 45267

Sharon A. Queally, Cold Spring Harbor Laboratory, Cold Spring Harbor, New York 11724

W. Rainey, The Department of Biological Sciences and The Genetics and Aging Centers, North Texas State University, Denton, Texas 76203

Steven P. Robertson, Section of Contractile Proteins, Department of Pharmacology and Cell Biophysics, University of Cincinnati College of Medicine, Cincinnati, Ohio 45267

Marschall Runge, Department of Cell Biology and Anatomy, The Johns Hopkins University School of Medicine, Baltimore, Maryland 21218

T. Sawada, The Department of Biological Sciences and The Genetics and Aging Centers, North Texas State University, Denton, Texas 76203

Jerry W. Shay, Department of Cell Biology, The University of Texas Health Science Center at Dallas, Dallas, Texas 75235

Katy Smith, Department of Biological Sciences, Columbia University, New York, New York 10027

P. Ross Smith, Department of Cell Biology, New York University School of Medicine, New York, New York 11016

R. John Solaro, Department of Pharmacology and Cell Biophysics and Department of Physiology, University of Cincinnati College of Medicine, Cincinnati, Ohio 45267

Marguerite Stauver, Department of Cell Biology, The University of Texas Health Science Center at Dallas, Dallas, Texas 75235

Bettie Steinberg, Department of Biological Sciences, Columbia University, New York, New York 10027; present address: Long Island Jewish Hospital, New Hyde Park, New York 11040

Peter Tseng, Department of Cell Biology and Anatomy, The Johns Hopkins University School of Medicine, Baltimore, Maryland 21218

Hitoshi Ueno, Department of Biology and McCollum–Pratt Institute, The Johns Hopkins University, Baltimore, Maryland 21218

Michael Verderame, Department of Biological Sciences, Columbia University, New York, New York 10027

Woodring E. Wright, Department of Cell Biology and Internal Medicine, The University of Texas Health Science Center at Dallas, Dallas, Texas 75235

Preface

The contributions to this volume were presented at a Symposium entitled "Current Topics in Muscle and Nonmuscle Motility" held in Dallas 19–21 November 1980 under the auspices of the A. Webb Roberts Center for Continuing Education, Baylor University Medical Center Dallas, and the University of Texas Health Science Center at Dallas. This very useful opportunity for a group of active investigators in motility to meet and discuss their latest findings was made possible in part by the income from an endowment fund established by a generous gift from Dr. Albert P. D'Errico in the Baylor University Medical Center. Dr. D'Errico was the first formally-trained neurosurgeon to practice in the Dallas area, the first Chief of Neurological Surgery, and a member of the Medical Board of the Baylor University Medical Center Dallas (1947–1964). The income from this fund is used to promote the dissemination of up-to-date information in the Neurosciences, to provide intellectual stimulation, to add to the fund of knowledge, and improve the skills of neurosurgeons, neurologists, internists, and others in specialized fields of medicine. We are all indebted for this generous gift that made this enriching educational experience possible. We are also grateful for support the Symposium received from Electron Microscopy Sciences, Forma Scientific, J.E.O.L. USA, Inc., Ladd Research Industries, M.J.O. Diatome Co., Organon Co., Upjohn Co., G. D. Searle & Co., and Smith, Kline and French.

Robert M. Dowben
Jerry W. Shay

Dallas

Contents

Chapter 1
Actin Organization as an in Vitro Assay for Tumorigenicity
*Robert Pollack, Nancy Nicholson, David Alcorta, Michael Verderame, Katy Smith,
and Bettie Steinberg*

Chapter 2
The Mechanism of Actin-Filament Assembly and Cross-Linking
Thomas D. Pollard, Ueli Aebi, John A. Cooper, Marshall Elzinga,
Walter E. Fowler, Linda M. Griffith, Ira M. Herman, John Heuser,
Gerhard Isenberg, Daniel P. Kiehart, Janelle Levy, Susan MacLean-Fletcher,
Pamela Maupin, Mark S. Mooseker, Marschall Runge, P. Ross Smith, and
Peter Tseng

Chapter 3
A Scanning and Transmission Electron Microscope Examination of
ACTH-Induced "Rounding up" in Triton X-100 Cytoskeleton
Residues of Cultured Adrenal Cells
J. Mrotek, W. Rainey, T. Sawada, R. Lynch, M. Mattson, and I. Lacko

Chapter 4
The Role of Tubulin in Steroidogenesis of Mouse Adrenal Y-1 Cells
and Rat Leydig CCL 43 Cells
Mike A. Clark and Jerry W. Shay

Chapter 10
Cytoskeletal Defects in Avian Muscular Dystrophy
Jerry W. Shay, John W. Fuseler, Ursula Neudeck, Gay Lorkowski,
Marguerite Stauver, and Howard Feit

Chapter 11
The Structure of Vertebrate Skeletal-Muscle Myosin Filaments
Frank A. Pepe

Chapter 20
Myosin Flexibility
Stephen C. Harvey and Herbert C. Cheung

1

Actin Organization as an in Vitro Assay for Tumorigenicity

Robert Pollack, Nancy Nicholson, David Alcorta, Michael Verderame, Katy Smith, and Bettie Steinberg

1. Introduction

The endpoints of *in vivo* and *in vitro* assays applied to cells after exposure to a potential oncogenic transforming agent are cellular tumorigenicity and transformation. Tumors are failures of *in vivo* growth control; transformations are failures of *in vitro* growth control. Many agents cause tumors *in vivo*, many agents transform normal cultured cells, and some agents do both. However, even when caused by a single agent, the *in vivo* and *in vitro* endpoint assays show only a partial overlap. That is, some but not all tumors will grow as transformed cells in culture, and some but not all *in vitro* transformants will be tumorigenic on injection into susceptible animals (Shin *et al.*, 1975). Recently, we have described a subset of *in vitro* phenotypic changes that correlate with *in vivo* tumorigenicity (Steinberg *et al.*, 1979; Barrett *et al.*, 1979; Pollack, 1981). In this chapter, we will describe recent studies on one of the *in vitro* changes linked to tumorigenicity, the disruption in organization of cytoskeletal actin.

1.1. Actin Organization and Oncogenic Transformation

The major intracellular protein of both normal and transformed cells is actin (Pollard and Weihing, 1974). The actin of a normal spread cell is par-

Robert Pollack, Nancy Nicholson, David Alcorta, Michael Verderame, Katy Smith, and Bettie Steinberg • Department of Biological Sciences, Columbia University, New York, New York 10027. Dr. Steinberg's present address is: Long Island Jewish Hospital, New Hyde Park, New York 11040.

titioned by cofactors into a low-molecular-weight globular (G)-actin complex and a set of macromolecular arrays that include long chains of filamentous (F)-actin (Spudich *et al.*, 1977; Heuser and Kirschner, 1980). These large arrays contain several components each, and are found in the cell as a microfilament gel, a set of actin-containing stress fibers or cables, and an isotropic matrix of single microfilaments that fills the cytoplasm, interacting with microtubules and intermediate filaments (Heuser and Kirschner, 1980). The gel is located just under the membrane. Microfilament cables contain polyactin, myosin, α-actinin, and tropomyosin. In normal cells, cables are located primarily at the adherent side of the cells just under the surface (Goldman *et al.*, 1975; Geiger, 1979). Thus, they share a localization with that part of the microfilament gel that also lies under the adherent surface.

In tumorigenic anchorage-transformed cells, the actin is repartitioned among these different macromolecular arrays. More of the actin is found in the microfilament gel and less of it in the cables (McNutt *et al.*, 1973; Heuser and Kirschner, 1980). This last event is most dramatic because the cables, mostly in one plane of focus, are easily visualized by immunofluorescence microscopy (Pollack *et al.*, 1975; Edelman and Yahara, 1976). Heuser and Kirschner (1980) have shown that this transition is accomplished by a shift of actin filaments in the cytoskeleton from bundles to a crudely interwoven structure. Apparently, then, transformation reweaves actin filaments.

Phalloidin is a compound derived from poisonous mushrooms that binds very tightly to F-actin (Lengsfeld *et al.*, 1974). To quantitate the pattern changes in actin-filament organization at the level of light microscopy, we have stained cells with fluorescent (Fl)-phalloidin (Wulf *et al.*, 1979; Verderame *et al.*, 1980). We have found two different changes in actin-cable size and number accompanying transformation and a human mutation leading to development of colonic neoplasms.

1.2. Actin Organization and Human Cancer

Most human cancers are of unknown etiology. The relative weight of environmental factors and host genetic susceptibility is thought usually to be tipped toward the environment. In rare cases, however, the disease occurs as the result of a host mutation. In such cases, it is reasonable to hope that all cells of an affected person might reveal an altered phenotype when appropriately examined. Detection of such an abnormality in easily cultured cells might permit prognosis of asymptomatic children of individuals affected by an inherited cancer.

Patients with the autosomal dominant mutation adenomatosis of the colon and rectum (ACR) develop colonic polyps and adenocarcinomas of the lower intestine by middle age. One half of their children carry the mutation as well. In 1977, we reported that cytoskeletal actin patterns in forearm-skin-biopsy fibroblasts from ACR patients and in some of their children were abnormal. Bundles of actin were replaced in many cells by a diffuse actin distribution (Kopelovich *et al.*, 1977). Recently, we reported that this change in actin patterns does not occur in skin fibroblasts from patients with other non-ACR, familial colonic neoplasms (Kopelovich *et al.*, 1980). We will consider here the quantitative distribution of F-actin patterns in skin cultures from patients with ACR and other inherited neoplasms and compare the distributions seen to those of normal and oncogenic transformed cell lines (Nicholson *et al.*, 1981).

2. Methods

2.1. Cells and Culture

A large number of cells from different sources were used in these studies. Established cells have been previously described (Steinberg and Pollack, 1979; Steinberg *et al.*, 1978; Pollack *et al.*, 1968). Human precrisis cells were received from Dr. M. Lipkin at the Memorial Sloan–Kettering Cancer Institute. All human and rat cells had been grown on plastic dishes in Dubecco's Modified Eagle medium (DME) (GIBCO H21) and 10% fetal calf serum (FCS) (Reheis) previous to assay. Mouse cells had been grown in DME and 10% calf serum (GIBCO). Culture conditions and procedures for passaging cells have been described (Steinberg and Pollack, 1979).

2.2. Anchorage Independence

Cells were plated in triplicate at 10^5, 10^4, and 10^3 cells/60-mm dish in 3 ml DME plus 10% FCS containing 0.33% agarose (Difco) over a 2-ml layer of 0.9% agarose in the same medium. Cultures were fed twice weekly with an additional 2 ml of the soft agarose–DME–10% FCS and cultured for 3 weeks. Large colonies greater than 0.2 mm in diameter were scored using a dissecting microscope. Total colony-volume increase in agar was determined as described by Steinberg and Pollack (1979).

2.3. Fixation and Staining for Localization of Actin by Antibody Immunofluorescence*

Cells were plated on coverslips at a density of $2–4 \times 10^3$ cells/cm^2 in 10% FCS. At 1 day later, medium was switched to 1% serum. After an additional day, coverslips were fixed in 10% formalin in phosphate-buffered saline (PBS), pH 7.1, kept in formalin for 4–8 days at 4°C, and then acetone-postfixed, stained sequentially with rabbit antiactin (1 : 80 in the PBS) and fluorescein-isothiocyanate-conjugated goat antiserum to rabbit IgG (1 : 20 in PBS), and mounted cell-side down in Aquamount (Pollack and Rifkin, 1975). Cells were scanned by fluorescence microscopy and were scored as positive for actin cables if fluorescent bands were seen to run the length of the

*The method described in this section is that of Kopelovich *et al.* (1977, 1980).

cell when the edge of the cell was in focus. More than 100 cells were scored on each coverslip, and all experiments were scored in ignorance of the origin of the cells examined.

2.4. Fixation and Staining for Localization of Actin by Fluorescent Phalloidin

Cells were plated on coverslips at a density of $2–4 \times 10^3$ cells/cm² in 10% FCS. At 1 day later, medium was switched to 1% serum. After an additional day, cells on coverslips were fixed in 10% formalin in PBS, for 20 min, rinsed with PBS, and extracted with 1% Nonidet P-40 in PBS for 20 min. Phalloidin concentrations were determined by absorbance spectroscopic measurements of concentrated stock solutions at 300 nm, which were corrected for the fluorescein absorbance by measurements at 492 nm. For Fl-phalloidin staining, 10 μl Fl-phalloidin (1 μg/ml in PBS, gift of T. Wieland, Göttingen West Germany) were incubated with cells at 37° C for 30 min. Coverslips were then rinsed three times in PBS and mounted on microscope slides with Aquamount.

2.5. Fluorescence Microscopy

Stained coverslips were examined with a Leitz Orthoplan microscope. A Zeiss Planapo 63× oil immersion objective was coupled with a Letiz L2 exciter–barrier filter cube to visualize fluorescein.

2.6. Specificity of Fluorescent Phalloidin Stain

Fl-phalloidin specificity for F-actin was determined in the following preincubation controls (Verderame *et al.*, 1980). G-actin, purified from acetone powder of chicken gizzard (J. Feramisco, personal communication), was converted to F-actin by the addition of KCl to a final concentration of 0.1 M. Actin preparations were mixed with Fl-phalloidin for 20 min at room temperature, then applied to coverslips of fixed 3T3 cells for 20 min at 37°C. Fl-phalloidin staining was completely blocked by 1 mg/ml of F-actin, not blocked by 2 mg/ml of G-actin, and only partially blocked by 10 mg/ml of G-actin. At this high concentration of G-actin, approximately 1%, or 0.1 mg/ml, is expected to be polymerized. Thus, staining with Fl-phalloidin was sensitive to the polymerized state of the actin. Staining by antibody to actin was completely blocked by 45-min preincubation with 2 mg/ml of G-actin.

Fl-phalloidin staining was blocked by preincubation of fixed cells with 1 mg/ml of unlabeled phalloidin solution or by coincubation with a mixture of 1 μg/ml of Fl-phalloidin and 1 mg/ml of phalloidin. Fixed cells preincubated with phalloidin at concentrations up to 100 μg/ml showed normal antibody staining with antiactin. Thus, phalloidin blocked specific binding of Fl-phalloidin, but not of actin antibody.

2.7. Photography

The photographs were taken by a Leitz Orthomat camera on Kodak Tri-X film developed in Microdol-X (1:3) at 22°C for 13 min. The prints were made on Kodak Polycontrast SC paper.

3. Results and Discussion

3.1. Fluorescent Phalloidin Permits Quantitation of Actin Patterns

Fl-phalloidin binds specifically to actin, and preferentially to F-actin (Wulf *et al.*, 1979; Verderame *et al.*, 1980). Well-spread fibroblasts stained with 1 µg/ml of Fl-phalloidin show many patterns of actin organization. In some cells, large cables predominate, while in others, fine cables fill the cytoplasm at the plane of focus of cell–substrate adhesion, but large cables are few or entirely lacking. Finally, some spread cells do not contain any cables, but rather show a diffuse fluorescence (Fig. 1).

To quantitate the overall degree of organization of actin in a cell population, we scored 200 random well-spread stained cells by placing each into one of the four categories typified by the cells in Fig. 1. For each line, some cells fell into each category. However, the distributions were reproducibly different for different cell lines (Verderame *et al.*, 1980; Nicholson *et al.*, 1981).

3.2. Tumorigenicity and Fluorescent-Phalloidin Patterns

To quantitate a possible relationship between actin organization and tumorigenicity, we scored phalloidin patterns in cultures from 15 rat and mouse cell lines with widely varying phenotypes of growth control (Table 1). These lines included precrisis mouse and rat fibroblasts (MEF, REF), postcrisis growth-controlled cell lines (3T3, Rat 1), fully transformed cell lines (SV101, SVR85, SVR87, 14B, MCA), and a set of cell lines of intermediate growth control isolated as revertants (FL SV101, 1–4, 3–8) and intermediate transformants (SVR42, SVR63, SVR13). Cellular tumorigenicity of cell lines was assayed in *nude* mice (Shin *et al.*, 1975). Anchorage independence correlated well quantitatively and qualitatively with tumorigenicity in these lines (Table 1), as had been seen previously for other sets of rodent cell lines (Barrett *el al.*, 1979; Kahn and Shin, 1979; Shin *et al.*, 1975).

In earlier studies using antibodies to actin, data on actin organization of a cell line were presented as the percentage of cells containing detectable cables of any sort (Pollack *et al.*, 1975; Pollack and Rifkin, 1975; Rifkin *et al.*, 1979). With Fl-phalloidin, distribution of cells into four categories permits us to characterize populations according to three different criteria: the percentage of cells filled with large cables (I), the percentage of cells containing at least two large cables (I+II), and the percentage of cells containing any detectable

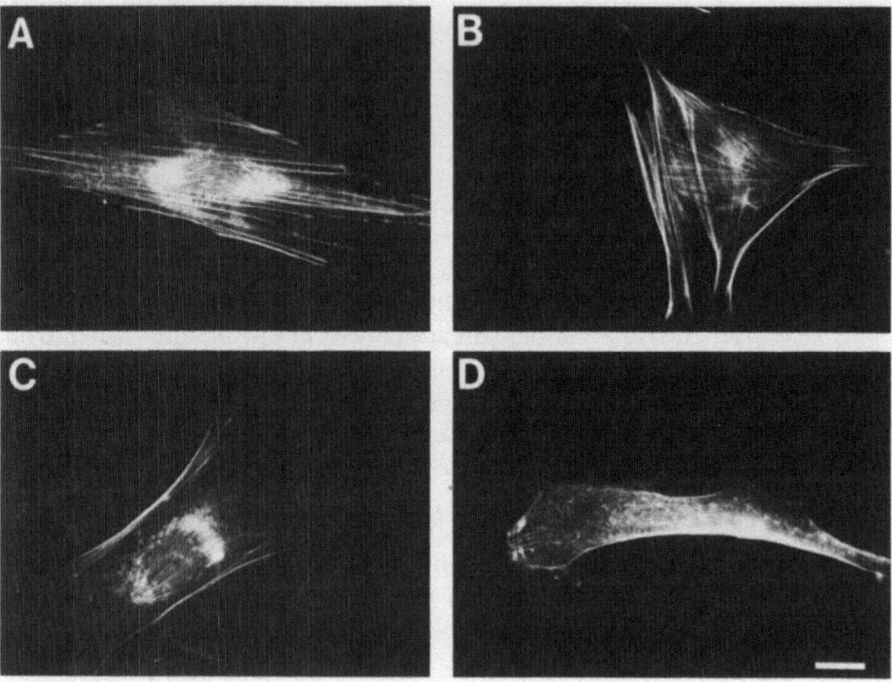

Figure 1. Four categories of F-actin distribution used for scoring cultures stained with Fl-phalloidin. (A) Category I: more than 90% of cell area filled with thick cables; (B) category II: at least two thick cables running under nucleus, rest of cell area filled with fine cables; (C) category III: no thick cables, but some fine cables present; (D) category IV: no cables visible in the central area of the cell. Scale bar: 20 μm. Adapted from Verderame *et al.* (1980).

Table 1. Properties of Rodent Fibroblastic Cells[a]

	Cell	Pre-/post-crisis	Transformed by	Subcloned from	Anchorage independence[b,d] RPE (%)	CVI	Tumorigenicity in *nude* mice[c,d]
Mouse	MEF	Pre-	—	—	≤0.001	ND	0/7
	3T3	Post-	—	—	≤0.001	ND	0/4
	SV101	Post-	SV40	3T3	27	ND	5/16
	FL SV101	Post-	—	SV101	0.01	ND	0/2
	SVR42	Post-	SV40	3T3	≤0.001	ND	ND
	SVR63	Post-	SV40	3T3	≤0.001	ND	0/2
	SVR13	Post-	SV40	3T3	0.2	ND	0/2
	SVR85	Post-	SV40	3T3	10.6	ND	ND
	SVR87	Post-	SV40	3T3	38.6	ND	3/4
Rat	REF	Pre-	—	—	<1	1.6	0/6
	Rat 1	Post-	—	—	7×10^{-3}	13	2/3
	14B	Post-	SV40	Rat 1	15	380	3/3
	1-4	Post-	—	14B	3×10^{-3}	21	0/4
	3-8	Post-	—	14B	4×10^{-3}	1	0/4
	MCA	Post-	—	1-4	5.4	4200	3/3

[a] Adapted from Verderame *et al.* (1980).
[b] Growth without anchorage is measured in two ways: Relative plating efficiency (RPE) measures colonies greater than 0.2 mm diameter in agar as a percentage of colonies on a plastic dish. Colony-volume increase (CVI) measures the total increase in cell number in agar culture. The latter measure is more sensitive to slow anchorage-independent cell growth (Steinberg and Pollack, 1979).
[c] *Nude* mice with tumors at 6 months/total animals injected with 10^7 cells (Kahn and Shin, 1979).
[d] (ND) Not done.

cables, whether large or fine (I+II+III). In Fig. 2, tumorigenicity of 15 cell lines (Table 1) is compared with the fraction of cells in categories I+II and with the fraction of cells in categories I+II+III.

According to both criteria of actin organization, actin structures are reduced in size and number in transformed cells compared with normal cells. Many normal cells, however, do not have large cables, but only fine ones. As a result, the fraction of cells in categories I+II is as low in these nontumorigenic lines as it is in the tumorigenic lines (Fig. 2). On the other hand, the percentage of cells with any detectable cables (I+II+III) is reproducibly higher for nontumorigenic lines than for tumorigenic ones, so that the fraction of cells in categories I+II+III is roughly inversely proportional to the tumorigenicity of a cell line (Fig. 2). The one "nontumorigenic line" with a low value for this measure of actin organization in Fig. 2 is the revertant rat line 1-4 (Steinberg

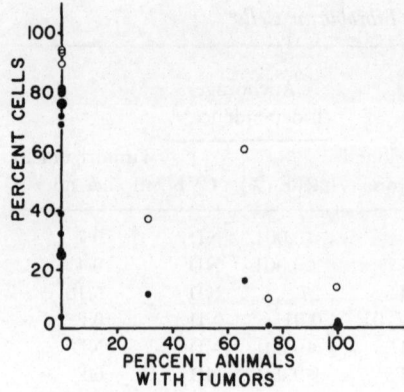

Figure 2. Correlation of disappearance of fine cables of F-actin in rodent cells with cellular tumorigenicity in *nude* mice. Rat and mouse cell lines with varying degrees of growth control (Table 1) were analyzed for F-actin organization by Fl-phalloidin stain. For a given cell line, the plots are of the fraction of cells in categories I+II (Fig. 1) vs. the tumorigenicity of that line (●) and of the fraction of cells in categories I+II+III vs. tumorigenicity (○). Tumorigenicity was measured by injection of cells into *nude* mice at 10^7 cells/mouse (Shin *et al.*, 1975; Steinberg *et al.*, 1979; Kahn and Shin, 1979).

et al., 1978, 1979). This line forms long-lasting benign nodules in *nude* mice (Steinberg *et al.*, 1979).

Thus, according to this last criterion, all tumorigenic lines fell below 65%, suggesting that a quantitative threshold for predicting tumorigenicity may be available with the assay. Apparently, the loss of all detectable cables is more directly related to the acquisition of tumorigenicity than is the shift from large to fine cables.

3.3. Organization of Actin in Human Skin Fibroblasts

Since 1977, we have been studying the actin organization of human skin fibroblasts from patients with inherited propensities to develop cancer. Table 2 summarizes the results obtained by scan of these cells with antibody to actin (Kopelovich *et al.*, 1977, 1980).

Six groups of subjects were examined for their actin cable patterns: (1) normal subjects, either from the general population of spouses from the cancer families; (2) patients diagnosed with adenomatosis of the colon and rectum (ACR); (3) asymptomatic children of ACR patients, who have a 50%

Table 2. *Actin Organization in Human Skin Fibroblasts as Determined with Antibody to Actin*[a]

| | Cells positive for actin cables | | | |
| | Kopelovich *et al.* (1977) | | Kopelovich *et al.* (1980) | |
Phenotype	Number of subjects	Positive (%)[b]	Number of subjects	Positive (%)[b]
Normal	9	77 (9.9)	16	76 (25)
ACR	11	32 (8.9)	7	37 (68.7)
CCP−	—	ND	7	81 (14.3)
CCP+	—	ND	6	67 (47.9)

[a] Adapted from Kopelovich *et al.* (1977, 1980).
[b] Positive cells contain at least two cables running the length of the cell. Results are expressed as mean percentage of cells (S.D.). (ND) Not done.

probability of developing the disease; (4) persons from colon-cancer-prone families who have colon cancer (CCP+); (5) persons from colon-cancer-prone families who have no symptoms but a 50% risk of developing colon cancer (CCP−); and (6) one patient with multiple primary colonic tumors. These published data with antiactin are based on a single threshold (+/−) score for the presence of cables (e.g., Pollack *et al.*, 1975).

In both earlier studies with antibody, we had found an ACR-specific reduction in the percentage of cells with cables (Table 2). However, since the percentage did not fall to zero, the possibility remained that cells from ACR patients had not undergone a change in actin similar to that in tumorigenic transformed cells, but rather had undergone a shift from larger to smaller cables.

Since we had already found that such a shift could occur in murine fibroblasts without acquisition of tumorigencity (Fig. 2), we reexamined these human skin fibroblasts with Fl-phalloidin (Nicholson *et al.*, 1981). Cells from ACR patients clearly show a dramatic loss of cells in categories I and II, compared with cells from other subjects (Table 3). In addition, a fraction of these cells lose all detectable cables (Table 3). Both changes are statistically significant for ACR by Student's *t* test (Table 4).

3.4. Asymptomatic Children of Patients with Adenomatosis of the Colon and Rectum

We examined cells from four children of ACR patients (Table 5). While these children were all free of symptoms of ACR at the time of their skin biopsies, we found that cells from one child (subject 4) had an Fl-phalloidin actin distribution much like that of cells from ACR patients (See Table 3). That is, cells at the category I threshold were not different, but subject 4 had fewer cells with sufficient cables for the I+II threshold or the I+II+III

Table 3. Actin Organization in Human Skin Fibroblasts Determined with Fluorescent Phalloidin

Phenotype	Number of subjects	Cells in categories (%)[a]		
		I	I+II	I+II+III
Normal	5	25 (28)	48 (26)	89 (10)
ACR[b]	7	4 (4)	15 (9)	68 (19)
CCP−[c]	2	54 (14)	66 (11)	88 (6)
CCP+[d]	2	26 (32)	53 (15)	88 (7)
MPT[e]	1	1 (2)	19 (21)	73 (9)

[a] Skin-fibroblast cultures were examined between the 5th and 12th passages. Categories of F-actin distribution in fixed cells are shown in Fig. 1. Briefly, category I cells have large F-actin cables, category II cells have a mixture of large and fine cables, category III cells have only fine cables, and category IV cells lack cables detectable with Fl-phalloidin. Results are expressed as mean percentage of cells (S.D.).
[b] Skin fibroblasts from patients with adenomatosis of the colon and rectum (Kopelovich *et al.*, 1980).
[c] Skin fibroblasts from asymptomatic colon-cancer-prone subjects.
[d] Skin fibroblasts from symptomatic colon-cancer-prone subjects.
[e] Skin fibroblasts from a subject with multiple primary colonic tumors.

Table 4. Differences among Human-Fibroblast Filamentous-Actin Distributions Significant by Student's T Test

Comparison	Cells in categories (Fl-phalloidin)		
	I	I+II	I+II+III
Normal vs. ACR	$0.2 > P > 0.1$	$0.01 > P > 0.001$	$0.01 > P > 0.001$
Normal vs. CCP−	$P > 0.9$	$P > 0.9$	$0.8 > P > 0.7$
Normal vs. CCP+	$P > 0.9$	$P > 0.9$	$P > 0.9$
Normal vs. MPT	$0.6 > P > 0.5$	$0.1 > P > 0.05$	$0.2 > P > 0.1$
CCP− vs. CCP+	$0.2 > P > 0.1$	$0.9 > P > 0.8$	$P > 0.9$

threshold than did the other three subjects. This distribution is significantly different from normal and also different from that of other non-ACR subjects or the other three asymptomatic children (Table 6).

All four children are under clinical observation, so it will be known in time whether any of them develop symptoms of ACR. Insofar as the Fl-phalloidin pattern distribution of subject 4 is not significantly different from the average distribution found in the cells of ACR patients (Table 6), it is reasonable to predict that subject 4 carries the ACR mutation and in time will develop symptoms of the disease.

3.5. Actin Organization vs. Age of Patients

Most patients with ACR present symptoms by early adulthood. Thus, an age bias may have arisen in the production of data of the sort presented above. If actin organization in fibroblasts increased with age of the individual, then our results might be explained as a consequence of the younger age of ACR patients and their children as compared with our normal control subjects.

To test this, we examined the skin fibroblasts of a group of normal children ranging in age from 3 months to 18 years. In Figs. 3 and 4, we show the actin organization of all persons examined so far, vs. their ages. Figure 3 shows the percentage of cells in categories I+II, while Fig. 4 shows the percentage of

Table 5. Percentages of Cells in Fluorescent-Phalloidin Categories from Asymptomatic Children of ACR Patients

Subject	Age	Sex	Cells in categories (%)[a]		
			I	I+II	I+II+III
1	20	F	13 (5)	48 (8)	91 (3)
2	10	F	7 (7)	51 (17)	92 (8)
3	19	M	33 (27)	64 (12)	97 (1)
4	8	M	10 (11)	20 (12)	40 (11)

[a] Results are expressed as mean percentage of cells (S.D.).

Table 6. *Differences among Fluorescent-Actin Distributions in Asymptomatic Children of ACR Patients Significant by Students t Test*[a]

Comparison	Cells in categories (Fl-phalloidin)		
	I	I+II	I+II+III
Subjects 1–3 avg. vs. Normal avg.	0.7 > P > 0.6	0.9 > P > 0.7	P > 0.9
Subjects 1–3 avg. vs. ACR avg.	0.2 > P > 0.1	0.3 > P > 0.2	0.01 > P > 0.001
Subject 4 vs. Normal avg.	0.7 > P > 0.6	0.4 > P > 0.3	0.01 > P > 0.001
Subject 4 vs. ACR avg.	0.5 > P > 0.4	0.6 > P > 0.5	0.05 > P > 0.02
Subjects 1–3 avg. vs. Subject 4	0.9 > P > 0.8	0.1 > P > 0.05	0.01 > P > 0.001

cells in categories I+II+III. Comparison of the two figures reveals that age does indeed have an effect on actin organization, but that this effect is largely on the fraction of cells with large cables (I+II) (Fig. 3). This fraction is quite low in young normal people under age 10, as low as it is in ACR patients.

However, the fraction of cells with any detectable cables (I+II+III) remains high at all ages (Fig. 4). By this criterion, actin cables in ACR cells are slightly but consistently less well organized, at all ages. In no case does the percentage of ACR cells in categories I+II+III fall below the threshold of 60%, which separates tumorigenic from nontumorigenic rodent fibroblasts (cf. Figs. 2 and 4).

4. Conclusions

From these data, we may reasonably conclude that:

1. The actin cytoskeleton of a normal cell can be disrupted in at least two distinct ways: loss of all detectable cables or loss of only large cables.

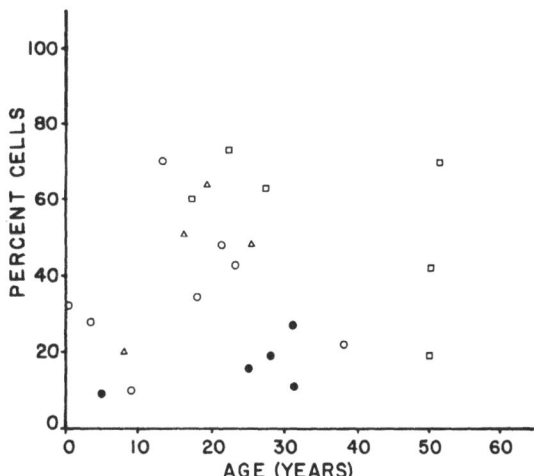

Figure 3. Percentage of human skin fibroblasts with large cables (categories I+II) from subjects of different ages. Skin fibroblasts from subjects with different diagnoses were stained with Fl-phalloidin and scored for the appearance of large cables (categories I+II in Fig. 1). For each culture, the fraction of cells in categories I+II is plotted vs. the age of the subjects. (●) patients with ACR; (○) normal subjects; (△) children of ACR patients; (□) CCP subjects.

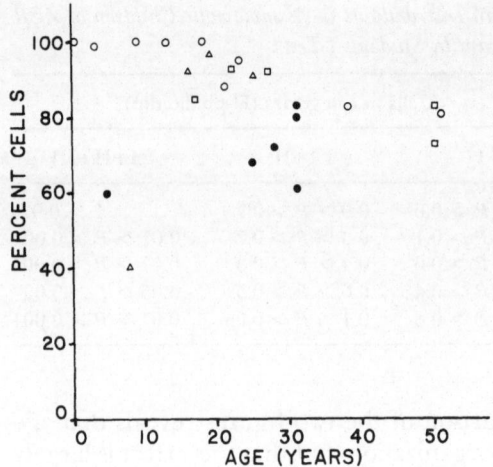

Figure 4. Percentage of human skin fibroblasts with detectable cables (categories I+II+III) from subjects of different ages. Skin fibroblasts from subjects with different diagnoses were stained with Fl-phalloidin and scored for the appearance of large and fine cables (categories I+II+III in Fig. 1). For each culture, the fraction of cells in categories I+II+III is plotted vs. the age of the subjects. For the key to the symbols, see the Fig. 3 caption.

2. Only the greater of these disruptions correlates with tumorigenicity.
3. The lesser disruption may be predictive for at least one inherited propensity to develop cancer.

Finally, Fl-phalloidin has permitted the detection of two different alterations in actin organization, and there is no reason to exclude the possibility of even finer distinctions existing among cell lines in culture. However, assaying the differences by eye has proven quite tedious. We plan in the near future to couple Fl-phalloidin fluorescence images to a vidicon-based digitalized image analyzer (Macagno *et al.*, 1979) to yield signals that can be analyzed and compared by computer. This will make possible detailed prospective studies of cultured skin-biopsy cells from persons at risk for cancer.

ACKNOWLEDGMENTS. We would like to thank Drs. Th. Wieland and A. Deboben for their gift of Fl-phalloidin, Peggy Monaghan for the initial cell cultures, and Marisa Bolognese for her excellent help with the manuscript. This work was supported by NIH grant CA-25066 and NRS Training Grant GM-07216.

References

Barrett, J. C., Crawford, B. D., Mixter, L. O., Schectman, L. M., Ts'o, P. O. P., and Pollack, R., 1979, Correlation of *in vitro* properties and tumorigenicity of Syrian hamster cell lines, *Cancer Res.* **39**:1504–1510.

Edelman, G., and Yahara, I., 1976, Temperature-sensitive changes in surface modulating assemblies of fibroblasts transformed by mutants of Rous sarcoma virus, *Proc. Natl. Acad. Sci. U.S.A.* **73**:2047–2051.

Geiger, B., 1979, A 130K protein from chicken gizzard: Its localization at the termini of microfilament bundles in cultured chicken cells, *Cell* **18**:193–205.

Goldman, R. D., Lazarides, E., Pollack, R., and Weber, K., 1975, The distribution of actin in non-muscle cells: The use of actin antibody in the localization of actin within the microfilament bundles of mouse 3T3 cells, *Exp. Cell Res.* **90**:333–344.

Heuser, J. E., and Kirschner, M. W., 1980, Filament organization revealed in platinum replicas of freeze–dried cytoskeletons, *J. Cell Biol.* **86**:212–234.

Kahn, P., and Shin, S., 1979, Cellular tumorigenicity in nude mice. II. Test of associations among loss of cell surface fibronectin, anchorage independence and tumor forming ability, *J. Cell Biol.* **82**:1–16.

Kopelovich, L., Conlon, S., and Pollack, R., 1977, Defective organization of actin in cultured skin fibroblasts from individuals with an inherited adenocarcinoma, *Proc. Natl. Acad. Sci. U.S.A.* **74**:3019–3022.

Kopelovich, L., Lipkin, M., Blattner, W. A., Fraumeni, J. F., Lynch, H. T., and Pollack, R., 1980, Organization of actin-containing cables in cultured skin fibroblasts from individuals at high risk of colon cancer, *Int. J. Cancer* **26**:301–308.

Lengsfeld, A. M., Low, I., Wieland, Th., Dancker, P., and Hasselbach, W., 1974, Interaction of phalloidin with actin, *Proc. Natl. Acad. Sci. U.S.A.* **71**:2803–2807.

Macagno, E., Levinthal, C., and Sobel, I., 1979, Three-dimensional computer reconstruction of neurons and neuronal assemblies, *Annu. Rev. Biophys. Bioeng.* **8**:323–351.

McNutt, N., Culp, L., and Black, P., 1973, Revertant cell lines isolated from SV40-transformed cells. IV. Microfilament distribution and cell shape in untransformed, transformed and revertant cells, *J. Cell Biol.* **56**:412–428.

Nicholson, N., Verderame, M., Lipkin, M., and Pollack, R., 1981, F-actin quantitated with Fl-phalloidin in skin fibroblasts of individuals genetically predisposed to colon cancer, in: *International Cell Biology, 1980—1981*, (H. Sweiger, ed.), pp. 331–335, Springer, Berlin.

Pollack, R., 1981, Hormones, anchorage and oncogenic cell growth, in: *Proceedings of 1980 International Symposium on Cancer*, Vol. 1 (J. Burchenal, ed.), pp. 501–517, Grune and Stratton, New York.

Pollack, R., and Rifkin, D., 1975, Actin-containing cables within anchorage-dependent rat embryo cells are dissociated by plasmin and trypsin, *Cell* **6**:495–506.

Pollack, R., Green, H., and Todaro, G., 1968, Growth control in cultured cells: Selection of sublines with increased sensitivity to contact inhibition and decreased tumor–producing activity, *Proc. Natl. Acad. Sci. U.S.A.* **60**:126–133.

Pollack, R., Osborn, M., and Weber, K., 1975, Patterns of organization of actin and myosin in normal and transformed non-muscle cells, *Proc. Natl. Acad. Sci. U.S.A.* **72**:994–998.

Pollard, T., and Weihing, R. R., 1974, Actin and myosin and cell movement, *CRC Crit. Rev. Biochem.* **2**:1.

Rifkin, D., Crowe, R., and Pollack, R., 1979, Tumor promotors induce changes in cytoskeletal organization in chick embryo fibroblasts, *Cell* **18**:361–368.

Shin, S., Freedman, V. H., Risser, R., and Pollack, R., 1975, Tumorigenicity of virus-transformed cells in *nude* mice is correlated specifically with anchorage independent growth *in vitro*, *Proc. Natl. Acad. Sci. U.S.A.* **72**:4435–4439.

Spudich, J. A., Mockrin, S. C., Brown, S. S., Rubenstein, P. A., and Levinson, A. 1977, The organization and interaction of actin and myosin in non-muscle cells, in: *Cell Shape and Architecture*, pp. 545–558. Alan R. Liss, New York.

Steinberg, B., and Pollack, R., 1979, Anchorage independence: Analysis of factors affecting the growth and colony formation of wild type and dl 54/59 mutant SV40 transformed line, *Virology* **99**:302–311.

Steinberg, B., Pollack, R., Topp, W., and Botchan, M., 1978, Isolation and characterization of T-antigen negative revertants from a line of transformed rat cells containing one copy of the SV40 genome, *Cell* **15**:19–32.

Steinberg, B., Rifkin, D., Shin, D., Boone, C., and Pollack, R., 1979, Tumorigenicity of revertants from an SV40-transformed line, *J. Supramol. Struct.* **11**:539–546.

Verderame, M., Alcorta, D., Egnor, M., Smith, K., and Pollack, R., 1980, Cytoskeletal F-actin patterns quantitated with Fl-phalloidin in normal and transformed cells, *Proc. Natl. Acad. Sci. U.S.A.* **77**:6624–6628.

Wulf, E., Deboben, A., Bautz, F. A., Faulstich, H., and Wieland, Th., 1979, Fluorescent phallotoxin, a tool for the visualization of cellular actin, *Proc. Natl. Acad. Sci. U.S.A.* **76**:4498–4502.

2

The Mechanism of Actin-Filament Assembly and Cross-Linking

Thomas D. Pollard, Ueli Aebi, John A. Cooper,
Marshall Elzinga, Walter E. Fowler, Linda M. Griffith,
Ira M. Herman, John Heuser, Gerhard Isenberg,
Daniel P. Kiehart, Janelle Levy, Susan MacLean-Fletcher,
Pamela Maupin, Mark S. Mooseker, Marschall Runge,
P. Ross Smith, and Peter Tseng

1. Introduction

Actin is one of the major proteins in eukaryotic cells, and actin filaments are the major structural element of both the contractile apparatus and the "cytoskeleton" of most cells. In their role as a contractile protein, filaments of actin are thought to interact with myosin to generate the force for cellular motility, much like they do in muscle contraction. The role of actin as a structural protein is less well defined, but probably no less important. The general idea is that the cytoplasm contains a three-dimensional network of actin filaments that can be cross-linked to form a gel. This network forms an internal scaffolding that traps the organelles (see Mast, 1926), distributes local contractile

Thomas D. Pollard, Ueli Aebi, John A. Cooper, Walter E. Fowler, Linda M. Griffith, Ira M. Herman, Gerhard Isenberg, Daniel P. Kiehart, Janelle Levy, Susan MacLean-Fletcher, Pamela Maupin, Marschall Runge, and Peter Tseng • Department of Cell Biology and Anatomy, The Johns Hopkins University School of Medicine, Baltimore, Maryland 21218. *Marshall Elzinga* • Biology Division, Brookhaven National Laboratories, Upton, New York 11973. *John Heuser* • Department of Physiology and Biophysics, Washington University School of Medicine, St. Louis, Missouri 63110. *Mark S. Mooseker* • Department of Biology, Yale University, New Haven, Connecticut 06510. *P. Ross Smith* • Department of Cell Biology, New York University School of Medicine, New York, New York 10016.

forces throughout the cytoplasm to the cell surface, and may also provide a scaffolding for certain enzyme systems (Clark and Masters, 1975).

An important feature of cytoplasmic actin is that its distribution changes considerably as the cell progresses through the cell cycle. For example, in cultured vertebrate cells, actin is spread throughout the cytoplasm in motile cells (Fig. 1a–d) (Herman *et al.*, 1981), concentrated in filament bundles called stress fibers in many stationary cells (Fig. 1e) (Lazarides and Weber, 1974), distributed in both the spindle and the surrounding cytoplasm of mitotic cells (Fig. 2a,b) (Herman and Pollard, 1979), and localized diffusely in cytokinetic cells (Fig. 2c,d) (Herman and Pollard, 1979). Although there is no direct

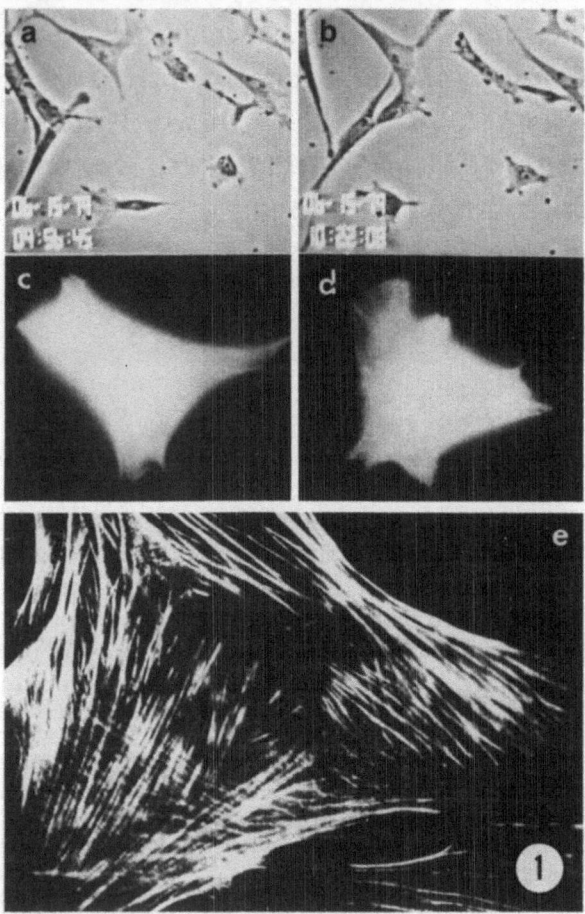

Figure 1. Fluorescent actin-antibody staining. (a,b) Videotape images of locomoting chick embryo cells taken 25 min apart. (c,d) Fluorescence micrographs of the cells on the lower left and lower right fixed immediately after frame (b) and then stained with fluorescent-antiactin. Note the diffuse staining of these two locomoting cells. (e) Stress fibers in a flat, nonmotile HeLa cell stained with fluorescent antiactin. Work of Ira Herman.

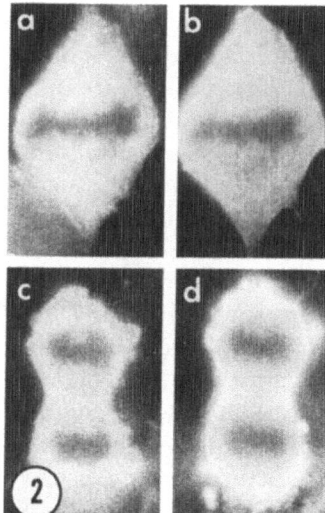

Figure 2. Double fluorescent-antibody staining of HeLa cells in mitosis (a,b) and cytokinesis (c,d). (a,c) Images of fluorescein-labeled antiactin. (b,d) Images of rhodamine-labeled antimyosin. Note the concentration of antiactin in the mitotic spindle fibers (a) and of antimyosin in the cortex (d) during cytokinesis. Work of Ira Herman.

evidence, it is thought that some of the actin filaments form and break down during these movements of the actin molecules from place to place in the cell.

Given these facts and speculations, it is clear why cell biologists and biochemists have made a considerable effort to understand the structure of the molecule, the mechanism of its assembly into filaments, and the way the filaments are linked together in the cell. As we describe in this chapter, much has been learned about these processes from studies on the purified actin molecule, and it is now established that cells have an interesting variety of proteins that regulate actin-filament formation, breakdown, and cross-linking. Although we emphasize molecular analysis here, our philosophy has been that complicated cellular processes can be understood only by a combination of biochemical work with morphological and physiological studies of intact cells.

Most of our work has been done with proteins from *Acanthamoeba,* a small, highly motile soil amoeba that can be grown in kilogram quantities in the laboratory. Aided by this abundance of material, we and Edward Korn's laboratory at the National Institutes of Health have been able to learn more about *Acanthamoeba* actin than about any other cytoplasmic actin. Although there are surely differences in the details, we expect the *Acanthamoeba* studies to have revealed general principles likely to be applicable to most other cell types.

2. Molecular Structure

Actin is one of the two or three most abundant proteins in eukaryotic cells. It has been estimated that except for the chloroplast protein D-ribulose-1,5-diphosphate carboxylase, there is more actin on earth than any other protein. Actin is also highly conserved. For example, about 95% of the amino acid sequences of rabbit-muscle actin (Collins and Elzinga, 1975) and *Acanthamoeba* actin are the same (Fig. 3). The differences are generally inconsequential substitutions of one amino acid by a similar amino acid; however, at position 228, *Acanthamoeba* actin has a histidine where other actins have a neutral residue. This may be the reason that the isoelectric point of *Acanthamoeba* actin is more basic than that of any other actin tested to date (Gordon *et al.*, 1977). *Acanthamoeba* actin is similar to other nonmuscle actins, which as a group (including vertebrate and *Physarum* actin) are distinct from muscle actins (Lu and Elzinga, 1976; Vanderkerckhove and Weber, 1978). Bovine-brain actin and *Acanthamoeba* actin are more closely related than bovine-brain and muscle actins. This led to the idea that muscle and nonmuscle actins have evolved independently under different selection pressures (Lu and Elzinga, 1976; Vanderkerckhove and Weber, 1978).

Actin has a molecular weight of 43,000 and has binding sites for an adenine nucleotide and a divalent cation (for a review, see Oosawa and Asakura, 1975). Actin has generally been assumed to be a globular protein, and recent structural studies have begun to reveal the features of the molecule in some detail.

We have studied the structure of two-dimensional crystalline sheets of actin by electron microscopy and image-processing techniques (Fig. 4 and 5) (Aebi *et al.*, 1980). These sheets form on dialysis of pure *Acanthamoeba* actin monomers (not actin filaments) against a molar excess of 6 : 1 or greater of the trivalent lanthanide Gd^{3+} (Fig. 4). Depending on the exact ionic strength and protein concentration, different polymorphic forms are observed, but all are built from the same two-dimensional "basic sheet," which is just one actin molecule thick. One of these polymorphic forms looks very similar to the crystalline tubes found earlier by exposure of rabbit-muscle actin to Gd^{3+} (DosRemedios and Dickens, 1978).

We have obtained a model of actin in projection to 1.5-nm resolution from negatively stained basic sheets (Fig. 5a). The unit cell is 5.6 × 6.5 nm and contains two actin molecules that are related to one another by a twofold axis of symmetry perpendicular to the sheet plane. From freeze–dried and metal-shadowed basic sheets, we determined their thickness to be 4.0 nm (Fig. 5b). Furthermore, these shadowed samples demonstrate that there is a significant asymmetry across the sheet plane (Fig. 5b): one surface appears smooth and textured with a near-rectangular right-handed lattice (5.5 × 6.4 nm), whereas the other surface looks coarse and untextured with no indication of a lattice. From these data obtained in projection, we conclude that actin is an elongated globular molecule (5.6 × 3.3 × 4.0 nm) with a pronounced asymmetry along the 4.0-nm axis. Some of our more recent reconstructions indicate that the molecule is divided into two lobes by a distinct cleft.

We have now collected tilted images of negatively stained sheets to ±60° under fairly low-dose conditions (1–5 $e^-/Å^2$). Some of these micrographs show sharp diffraction spots to $(1.1\ nm)^{-1}$, so we are confident that by late 1981, we shall have a three-dimensional model of actin to 1.5-nm resolution or better.

(Ac-Asp-Glu-Asp-Glu-Thr-Thr-Ala-Leu-Val-Cys-Asp-Asn-Gly-Ser-Gly)$^{Met}_{Leu}$$^{Cys}_{Val}$-Lys-Ala-Gly-^{20}

Phe-Ala-Gly-Asp-Asp-Ala-Pro-Arg-Ala-Val-Phe-Pro-Ser-Ile-Val-Gly-Arg-Pro-Arg-His-^{40}

Gln-Gly-Val-Met-Val-Gly-Met-Gly-Gln-Lys-Asp-Ser-Tyr-Val-Gly-Asp-Glu-Ala-Gln-Ser-^{60}

Lys-Arg-Gly-Ile-Leu-Thr-Leu-Lys-Tyr-Pro-Ile-Glu-TMH-Gly-Ile-$^{Val}_{Ile}$-Thr-Asn-Trp-Asp-^{80}

Asp-Met-Glu-Lys-Ile-Trp-His-His-Thr-Phe-Tyr-Asn-Glu-Leu-Arg-Val-Ala-Pro-Glu-Glu-^{100}

His-Pro-$^{Val}_{Thr}$-Leu-Leu-Thr-Glu-Ala-Pro-Leu-Asn-Pro-Lys-Ala-Asn-Arg-Glu-Lys-Met-Thr-^{120}

Gln-Ile-Met-Phe-Glu-Thr-Phe-Asn-$^{Thr}_{Val}$-Pro-Ala-Met-Tyr-Val-Ala-Ile-Gln-Ala-Val-Leu-^{140}

Ser-Leu-Tyr-Ala-Ser-Gly-Arg-Thr-Thr-Gly-Ile-Val-Leu-Asp-Ser-Gly-Asp-Gly-Val-Thr-^{160}

His-Asn-Val-Pro-Ile-Tyr-Glu-Gly-Tyr-Ala-Leu-Pro-His-Ala-Ile-Met-Arg-Leu-Asp-Leu-^{180}

Ala-Gly-Arg-Asp-Leu-Thr-Asp-Tyr-Leu-Met-Lys-Ile-Leu-Thr-Glu-Arg-Gly-Tyr-Ser-Phe-^{200}

Val-Thr-Thr-Ala-Glu-Arg-Glu-Ile-Val-Arg-Asp-Ile-Lys-Glu-Lys-Leu-Cys-Tyr-Val-Ala-^{220}

Leu-Asp-Phe-Glu-Asn-Glu-Met-$^{His}_{Ala}$-Thr-Ala-Ala-Ser-Ser-Ser-Ser-Leu-Glu-Lys-Ser-Tyr-^{240}

Glu-Leu-Pro-Asp-Gly-Gln-Val-Ile-Thr-Ile-Gly-Asn-Glu-Arg-Phe-Arg-Cys-Pro-Glu-Thr-^{260}

Leu-Phe-Gln-Pro-Ser-Phe-$^{Leu}_{Ile}$-Gly-Met-Glu-Ser-Ala-Gly-Ile-His-Glu-Thr-Thr-Tyr-Asn-^{280}

Ser-Ile-Met-Lys-Cys-Asp-$^{Val}_{Ile}$-Asp-Ile-Arg-Lys-Asp-Leu-Tyr-$^{Gly}_{Ala}$-Asn-$^{Val}_{Asn}$-Val-$^{Leu}_{Met}$-Ser-^{300}

Gly-Gly-Thr-Thr-Met-$^{Phe}_{Tyr}$-Pro-Gly-Ile-Ala-Asp-Arg-Met-Gln-Lys-Glu-$^{Leu}_{Ile}$-Thr-Ala-Leu-^{320}

Ala-Pro-Ser-Thr-Met-$^{MeK}_{Lys}$-Ile-Lys-Ile-Ile-Ala-Pro-Pro-Glu(Arg-Lys-Tyr-Ser-Val-Trp-^{340}

Ile-Gly-Gly-Ser-Ile-Leu-Ala-Ser-Leu-Ser-Thr-Phe-Gln-Gln-Met)Trp-Ile-$^{Ser}_{Thr}$-Lys-Gln-^{360}

Glu-Tyr-Asp-Glu-$^{Ser}_{Ala}$-Gly-Pro-Ser-Ile-Val-His-Arg-Lys-Cys-Phe

Figure 3. Comparison of the amino acid sequences of actins from rabbit skeletal muscle and *Acanthamoeba*. The sequences in parentheses have not been determined for *Acanthamoeba*. Where alternate residues are indicated, the upper is that found in *Acanthamoeba,* and the lower is the one present in rabbit. TMH (at 73) is *N*-methyl histidine; MeK (at 326) is methyl lysine. The sequence contains 375 residues, reflecting the presence of a Ser at 234 that was overlooked in the original sequence studies on actin. Work of Marshall Elzinga.

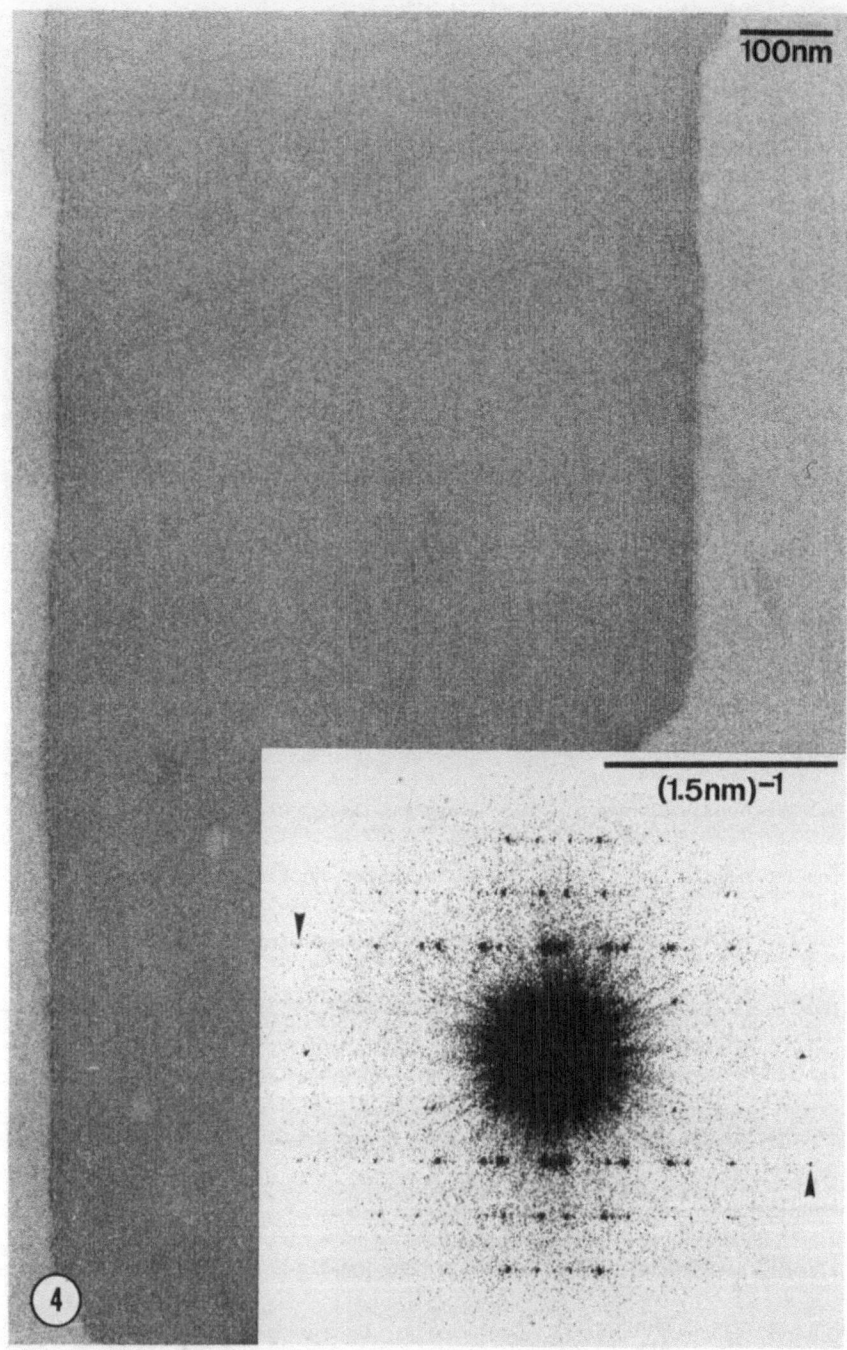

Figure 4. Low-dose (1–5 e⁻/Å²) electron micrograph recorded from a negatively stained (0.75% uranyl formate, pH 5.25), rectangular-type *Acanthamoeba* actin sheet. This polymorphic variant is two actin molecules thick (≈8.0 nm) and can be explained as a parallel association of two basic sheets, one atop the other, in an apolar fashion. The inset shows an optical-diffraction pattern obtained from a circular area of this sheet containing about 1000 5.6 × 6.5 nm unit cells. Note that the highest-order diffraction spots visible (indicated by arrows) correspond to structural detail of 1.25 nm. Work of U. Aebi and W. Fowler.

3. Filament Structure

Actin filaments are approximately 6-nm-wide double-helical arrays of actin molecules (Fig. 6) (Huxley, 1963; Moore *et al.*, 1970). We do not know exactly how our new model of the actin molecule fits into the filament, but it is virtually certain from the dimensions of the filament that the 5.6-nm axis of the molecule is aligned with the filament axis and that the 3.3-nm axis of the molecule is close to radial in the filament. Note, however, that the orientation of adjacent molecules must be different in the filaments and the sheets. In the sheets, the adjacent longitudinal rows of molecules are antiparallel, while the molecules in the two helical strands of the filament have the same polarity, as revealed by decoration with myosin heads (Huxley, 1963).

4. Filament Formation

It is generally believed that the formation of an actin filament from subunits involves at least two (and probably three) steps (Oosawa and Asakura, 1975). Initially, there is a slow nucleation step during which three or four actin molecules get together to form a tiny fragment of a filament referred to as a nucleus (Fig. 7). This slow step is thought to account for the lag usually observed at the onset of actin polymerization from monomers (Fig. 8). Then, more actin molecules add rapidly to the ends of the nucleus to elongate the filament (Fig. 7). This elongation step is responsible for the rapid increase in viscosity illustrated in Fig. 8. Also, two short filaments can anneal end to end to form a longer filament. Each of these three reactions is reversible. Filaments and nuclei can become shorter by dissociation of molecules from their ends, and long filaments can be broken into smaller filaments. Most of what we know about these processes comes from studies of muscle actin, but comparative studies of the polymerization of cytoplasmic actins (e.g., Gordon *et al.*, 1977) have revealed only minor differences.

To understand the mechanisms of these reactions, the steps involved in each reaction must be identified and their rate constants measured. In the case of nucleation, essentially nothing is known about either the nature or the rates of the steps.

Much more has been learned about the elongation reaction. It is assymetrical; that is, subunits add more rapidly at one end than at the other. The ends can be identified by decoration of the filament with myosin fragments that form arrowhead-shaped complexes along the filament. The rapidly polymerizing end has the barbs of the arrowheads (Fig. 9) (Woodrum *et al.*, 1975). The pointed end is slow. At some point during the elongation process, the ATP molecule bound to each free actin molecule is dephosphorylated, because each actin molecule in the filament has a bound ADP (Oosawa and Askura, 1975). However, it is not known how tightly the dephosphorylation is coupled to subunit addition. Some experiments (e.g., Brenner and Korn, 1980) seem to show a lag in phosphorylation with respect to polymerization.

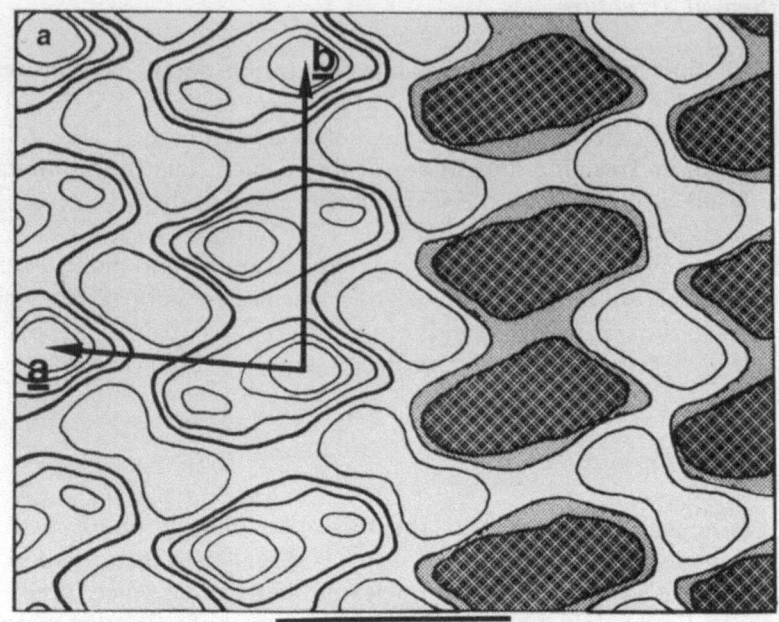

5nm

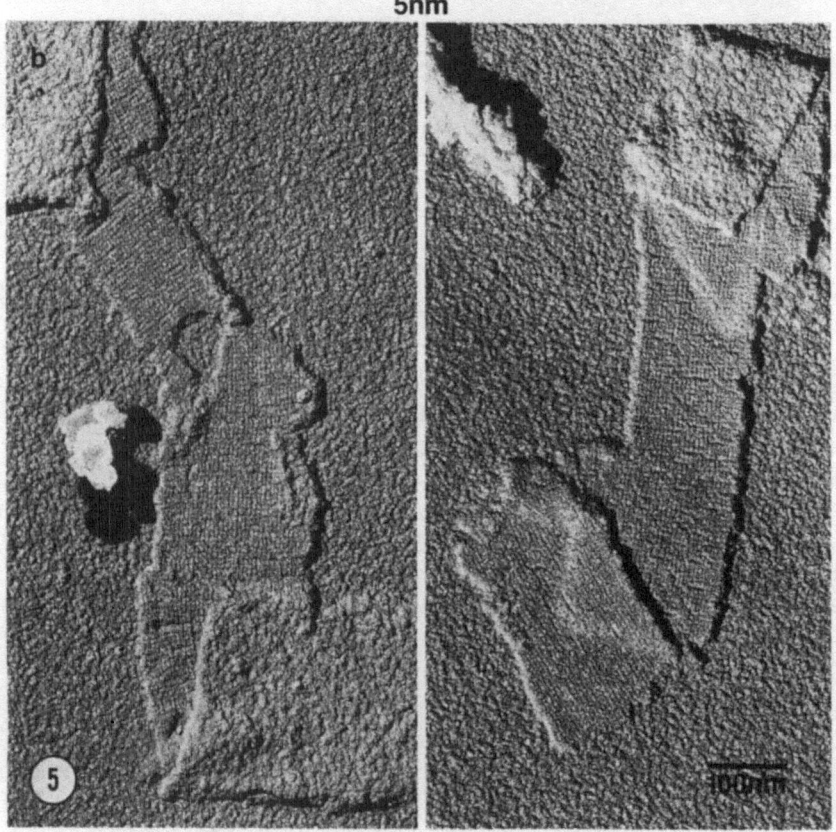

100nm

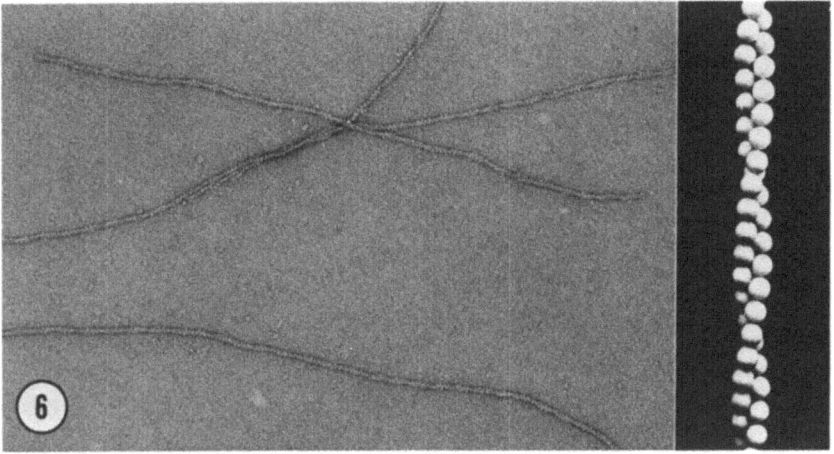

Figure 6. Electron micrograph of actin filaments negatively stained with uranyl acetate. × 80,000. *Inset:* Model of the filament.

The elongation reaction can be described by a simple kinetic equation

$$\frac{dl}{dt} = k_+ \cdot c_1 - k_-$$

where dl/dt is the change in the length of the filament with time, k_+ is a second-order association-rate constant, c_1 is the concentration of actin monomers, and k_- is a first-order dissociation-rate constant (Oosawa and Asakura, 1975). What this equation says is that the rate at which monomers add to the end is equal to the product of the monomer concentration and a constant. At the same time, monomers are lost from the end at a constant rate given by k_-. When $k_+ \cdot c_1 = k_-$, the rates of association and dissociation are balanced, and the length does not change. The apparent equilibrium constant at an end is given by

$$K_{eq} = \frac{k_+}{k_-} = \frac{1}{c_1^0}$$

This concentration of monomers, c_1^0, in equilibrium with polymer is called the "critical concentration."

Since the net rates of subunit addition to the two ends of actin filaments are unequal, the rate constants must be different at the two ends, although the ratio k_+/k_- (the critical concentration) may or may not differ at the two ends. The equations for the elongation rates at the two ends are

---

Figure 5. (a) Contour plot of the computer-filtered actin basic sheet. The unit cell containing two molecules is indicated by the lattice vectors a and b. The actin monomers and dimers are indicated by the light- and dark-shaded regions at right. (b) Electron micrographs of freeze-dried, shadowed actin basic sheets illustrating the distinct surface topographies of the two sides. The specimens were shadowed unidirectionally with platinum. (a) From Aebi *et al.* (1980). (b) Work of U. Aebi and P. R. Smith.

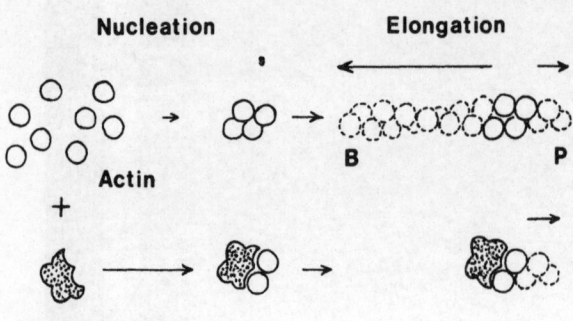

Figure 7. Schematic model of actin polymerization showing the slow nucleation step and the rapid bidirectional elongation step. (B) Barbed end; (P) pointed end. A speculation regarding the mechanism of action of capping protein is shown at the bottom.

$$\frac{dl^B}{dt} = k_+^B \cdot c_1 - k_-^B \quad \text{and} \quad \frac{dl^P}{dt} = k_+^P \cdot c_1 - k_-^P$$

where the superscripts B and P denote the barbed and the pointed end. An additional complication is that monomers bearing either ATP or ADP may associate or dissociate from either end. We will indicate the bound nucleotide by the superscript T or D for ATP or ADP (Fig. 10).

These equations have the form $y = mx + b$, so that the two unknowns (k_+ and k_-) can be evaluated by plotting the elongation rate vs. monomer concentration. This was done for microtubules (Bergen and Borisy, 1980). It is necessary to observe the growth process by electron microscopy, because it is impossible to determine the elongation rates at the two ends by measurement of the overall growth rate.

Pollard and Mooseker (1981) measured the growth rates of muscle-actin filaments by electron microscopy using the bundle of cytoplasmic actin fila-

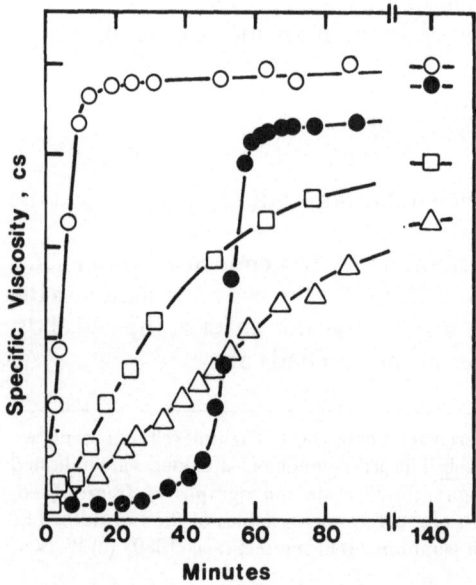

Figure 8. Effect of filament capping protein on the polymerization kinetics of pure actin. Polymerization was monitored by measuring the high-shear viscosity in an Ostwald capillary viscometer. Gel-filtered actin monomer (0.5 mg·ml⁻¹) (●) was mixed with 2.2 μg·ml⁻¹ capping protein (△), 30 μg·ml⁻¹ actin filaments (○), or 2.2 μg·ml⁻¹ capping protein plus 30 μg·ml⁻¹ actin filaments (□) and polymerized in 20 mM KCl, 10 mM imidazole, 0.7 mM tris, 0.07 mM ATP, 0.2 mM dithiothreitol (DTT), and 0.07 mM CaCl₄, pH 7.5, at 25°C. From Isenberg *et al.* (1980).

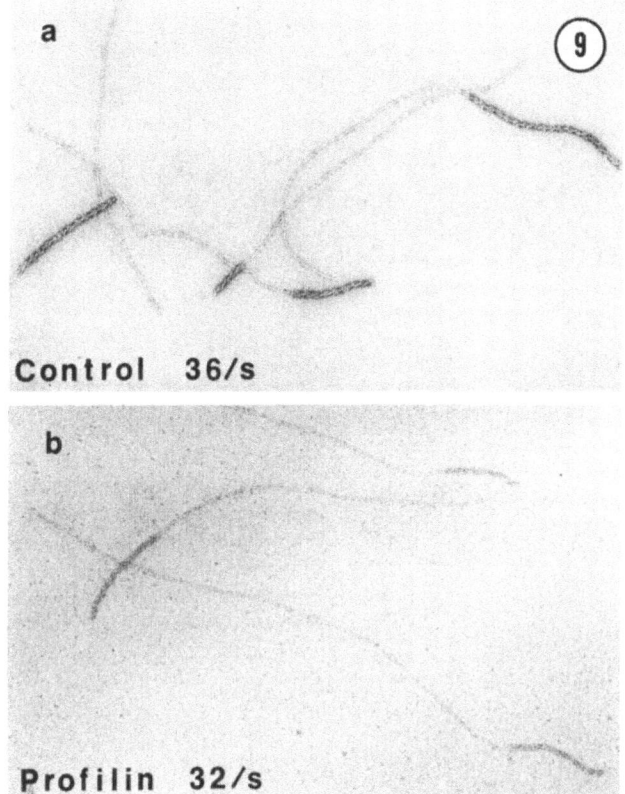

Figure 9. Electron micrographs of actin polymerization at the barbed end of actin-filament nuclei (decorated with myosin subfragment-1) in the absence (a) or the presence (b) of an excess of *Acanthamoeba* profilin. (36/s, 32/s) Average growth rates (molecules/sec.) ×33,000. Work of M. Runge.

ments isolated from the microvilli of intestinal epithelial cells as a morphologically distinct nucleus (Fig. 11). The length of the filaments grown from these nuclei is proportional to the time of incubation of the nuclei with actin monomers. Plots of the growth rates vs. ATP-monomer concentration are linear (Fig. 12), as predicted by the theory presented above. The apparent rate constants are given in Table 1. Note that the dissociation-rate constants given in Table 1 include both the ATP- and ADP-monomer dissociation-reactions, since the state of the nucleotide on the dissociating molecules was not known.

The apparent association-rate constant for ATP-monomers at the barbed end ($k^{B,T}$) is very large, suggesting that the process is diffusion-limited, rather than limited by some subsequent chemical reaction. In contrast, the apparent association-rate constant at the pointed end is only about 20% as fast, so that the rate of subunit association at that end may actually be determined by a slower process that follows the formation of the collision intermediate. One

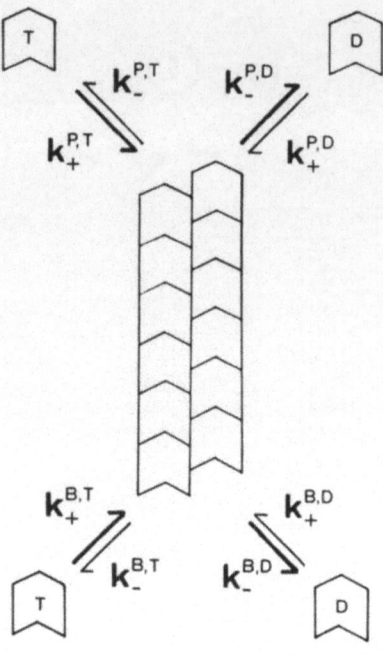

Figure 10. Model of actin polymerization. The symbols are defined in the text. The generally accepted pathways of association and dissociation are indicated by the heavier arrows. The polarity of the subunits in the double-helical actin filament is indicated by a chevron shape. From Pollard and Mooseker (1981).

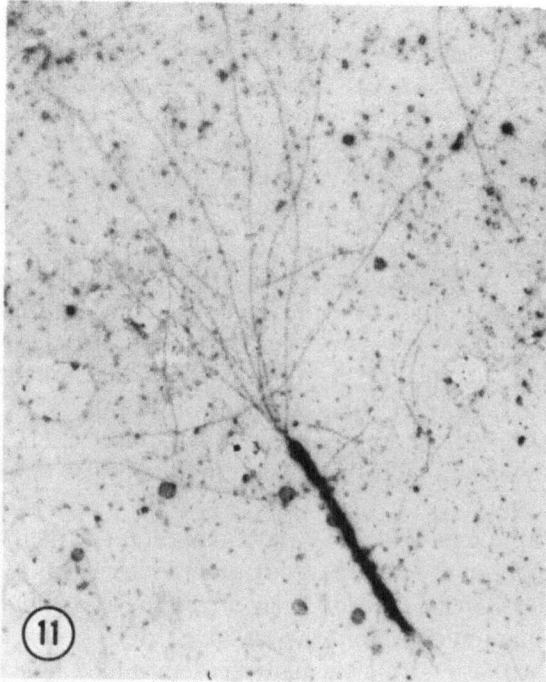

Figure 11. Electron micrograph of a negatively stained microvillus core used to nucleate polymerization of 5.7 μM actin for 20 sec. Longer filaments have grown from the upper (barbed) end than from the pointed end. ×13,100. From Pollard and Mooseker (1981).

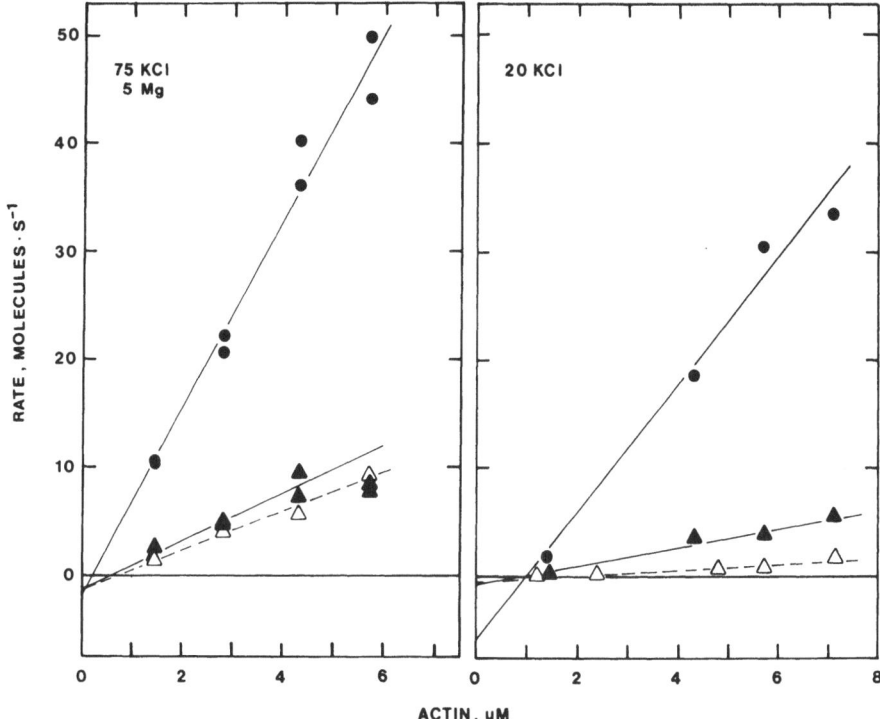

Figure 12. Dependence of actin-filament growth rates on actin-monomer concentration in either the 75 mM KCl–5 mM MgSO₄ buffer (*left*) or the 20 mM KCl buffer (*right*). (●) Barbed end; (▲) control pointed end; (△) pointed end with 2 μM cytochalasin B. No growth was observed at the barbed end in cytochalasin B. In the KCl–Mg experiment, growth rates at two time points for each actin concentration are plotted. In the 20 mM KCl experiment, mean growth rates measured from three or more time points are plotted. Correlation coefficients were greater than 0.98 except for the control pointed end in KCl–Mg (▲) (0.96 omitting the 5.6 μM actin data as shown; 0.90 including all the data) and the pointed end in 20 mM KCl with cytochalasin B (△) (0.86). From Pollard and Mooseker (1981).

such process could be ATP splitting. This hypothesis leads to the prediction that ATP splitting is tightly coupled to monomer association at the pointed but not at the barbed end. Consequently, a number of internal subunits at the barbed end may have bound ATP, accounting for the lag in ATP splitting compared with subunit incorporation into the growing filament (Brenner and Korn, 1980).

The rate constants can also be used to calculate the efficiency of the polymerization reaction. In a typical laboratory experiment with 20 μM actin monomer, the filament will grow bidirectionally. In 1 sec, 180 subunits will add and 2 will be lost from the barbed end, an efficiency of approximately 98%. At the pointed end, the efficiency is about 95%.

The events taking place once a sample of actin monomers and filaments has reached a steady state depend on the composition of the buffer. In 20 mM KCl, the critical concentrations (*x* intercept on the graphs in Fig. 12) are the same at both ends. Although both subunit association and dissociation con-

Table 1. Rate Constants for Actin Polymerization[a]

Conditions	Experiment	$k_+^{B,T}$	$k_+^{P,T}$	k_-^B	k_-^P
20 mM KCl	Control	5.9	0.8	6.0	0.7
	2 μM Cytochalasin B	≈0	0.3	≈0	0.6
75 mM KCl	Control	8.8	2.2	2.0	1.4
5 mM MgSO₄	2 μM Cytochalasin B	≈0	1.8	≈0	1.3

[a] From Pollard and Mooseker, 1981. These rate constants were obtained from the plots shown in Fig. 12. Units: k_+ = molecules·sec⁻¹·μM⁻¹; k_- = sec⁻¹.

tinue at steady state and the rate of these reactions is greater at the barbed than at the pointed end, neither end changes in length because the ratio, k_+/k_-, is the same at both ends. In contrast, in the KCl–MgSO₄ buffer, the critical concentrations appear to be slightly different at the two ends (Fig. 12). Consequently, a steady state is achieved at a concentration of free monomer where the filaments maintain a constant length, but there is net addition of subunits at the barbed end balanced by net loss of subunits at the pointed end. This causes a flux of monomers through the filament from the barbed to the pointed end. Such a flux was predicted by Wegner (1976) from theoretical considerations and supported by his experiment on the time–course of the equilibrium of labeled subunits with a steady-state polymer solution in 2 mM CaCl₂. In our KCl–Mg buffer, the rate of this flux is about 1 subunit/sec, so it takes less than 10 min for a molecule adding at the barbed end of a 1-μm filament to pass through the filament and to be lost at the pointed end. The efficiency of this process is about 20%, so that about 1 of 5 molecules associating with the filament ends contributes to this flux, while the other 4 are lost from the same end by dissociation. The rate of flux calculated from the rate constants is high enough to account for the rapid equilibration of radioactive actin monomers with nonradioactive actin filaments observed by Simpson and Spudich (1980).

5. Regulation of Actin Polymerization

The fungal metabolite cytochalasin and several cellular proteins can modify the actin-filament-assembly reactions. It would appear likely that these proteins (and others that are yet to be identified) are responsible for controlling actin-filament formation in the cell. In this section, we will summarize our work on the mechanism of action of cytochalasin and of two actin-regulatory proteins, profilin and capping protein.

Profilin is a small protein originally isolated from lymphoid tissues as a complex with actin (Carlsson *et al.*, 1977). This complex does not polymerize, so it was thought that profilin blocks polymerization. However, it was found later that profilin purified from *Acanthamoeba* (Reichstein and Korn, 1979) and platelets (Grumet and Lin, 1980a) inhibits the initiation but not the extent of polymerization. This was attributed to a reduction in the rate of nucleation.

We have confirmed these observations on *Acanthamoeba* profilin and have also demonstrated by a nonequilibrium dialysis technique that profilin binds weakly ($k_d \sim 10^{-5}$ M) to actin monomers. It does not bind to actin filaments in either polymerizing or depolymerizing buffers. Consequently, profilin must dissociate from the monomers as they add to the ends of the filaments. Remarkably, the actin–profilin complex adds to the ends of decorated nuclei as fast as actin monomers (see Fig. 9). No mechanism unifying these effects of profilin has been proposed, but taken together the evidence suggests that profilin might be used to suppress spontaneous nucleation of actin in the cell.

It has been known for more than a decade that the various cytochalasins will reversibly inhibit many cellular movements, especially cytokinesis (Carter, 1967). Beginning with the observation by Lin and Lin (1979) that cytochalasin binds to a crude actin–spectrin complex from erythrocytes and inhibits the growth of actin filaments from this complex, several laboratories have focused on the inhibition of actin polymerization by cytochalasins. Independently, Brenner and Korn (1979), Brown and Spudich (1979), Lin *et al.* (1980), and MacLean-Fletcher and Pollard (1980b) found that cytochalasin inhibits the growth of actin filaments from actin-filament nuclei in a concentration-dependent fashion. All the cytochalasins act substoichiometrically, and their potencies parallel their affinity for binding to a small number of high-affinity, cytochalasin-binding sites on actin filaments. This site has not been identified directly, but it is likely to be at the barbed end of the filament. MacLean-Fletcher and Pollard (1980b) found that cytochalasin B blocks the growth of actin filaments from the barbed end of short actin filaments decorated with myosin heads. Mechanistically, this is due to a reduction of both the association- and dissociation-rate constants at the barbed end to approximately zero (Table 1) (Pollard and Mooseker, 1981). Kinetically, the barbed end is thus completely blocked. The rate constants at the pointed end may also be reduced to a small extent, but the major effect is on the barbed end. This blocking of the barbed end stops the rapid equilibration of actin monomers with a steady-state solution of actin filaments (Simpson and Spudich, 1980), which is attributable to the flux of molecules through the filaments. It is not yet clear how these effects of cytochalasin on actin polymerization might be responsible for the inhibition of cellular movements.

Acanthamoeba contains a cytochalasinlike protein that we purified and named capping protein (Isenberg *et al.*, 1980). It consists of two polypeptides with molecular weights of 28,000 and 31,000. Antibodies that react exclusively with the individual capping-protein polypeptides precipitate both when reacted with the native protein. The Stokes radius and sedimentation co-efficient of capping protein yield a native molecular weight of 72,000 with one copy of each polypeptide. Like cytochalasin capping protein blocks the addition of actin monomers to the barbed end of actin filaments, but also seems to be capable of nucleating actin polymerization (see Fig. 8). A simple mechanism might be that it binds two actin molecules at their barbed poles, forming a stable complex to which more monomers can add at the pointed end (see Fig. 7). Further evidence for binding of capping protein to the barbed end is that it is a competitive inhibitor of cytochalasin binding to actin filaments (unpublished observation by M. Grumet, D. Cribbs and S. Lin), like a similar partially purified cytochalasinlike molecule from platelets (Grumet and Lin, 1980b). Although the concentration of the capping protein in the cell is low, about 40 $\mu g/ml$, this is enough capping protein to block the barbed ends of all the actin filaments in the cell, assuming that most of the actin is in 1-μm-long filaments. In binding to the actin filaments, this remarkable protein is in a position to determine the site and direction of actin-filament polymerization. In this way, it may also regulate the number and length of actin filaments in the cell.

6. Actin-Filament Networks

Although it has been known for years that cytoplasm is gelatinous, the molecular basis of this property of cytoplasm has been revealed only recently. Dujardin (1835) speculated that a common substance, sarcode, is responsible for both the high viscosity of cytoplasm and its contractibility. A connection between these cellular processes at the molecular level was made by Pollard and Ito (1970), who found that gelation of an extract of *Amoeba proteus* is correlated with the formation of a network of actin filaments. Since then, work on extracts of a number of different organisms has established that actin-filament gels are found in most cells (see Taylor and Condeelis, 1979).

The gelation reaction in extracts of *Acanthamoeba* has been characterized in detail using a low-shear, falling-ball viscometer to follow the time–course of the reaction (Fig. 13) (MacLean-Fletcher and Pollard, 1980a). The conditions required for gelation of a cold extract (Table 2) are likely to be found in the cytoplasm, so we expect that such a gel accounts for the high viscosity of the cytoplasm. Of particular interest is the requirement for MgATP in concentrations higher than those needed for actin polymerization. Nonhydrolyzable analogues of ATP (e.g., AMP-P-N-P) will substitute for ATP in actin polymerization, but not in the gelation reaction. These facts suggest that an ATP-requiring enzyme is involved with gelation. It could be a kinase, because many of the proteins in the extract are phosphorylated during the gelation reaction. The fact that micromolar Ca^{2+} inhibits gelation in *Acanthamoeba* (Fig. 13) (Pollard, 1976; MacLean-Fletcher and Pollard, 1980a) and in other cells such as *Dictyostelium* (Condeelis and Taylor, 1977) makes Ca^{2+} the major candidate for regulating gelation.

It is almost certain that the actin gels are constructed of filaments held together in a three-dimensional network by physical connections (cross-links). This is true for other polymeric gels (Flory, 1953). Most investigators have assumed that these cross-links are formed by protein molecules that bind to actin, and such gelation factors have been isolated from a variety of cells (Table 3). We will return to our work on *Acanthamoeba* gelation factors, but first it is necessary to consider another type of cross-link that contributes to the stability of the actin networks: direct binding of actin filaments to each other.

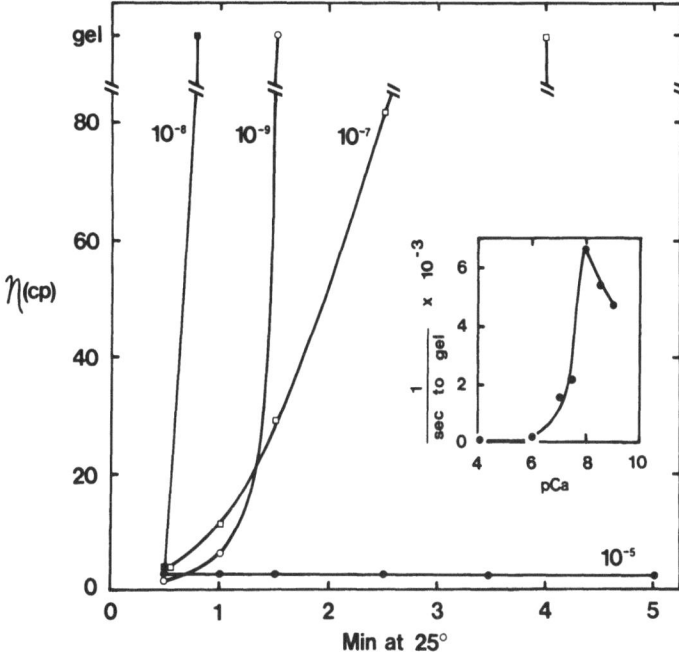

Figure 13. Dependence of the time–course of gelation of desalted *Acanthamoeba* extract on the free calcium ion concentration. We preincubated samples containing 9.2 mg/ml protein in 0.34 M sucrose, 10 mM imidazole (pH 7), 2 mM $MgCl_2$, 1 mM ATP, mM ethyleneglycol-bis (β-aminoethyl ether)N,N'-tetracetic acid (EGTA), 0.5 mM DTT, and various amounts of $CaCl_2$ for 5 min at 0°C before warming them to 25°C and measuring the time–courses of viscosity changes. Free calcium ion concentrations: 10^{-9} M (○), 10^{-8} M (■), 10^{-7} M (□), 10^{-5} M (●). *Inset:* Plot of the reciprocal of the time to gel vs. pCa from another experiment. From MacLean-Fletcher and Pollard (1980a).

Anyone who has purified actin from muscle or nonmuscle cells has noticed that solutions containing high concentrations of actin filaments so-lidify at rest. These actin gels are weak, so that stirring disrupts the network and yields a viscous solution. This tendency of purified muscle-actin filaments to self-associate has been documented by sensitive viscometric techniques by Maruyama *et al.* (1974), who found that the absolute viscosity is inversely proportional to the shear rate. As a result, the viscosity approaches infinity at low shear rates like those in living cells.

It was possible that this apparent self-association of muscle-actin fila-

Table 2. *Conditions for the Gelation of Acanthamoeba Extracts[a]*

Temperature	15–35°C
Actin concentration	$\geq 10\ \mu M$
$MgCl_2$ concentration	$\geq 1.5\ \mu M$
ATP concentration	$\geq 1.0\ \mu M$
Ca^{2+} concentration	$< 1.0\ \mu M$
Cytochalasin B	$< 0.5\ \mu M$

[a] From MacLean-Fletcher and Pollard (1980a).

Table 3. Examples of Actin-Cross-Linking Gelation Proteins

Source	Name	Molecular subunit	Calcium sensitive	Reference
Acanthamoeba	Gelation factor-23	23,000	No	Maruta and Korn (1977)
	Gelation factor-29	29,000	No	
	Gelation factor-33	33,000	No	
	Gelation factor-38	38,000	No	
	Gelation factor-85	85,000	Yes	Pollard (1981)
Ascites-tumor cells	Actinogelin	110,000	Yes	Mimura and Asano (1980)
Sea urchin egg	Fascin	58,000	No	Bryan and Kane (1978)
		220,000	No	
Smooth muscle Vertebrate nonmuscle	Filamin	250,000	No[a]	Wang and Singer (1977)
cells	Actin-binding protein	250,000	No[a]	Brotschi *et al.* (1978)

[a] Unless an additional protein, "gelsolin," is present (Yin and Stossel, 1979).

ments was due to a low concentration of cross-linkers contaminating the actin, but attempts to remove such hypothetical cross-linkers have led to actin with a stronger tendency to self-associate, rather than the opposite (Fig. 14) (MacLean-Fletcher and Pollard, 1980c). This is due to the separation of a cytochalasinlike protein from the actin during further purification by gel filtration. When added back to the actin, this unpurified cytochalasinlike protein reduces the low shear viscosity of the actin.

Actin-filament self-association is promoted by a variety of factors (Griffith and Pollard, 1982a). KCl (Fig. 14), K glutamate, $CaCl_2$, $MgCl_2$, $CrCl_2$, $CoCl_2$, $FeCl_2$, $CdSO_4$, $CuSO_4$, and $Pb(NO_3)_2$ all cause the low shear viscosity of pure actin to be high. In fact, millimolar concentrations of Cu^{2+}, Fe^{2+}, Cd^{2+}, Co^{2+}, or Pb^{2+} and acid pH all cause pure actin to gel. Consequently, it is clear that actin filaments themselves can form networks with some rigidity under conditions that might be found in cells, and this self-association must be considered as a potentially important component of actin-gel formation.

In addition to actin–actin bonds, actin networks can also be stabilized by molecules that cross-link the filaments. A number of such protein molecules have been isolated from nonmuscle and smooth-muscle cells (Table 3). All these proteins raise the low shear viscosity (or any other parameter of gel formation) in a concentration-dependent fashion (Figs. 15–17). As first shown for actin-binding protein and filamin (Brotschi *et al.*, 1978), the curve relating network formation to cross-linker concentration is hyperbolic, as predicted (Flory, 1953) for any system of cross-linked polymers. The explanation for the abrupt transition from a viscous liquid to a gel at a "critical gelling concentration" is that the gel forms only when all the polymers are cross-linked to at least one other filament to form a continuous network. Below the critical gelling concentration of cross-linker, the viscosity is relatively low because the individual groups of cross-linked polymers are small.

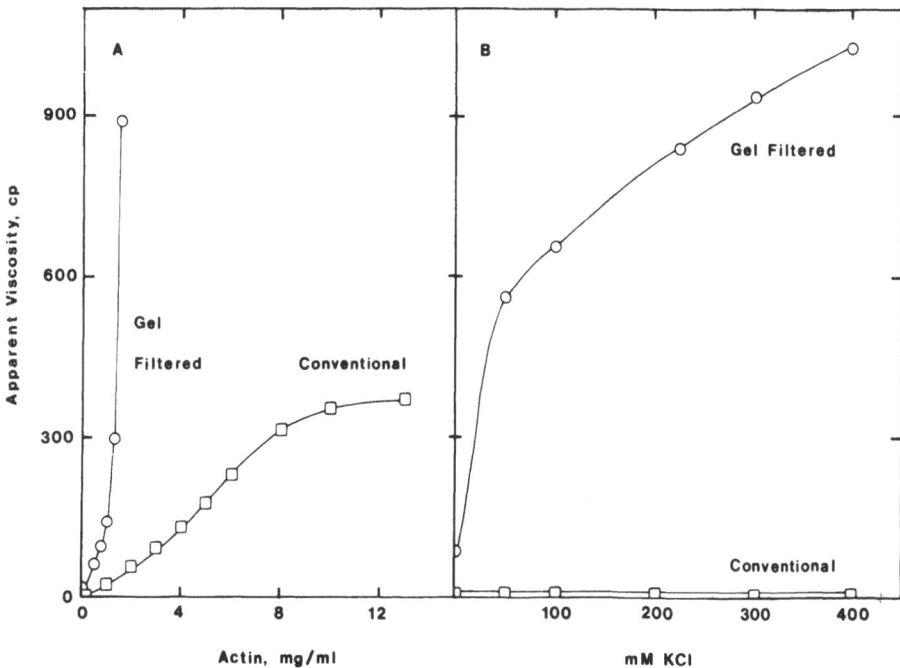

Figure 14. (A) Dependence of low-shear apparent viscosity on actin concentration. Conditions: 2 mM $MgCl_2$, 1 mM ATP, 1 mM EGTA, and 10 mM imidazole, pH 7, 10 min at 25°C. (B) Dependence of low-shear apparent viscosity on KCl concentration. Conditions: As in (A); actin: 1 mg/ml ($24\mu M$). (O) Gel-filtered actin; (□) conventional actin. From Griffith and Pollard (1982a).

The molecules that can cross-link actin are quite varied (Table 3). Most have been isolated from cells using assays for actin binding or gelation, and it has been assumed that these proteins function as actin-filament cross-linkers in cells. However, a number of basic macromolecules, which are unlikely to have access to actin in the cell, can also cross-link actin filaments (Table 4) (Griffith and Pollard, 1982a). For example, ribonuclease is sequestered in zymogen granules and lysosomes and is unlikely to associate with actin in the cytoplasm. It could be released during cell fractionation and isolated as a gelation factor even though it is irrelevant to actin-network formation in the cytoplasm. Consequently, additional criteria are needed to prove that any of the isolated gelation factors are part of the cytoplasmic actin network in the cell.

It is difficult to establish the physiological relevance of a gelation protein with purely biochemical experiments. For example, both nonspecific (polylysine) and presumably specific [gelation protein-85 (GP-85)] gelation factors can have high specific activities (e.g., will form a gel at low concentrations). Another sort of biochemical criterion, which may be more helpful, is the analysis of how solution conditions affect gelation. For example, it is well established that micromolar Ca^{2+} inhibits gelation in crude extracts (see Fig. 13), and purified gelation factors might be expected to have this property if they were physiologically relevant. In fact, both actinogelin (Mimura and Asano, 1980) and *Acanthamoeba* GP-85 (Fig. 15) (Pollard, 1981) are calcium-

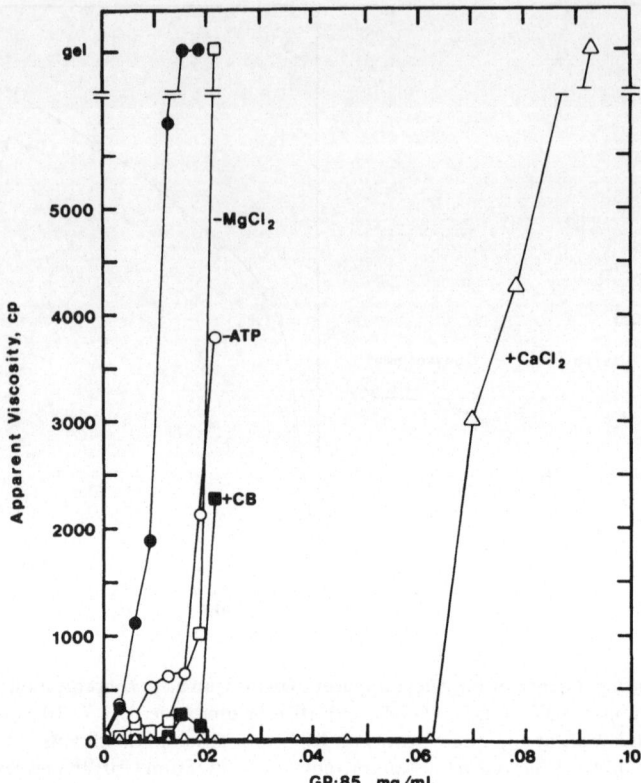

Figure 15. Low-shear, falling-ball viscometry of actin with *Acanthamoeba* gelation protein-85 (GP-85). Conditions: 10 mM imidazole, pH 7.0, 2 mM $MgCl_2$, 1 mM ATP, 1 mM EGTA; 0.5 mg/ml actin; 10-min incubation at 25°C. The GP-85 concentration is varied as indicated. (●) Complete buffer; (□) buffer with 50 mM KCl substituted for $MgCl_2$; (○) buffer minus ATP; (■) buffer plus 2 μM cytochalasin B; (△) buffer with 100 μM $CaCl_2$ substituted for EGTA. From Pollard (1981).

sensitive. On the other hand, the lack of direct calcium sensitivity does not mean that a gelation protein is not physiologically important. For example, macrophage actin-binding protein is not Ca^{2+}-sensitive itself, but in combination with the protein gelsolin, it can form Ca^{2+}-sensitive actin gels (Yin and Stossel, 1979). Perhaps the best evidence that gelation is required for the cell to maintain its cytoplasmic gel would be provided by the microinjection of antibodies to the gelation proteins into living cells. Detailed localization of the gelation proteins in the cell may also be helpful in establishing their physiological function.

In principle, the formation of actin-filament gels could be controlled directly at the level of the cross-linking proteins or of the actin filaments themselves by modulating the concentration, length, and degree of self-association of the filaments. One example of regulation of gelation by action on the actin filaments is provided by gelsolin (Yin and Stossel, 1979), a Ca^{2+}-sensitive protein that can inhibit gelation by shortening actin filaments. Both

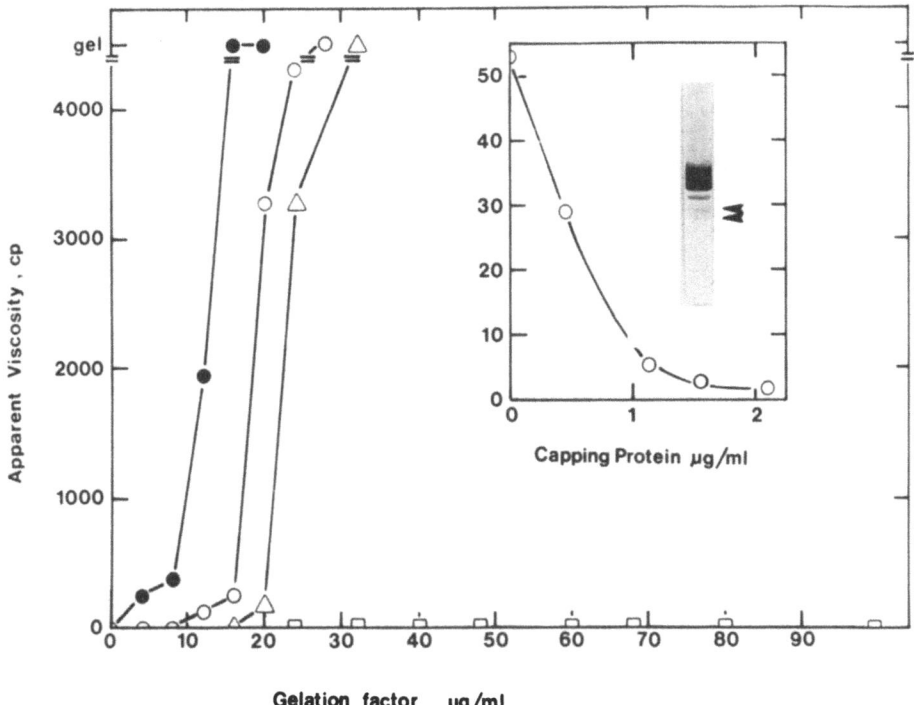

Gelation factor µg/ml

Figure 16. Effect of filament capping protein on actin-gel formation and on the low-shear viscosity of actin filaments. Samples consisting of 0.5 mg·ml⁻¹ actin with 0 (●), 0.44 µg·ml⁻¹ (○), 0.88 µg·ml⁻¹ (△), or 1.1 µg·ml⁻¹ (□) capping protein were mixed with various concentrations of GP-85 from *Acanthamoeba* and polymerized in capillary tubes for 10 min in 2 mM MgCl₂, 1 mM ATP, 1 mM EGTA, and 10 mM imidazole, pH 7. The low-shear apparent viscosity was measured with the falling-ball device. *Inset:* Actin alone or with several concentrations of capping protein was polymerized in capillary tubes for 10 min, after which the apparent viscosity was measured with the falling-ball device. Incubation for up to 40 min did not change the viscosity of actin plus 2.2 µg·ml⁻¹ capping protein. The electrophoretic gel illustrates the composition of a sample with 0.5 mg·ml⁻¹ actin with 1.1 µg·ml⁻¹ capping protein. From Isenberg *et al.* (1980).

cytochalasins (Hartwig and Stossel, 1979; MacLean-Fletcher and Pollard, 1980b) and capping protein (Isenberg *et al.,* 1980) also inhibit gelation of actin by a variety of cross-linking proteins by direct action on the filaments. Both these molecules raise the concentration of any gelation protein required to form a gel (Figs. 15–17). The resulting critical gelling concentration depends on the concentration of cytochalasin or capping protein as though they were competing with the gelation protein. This, however, is unlikely, because the nature of the gelation factor is irrelevant. For example, cytochalasin inhibits gelation of actin by all molecules tested, including filamin, actin-binding protein (Hartwig and Stossel, 1979), polylysine, microtubule-associated proteins (MAPs) (MacLean-Fletcher and Pollard, 1980b), *Acanthamoeba* GP-85 (Pollard, 1981), aldolase, histone, lysozyme, and ribonuclease (Griffith and Pollard, 1982a). Consequently, the effect of cytochalasin and capping protein must be explained by their effect directly on actin. The inhibition of gelation

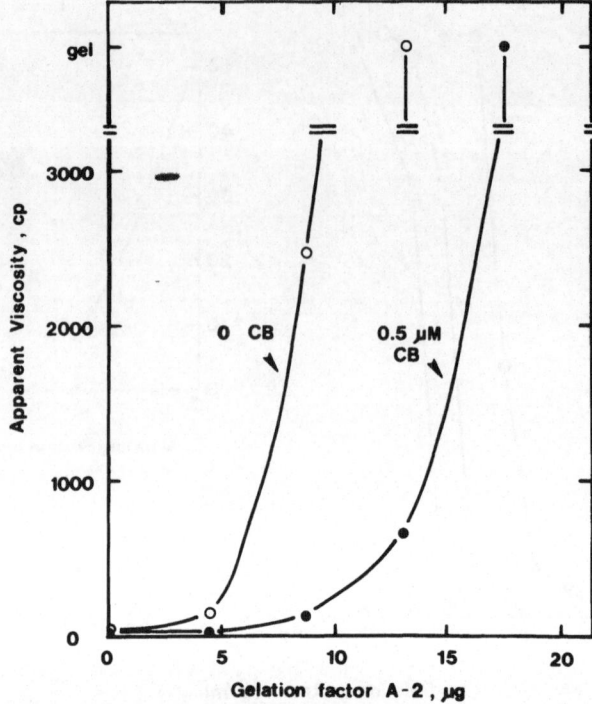

Figure 17. Cytochalasin B (CB) inhibition of the low-shear viscosity of mixtures of actin and purified *Acanthamoeba* gelation factor-33. Actin, $12\mu M$, with various concentrations of gelation factor was incubated for 10 min at 25°C in 10 mM imidazole (pH 7), 2 mM $MgCl_2$, 1 mM ATP, 1 mM EGTA, and 0.005% dimethylsulfoxide (DMSO) with 0 (○) or 0.5 μM (●) CB. The inset on the top left shows Sodium dodecyl sulfate gel electrophoresis of the purified gelation factor. The major band had a molecular weight of 33,000. From MacLean-Fletcher and Pollard (1980b).

is probably the consequence of the loss of the actin-filament self-associations that normally constitute a major part of the total cross-links that stabilize the filament network of the gel, because both cytochalasin (MacLean-Fletcher and Pollard, 1980b) and capping protein (Fig. 16) (Isenberg *et al.*, 1980) inhibit gelation at the same concentrations at which they inhibit the low-shear viscosity of pure actin. It is important to note that there is only about one molecule of either cytochalasin or capping protein bound to each filament when the low-shear viscosity is inhibited 90%.

The mechanisms by which cytochalasin and capping protein inhibit actin-filament low-shear viscosity are controversial. Hartwig and Stossel (1979) proposed that cytochalasin does so by reducing the length of the filaments. Given a constant amount of total polymer, this would mean that there are more, but shorter, filaments in the presence of cytochalasin. According to the theory of Flory (1953), this would raise the concentration of cross-linker required to form a gel. Using three different assays, we could not find convincing evidence that actin filaments are shorter in 2 μm cytochalasin B, which inhibits the low-shear viscosity of actin filaments more than 95% (MacLean-Fletcher and Pollard, 1980b). Similarly, a concentration of capping protein

Table 4. Interaction of Macromolecules with Actin Filaments[a]

Macromolecule	pI	Critical gelling concentration (mg/ml)[b]	Fraction binding to actin filaments[c]
Polylysine	10.5	0.003	—
Histone	10.5–11.0	0.030	0.96
Lysozyme	11.0	0.070	0.82
MAPs	—	0.10	0
Ribonuclease A	9.45	0.12	0.34
Aldolase	8.2–8.6	0.30	1.00
Myosin	4.8–6.2	0.60	1.00
Serum albumin	4.9	>10	0.05
tRNA	≈2.5	>8.2	0
Polyadenylic acid	≈2.5	>10	0

[a] From Griffith and Pollard (1982a). Conditions: 0.5mM $MgCl_2$ (0.1 M methylethanesulfonate, pH 6.4, 1 mM EGTA, plus 0.5 M KCl for myosin only).

[b] The minimal concentration of the added macromolecule required to give 0.5 mg/ml actin an apparent viscosity of more than 12,000 cp in 20 min at 37°C. All samples with serum albumin, transfer RNA (tRNA), and polyadenylic acid had the same viscosity as actin alone.

[c] Samples of 1 ml containing 0.25 mg histone, 0.10 mg lysozyme, 0.90 mg microtubule-associated proteins (MAPS), 0.20 mg ribonuclease A, 0.50 mg aldolase, 0.5 mg myosin, 2 mg serum albumin, 2 mg tRNA, or 2 mg polyadenylic acid with or without 1 mg actin were incubated 30 min at 37°C and then centrifuged at $100,000g$ for 1 hr at 25°C. The fraction of the molecule that pelleted with actin was determined by quantitative gel electrophoresis or UV absorption.

that inhibits the low-shear viscosity of actin by 90% does not alter the length distribution of actin filaments observed by electron microscopy (Isenberg *et al.*, 1980).

Although our work cannot rule out the mechanism proposed by Hartwig and Stossel (1979), we think that it is important to consider alternate mechanisms. The one that we favor at present is based on the fact that both cytochalasin and capping protein bind to the barbed end of the filament, and both inhibit subunit addition there at the same concentrations that inhibit the low-shear viscosity. This suggests that a common binding site at the barbed end of the filament is responsible for the effects on both polymerization and network formation. Moreover, there is only one or a few molecules of either cytochalasin or capping protein bound to each filament in the samples with reduced low-shear viscosity, so the mechanism must focus on the barbed end of the filament. We suggested (MacLean-Fletcher and Pollard, 1980b; Isenberg *et al.*, 1980) that actin-filament self-association is due, in part, to the binding of the barbed end of filaments to the side of other actin filaments. It is proposed that the formation of these T-junctions is inhibited by either cytochalasin or capping protein binding to the barbed end of the actin filaments. The elimination of this class of actin-filament self-association would increase the concentration of protein cross-links necessary to form a gel.

If this mechanism is true, one of the functions of capping protein may be to inhibit the self-association of the actin filaments in the cell. There would be at least two consequences: (1) the viscosity attributable to actin alone would not be as high as predicted for pure actin at the low shear rates found in cells

(Maruyama *et al.*, 1974) and (2) the formation of cytoplasmic actin gels would then be determined entirely by cross-linking proteins. Both would appear to offer advantages for the cell.

7. Structure of Actin-Filament Gels

We have used both polarized-light microscopy and electron microscopy to study the structure of these actin gels. The gels are optically isotropic (Fig. 18a), showing that the constituent filaments are randomly arranged. This is the same impression given by thin sections (Fig. 18c) and shadowed specimens prepared by rapid freezing and etching (Fig. 18d). The most remarkable feature of the electron micrographs is the absence of any organized structure. In fact, it is impossible to distinguish shadowed specimens of pure actin filaments (apparent viscosity ≈150 cp) from actin filaments in cytochalasin (<10 cp) or a gel of actin and GP-85 (>12,000 cp). There are many side-to-side contacts in all these specimens, but they must not contribute substantially to the viscosity. In the presence of cross-linkers, there must be a few, widely spaced bonds between the filaments that prohibit them from sliding past each other. This is obvious from the high viscosity and can be demonstrated visually by polarization microscopy. When actin filaments are stirred with a microneedle, the flow birefringence dissipates in a few seconds, but the birefringence induced in a gel is stable (Fig. 18b). These properties of reconstituted gels are entirely consistent with the optical properties of the gelled hyaline ectoplasm of living cells (Allen *et al.*, 1965) and with the random arrangement of actin filaments in the cortex of fixed, thin-sectioned cells (Ishikawa *et al.*, 1969; Pollard *et al.*, 1970) and in quick-frozen, shadowed cells (Heuser and Kirschner, 1980).

8. Actin–Microtubule Interactions

Actin filaments can also form gels by virtue of weak interactions with the MAPs found on the surface of the microtubules (Fig. 19) (Griffith and Pollard, 1978). Microtubules without MAPs do not form a high-viscosity complex (Fig. 19). The actin–microtubule gels consist of a random, three-dimensional network of the two fibers, with a number of actin filaments in parallel association with the microtubules over short distances (Fig. 20). The mechanism of this binding has been studied by isolating the MAPs and attempting to identify which of the several proteins are responsible for actin interaction. We found that a subset of the high-molecular-weight MAPs (Molecular weight ≈250,000) and two tau (molecular weight 60,000–70,000) fractions could cross-link actin; however, we have not yet succeeded in using either fraction alone to reconstitute an actin–microtubule gel with pure tubulin (Griffith and Pollard, 1982b).

These actin–microtubule interactions could contribute to the structure of

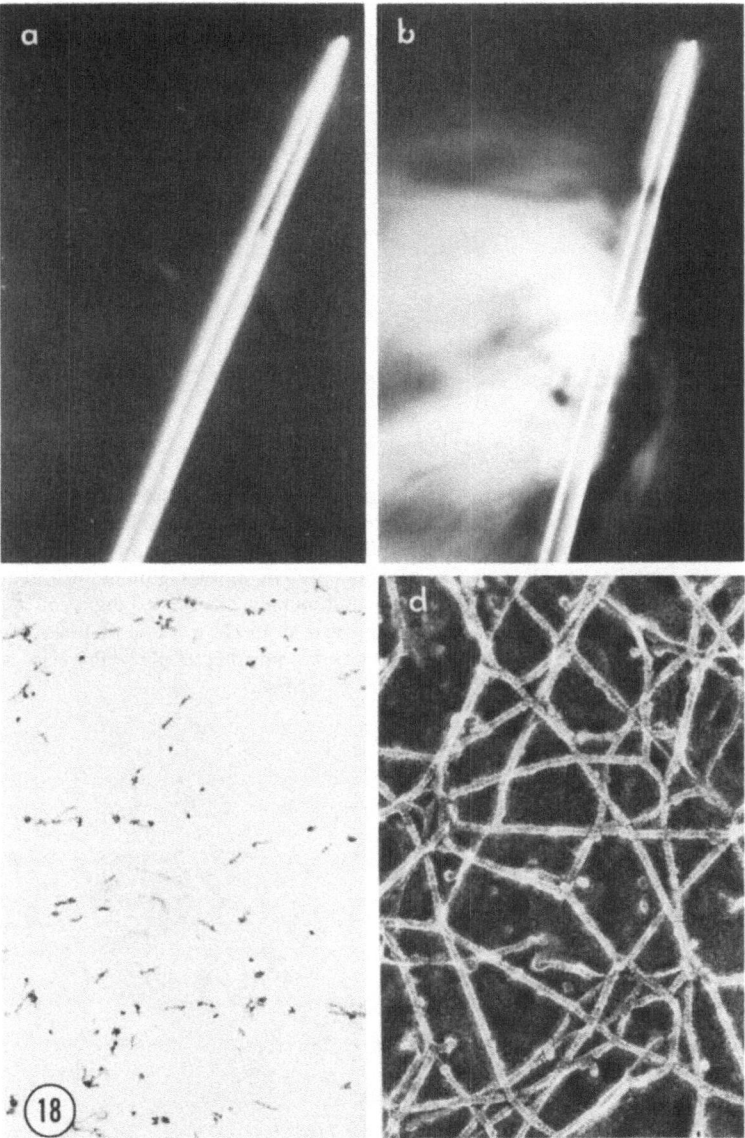

Figure 18. Structure of actin gels. (a,b) Polarized-light microscopy of a mixture of 1 mg/ml of actin with 90 μg/ml of *Acanthamoeba* GP-85. The gel is optically isotropic (a), but stable birefringence can be induced (b) by moving the microneedle through the gel. (c,d) Electron micrographs of actin gels. (c) The specimen contained gelation factor-33 (see Fig. 17) and was prepared by tannic acid–glutaraldehyde fixation, embedded, and sectioned. ×40,000. (d) The specimen contained GP-85 and was prepared by rapid freezing, etching, and rotary shadowing with platinum. The exposed filament network is seen above the dark ice table. ×122,000. (a,b) Work of D. Kiehart. (c) Work of P. Maupin. (d) Work of J. Heuser.

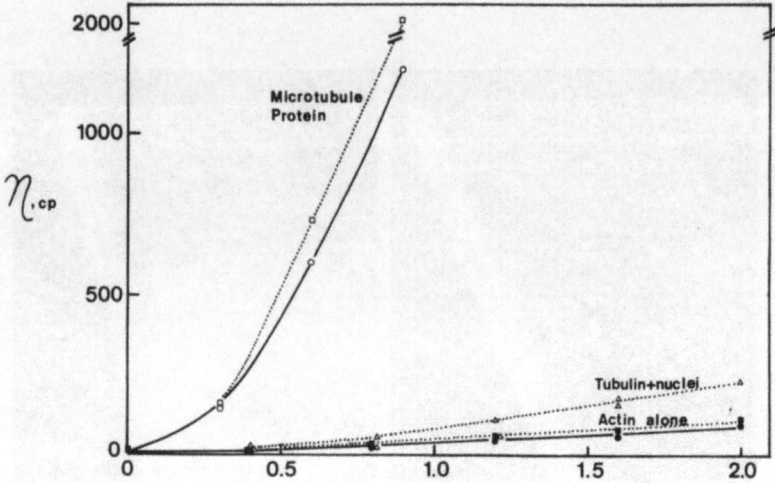

Figure 19. Comparison of the low-shear apparent viscosity (η, cp) of mixtures of actin polymerized with microtubule protein or nucleated purified tubulin. Solutions added to actin had viscosities of 15 cp. The concentrations of protein added to actin were 3.2 mg/ml of microtubule protein in 10% DMSO, or 5.0 mg/ml of microtubule protein, or 1.9 mg/ml of nucleated purified tubulin with 10% DMSO. Each sample of nucleated tubulin contained 0.3 mg/ml of fragmented microtubules. Nuclei constituted 12% of total protein at 15 cp. (O) Microtubule protein with actin; (□) microtubule protein with actin and DMSO; (△) purified tubulin with nuclei, actin, and DMSO; (●) actin; (■) actin with DMSO. Work of L. Griffith.

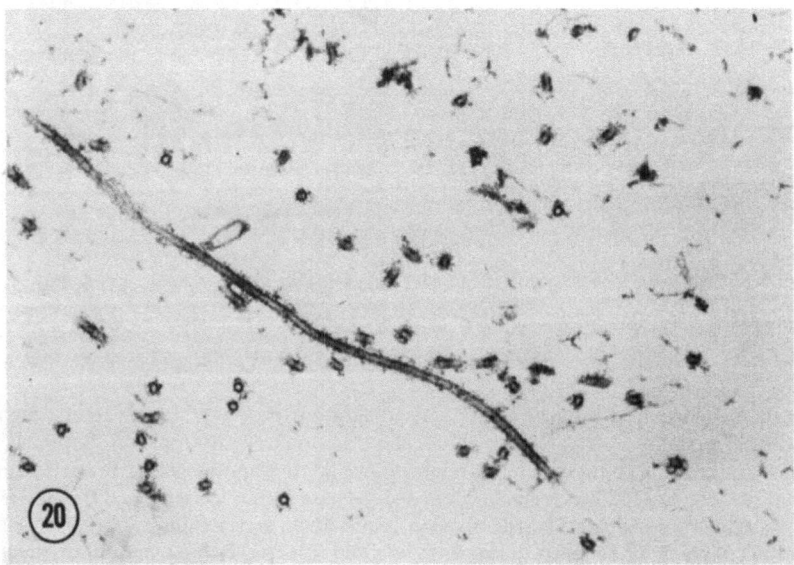

Figure 20. Electron micrograph of a thin section of an actin filament–microtubule gel fixed *in situ* with glutaraldehyde and tannic acid. ×48,000. Work of P. Maupin and L. Griffith.

the cytoplasm, but it is attractive to consider that they may also be essential for some microtubule-dependent movements powered by contractile proteins. There is no direct evidence that such a class of movements exists, but the continued difficulty in isolating an energy-transducing enzyme associated with cytoplasmic microtubules forces us to consider such a mechanism for fast axonal transport and other saltatory movements. Although powered by contractile proteins, they could be dependent on microtubules, just as muscles are dependent on bones for support and transmission of forces.

9. Perspectives

It is hoped that the work for the immediate future will provide further insight into the mechanisms of actin-filament assembly and cross-linking. Improved chromatography methods should make it possible to isolate the *Acanthamoeba* actin peptides that have not been sequenced, so that the primary structure can be completed.

The overall shape of the *Acanthamoeba* actin molecule will soon be known from the three-dimensional image recontruction of the crystalline sheets. The unpredictable part of this work is the resolution limit of state-of-the-art electron microscopy and image-processing techniques. We already know that radiation damage is a major limiting factor in recording structural information from negatively stained sheets. We are optimistic that reducing the electron dose further, using unstained specimens, and developing specific heavy metal or antibody labels will allow us to push the resolution below 1.0 nm and to trace parts of the polypeptide chain. A high-resolution structure of the actin filament will make it possible to orient the model of the actin molecule in the filament.

To better understand the mechanism of actin polymerization, we intend to measure all the rate constants in Fig. 10 under a variety of conditions. This work has been limited to some extent by difficulty in measuring polymerization rates at the slow end of the filament in the face of rapid polymerization at the fast end. The availability of capping protein to block the fast end may solve this problem. This work should also provide further insight into the mechanism of action of molecules that modify actin polymerization, such as cytochalasin, phalloidin, and profilin.

Three major questions regarding actin gelation in *Acanthamoeba* are the nature of actin-filament self-association, the mechanisms of the cross-linking proteins, and the relative physiological importance of the various molecules that might control this process in the living cell. Answers will come from further analysis of the biochemistry and structure of the gels, but events in the living cell will have to be probed there. We hope that microinjection of specific inhibitors will be informative.

ACKNOWLEDGMENTS. This work was supported by research grants from the NIH to Drs. Pollard, Aebi, Elzinga, Heuser, Mooseker, and Smith and from the Muscular Dystrophy Association of America to Drs. Aebi, Herman, Heuser, and Kiehart.

References

Aebi, U., Isenberg, G., Pollard, T. D., and Smith, P. R., 1980, Structure of crystalline actin sheets, *Nature (London)* **288:**296.

Allen, R. D., Francis, D. W., and Nakajima, H., 1965, Cyclic birefringence changes in pseudopods of *Chaos carolinensis* revealing the localization of the motive force in pseudopod extension, *Proc. Natl. Acad. Sci. U.S.A.* **54:**1153.

Bergen, L. G., and Borisy, G. G., 1980, Head-to-tail polymerization of microtubules *in vitro:* Electron microscope analysis of seeded assembly, *J. Cell Biol.* **84:**141.

Brenner, S. L., and Korn, E. D., 1979, Substoichiometric concentrations of cytochalasin D inhibit actin polymerization, *J. Biol. Chem.* **254:**9982.

Brenner, S. L., and Korn, E. D., 1980, The effects of cytochalasins on actin polymerization and actin ATPase provide insights into the mechanism of polymerization, *J. Biol. Chem.* **255:**841.

Brotschi, E. A., Hartwig, J. H., and Stossel, T. P., 1978, Gelation of actin by actin-binding protein, *J. Biol. Chem.* **253:**8988.

Brown, S. S., and Spudich, J. A., 1979, Cytochalasin inhibits the rate of elongation of actin filament fragments, *J. Cell Biol.* **83:**657.

Bryan, J., and Kane, R. E., 1978, Separation and interaction of major components of sea-urchin actin gel, *J. Mol. Biol.* **125:**207.

Carlsson, L., Nystrom, L. E., Sundkvisk, I., Markey, F., and Lindberg, U., 1977, Actin polymerizability is influenced by profilin, a low molecular weight protein in non-muscle cells, *J. Mol. Biol.* **115:**465.

Carter, S. B., 1967, Effects of cytochalasins on mammalian cells, *Nature (London)* **213:**261.

Clark, F. M., and Masters, C. J., 1975, On the association of glycolytic enzymes with structural proteins of skeletal muscle, *Biochim. Biophys. Acta* **381:**37.

Collins, J. H., and Elzinga, M., 1975, The primary structure of actin from rabbit skeletal muscle: Completion and analysis of the amino acid sequence, *J. Biol. Chem.* **250:**5915.

Condeelis, J. S., and Taylor, D. L., 1977, Contractile basis of ameboid movement. 5. Control of gelation, solation, and contraction in extracts from *Dictyosteliumdiscoideum, J. Cell Biol.* **74:**901.

DosRemedios, C. G., and Dickens, M. J., 1978, Actin microcrystals and tubes formed in presence of gadolinium ions, *Nature (London)* **276:**731.

Dujardin, F., 1835, Recherches sur les organismes inférieurs, *Ann. Sci. Nat. Zool.* **4:**343.

Flory, P. J., 1953, *Principles of Polymer Chemistry,* Cornell University Press, New York.

Gordon, D. S., Boyer, J. L., and Korn, E. D., 1977, Comparative biochemistry of non-muscle actins, *J. Biol. Chem.* **252:**8300.

Griffith, L. M., and Pollard, T. D., 1978, Evidence for actin filament–microtubule interaction mediated by microtubule-associated proteins, *J. Cell Biol.* **78:**958.

Griffith, L. M., and Pollard, T. D., 1982a, Crosslinking of actin filament networks by self-association and actin binding macromolecules, *J. Biol. Chem.* (submitted).

Griffith, L. M., and Pollard, T. D., 1982b, The interaction of actin filaments with microtubules and microtubule associated proteins, *J. Biol. Chem.* (submitted).

Grumet, M., and Lin, S., 1980a, Reversal of profilin inhibition of actin polymerization *in vitro* by erythrocyte cytochalasin-binding complexes and cross-linked actin nuclei, *Biochem. Biophys. Res. Commun.* **92:**1324.

Grumet, M., and Lin, S., 1980b, A platelet inhibitor protein with cytochalasin-like activity against actin polymerization *in vitro, Cell* **21:**439.

Hartwig, J. H., and Stossel, T. P., 1979, Cytochalasin B and the structure of actin gels, *J. Mol. Biol.* **134:**539.

Herman, I. M., and Pollard, T. D., 1979, Comparison of purified anti-actin and fluorescent-heavy meromyosin staining patterns in dividing cells, *J. Cell Biol.* **80:**509.

Herman, I., Crisona, N., and Pollard, T. D., 1981, Relation between cell activity and the distribution of cytoplasmic actin and myosin, *J. Cell Biol.* **90**:84.

Heuser, J. E., and Kirschner, M. W., 1980, Filament organization revealed in platinum replicas of freeze–dried cytoskeletons, *J. Cell Biol.* **86**:212.

Huxley, H. E., 1963, Electron microscopic studies on the structure of natural and synthetic protein filaments from striated muscle, *J. Mol. Biol.* **7**:281.

Isenberg, G., Aebi, U., and Pollard, T. D., 1980, A novel actin binding protein from *Acanthamoeba* which regulates actin filament polymerization and interactions, *Nature (London)* **288**:455.

Ishikawa, H., Bischoff, R., and Holtzer, H., 1969, Formation of arrowhead complexes with heavy meromyosin in a variety of cell types, *J. Cell Biol.* **43**:312.

Lazarides, E., and Weber, K., 1974, Actin antibody: The specific visualization of actin filaments in non-muscle cells, *Proc. Natl. Acad. Sci. U.S.A.* **71**:2268.

Lin, D. C. and Lin, S., 1979, Actin polymerization induced by a motility-related high affinity cytochalasin binding complex from human erythrocyte membrane, *Proc. Natl. Acad. Sci. U.S.A.* **76**:2345.

Lin, D. C., Tobin, K. D., Grumet, M., and Lin, S., 1980, Cytochalasins inhibit nuclei-induced actin polymerization by blocking filament elongation, *J. Cell Biol.* **84**:455.

Lu, R., and Elzinga, M., 1976, Comparison of amino acid sequences of actins from bovine brain and muscles, in: *Cell Motility* (R. Goldman, T. Pollard, and J. Rosenbaum, eds.), pp. 487–492, Cold Spring Harbor Laboratory, New York.

MacLean-Fletcher, S., and Pollard, T. D., 1980a, Viscometric analysis of the gelation of *Acanthamoeba* extracts and purification of two gelation factors, *J. Cell Biol.* **85**:414.

MacLean-Fletcher, S., and Pollard, T. D., 1980b, Mechanism of action of cytochalasin B on actin, *Cell* **20**:329.

MacLean-Fletcher, S., and Pollard, T. D., 1980c, Identification of a factor in conventional muscle actin preparation which inhibits actin filament self-association, *Biochem. Biophys. Res. Commun.* **96**:18.

Maruta, H., and Korn, E. D., 1977, Purification of *Acanthamoeba castellanii* of proteins that induce gelation and syneresis of F-actin, *J. Biol. Chem.* **252**:399.

Maruyama, K., Kaibara, M., and Fukada, E., 1974, Rheology of F-actin. I. Network of F-actin in solution, *Biochim. Biophys. Acta* **271**:20.

Mast, S. O., 1926, Structure, movement, locomotion and stimulation of amoeba, *J. Morphol. Physiol.* **41**:347.

Mimura, N., and Asano, A., 1980, Ca^{2+}-sensitive gelation of actin-filaments by a new protein factor, *Nature (London)* **282**:44.

Moore, P. B., Huxley, H. E., and DeRosier, D. J., 1970, Three-dimensional reconstruction of F-actin, thin filaments and decorated thin filaments, *J. Mol. Biol.* **50**:279.

Oosawa, F., and Asakura, S., 1975, *Thermodynamics of the Polymerization of Protein*, Academic Press, New York.

Pollard, T. D., 1976, The role of actin in the temperature dependent gelation and contraction of extracts of *Acanthamoeba*, *J. Cell Biol.* **68**:579.

Pollard, T. D., 1981, Purification of a calcium-sensitive actin gelation protein from *Acanthamoeba*, *J. Biol. Chem.* **256**:7666.

Pollard, T. D., and Ito, S., 1970, Cytoplasmic filaments of *Amoeba proteus*. I. The role of filaments in consistency changes and movement, *J. Cell Biol.* **46**:267.

Pollard, T. D., and Mooseker, M. S., 1981, Direct measurement of actin polymerization rate constants by electron microscopy of actin filaments nucleated by isolated microvillus cores, *J. Cell Biol.* **88**:654.

Pollard, T. D., Shelton, E., Weihing, R. R., and Korn, E. D., 1970, Ultrastructural characterization of F-actin isolated from *Acanthamoeba castellanii* and identification of cytoplasmic filaments as F-actin by reaction with rabbit muscle heavy meromyosin, *J. Mol. Biol.* **50**:91.

Reichstein, E., and Korn, E. D., 1979, *Acanthamoeba* profilin—protein of low-molecular weight from *Acanthamoeba castellanii* that inhibits actin nucleation, *J. Biol. Chem.* **254**:6174.

Simpson, P. A., and Spudich, J. A., 1980, ATP-driven steady-state exchange of monomeric and filamentous actin from *Dictyostelium discoideum*, *Proc. Natl. Acad. Sci. U.S.A.* **77**:4610.

Taylor, D. L., and Condeelis, J. S., 1979, Cytoplasmic structure and contractility in ameboid cells, *Int. Rev. Cytol.* **56:**57.

Vanderkerckhove, J., and Weber, K., 1978, Amino-acid sequence of *Physarum* actin, *Nature (London)* **276:**720.

Wang, K., and Singer, S. J., 1977, Interaction of filamin with F-actin in solution, *Proc. Natl. Acad. Sci. U.S.A.* **74:**2021.

Wegner, A., 1976, Head to tail polymerization of actin, *J. Mol. Biol.* **109:**139.

Woodrum, D. T., Rich, S. A., and Pollard, T. D., 1975, Evidence for the biased bidirectional polymerization of actin using heavy meromyosin produced by an improved method, *J. Cell Biol.* **67:**231.

Yin, H. L., and Stossel, T. P., 1979, Control of cytoplasmic actin gel-sol transformation by gelsolin, a calcium-dependent regulatory protein, *Nature (London)* **281:**583.

3

A Scanning and Transmission Electron Microscope Examination of ACTH-Induced 'Rounding up' in Triton X-100 Cytoskeleton Residues of Cultured Adrenal Cells

J. Mrotek, W. Rainey, T. Sawada, R. Lynch, M. Mattson, and I. Lacko

1. Introduction

Several investigators have used cultured Y-1 mouse adrenal tumor cells as a model to study the action of the pituitary hormone adrenocorticotropin (ACTH) in controlling glucocorticoid synthesis (Kowal, 1970; Mrotek and Hall, 1975, 1977, 1978).

In most steroid-producing cells, the enzymes of the steroidogenic pathway have been identified in the mitochondrial and microsomal subcellular fractions (Schulster *et al.*, 1976). In addition, the steroid intermediates of this pathway have also been identified in these fractions (Kowal, 1970; Tait and Tait, 1979). Although ACTH is known to accelerate the production of mitochondrial steroid from cholesterol, (Hall and Koritz, 1964, 1965; Karaboyas and Koritz, 1965) the ACTH-controlled intracellular mechanisms by which cholesterol conversion takes place are not known.

When cytochalasin B, an inhibitor of microfilament assembly (MacLean *et al.*, 1978; Lin *et al.*, 1980) (also see Chapter 2), and ACTH were added to Y-1

J. Mrotek, W. Rainey, T. Sawada, R. Lynch, M. Mattson, and I. Lacko • The Department of Biological Sciences and The Genetics and Aging Centers, North Texas State University, Denton, Texas 76203.

adrenal tumor cells, the normal ACTH-related increase in mitochondrial cholesterol did not occur (Mrotek and Hall, 1975, 1977), suggesting that the amount of cholesterol available to the mitochondria is the rate-limiting step in steroid synthesis. It has also been suggested that microfilaments are responsible for transporting cholesterol to the mitochondria in response to ACTH stimulation. Several other investigators, using isolated rat adrenal cells (Crivello and Jefcoate, 1978), and ovine corpora lutea (Silavin, 1979) confirmed the observation that cytochalasin B inhibited steroid synthesis.

In addition, the following indirect evidence also suggests that Y-1 adrenal cell microfilaments may undergo changes in the response to ACTH stimulation. When cultured Y-1 adrenal tumor cells were incubated with either ACTH or cyclic 3′,5′-adenosine monophosphate (cAMP), steroid-hormone production increased simultaneously with the transformation of the cell from a flattened to a rounded state (Yasumura *et al.,* 1966; Masui and Garren, 1971; Temple and Wolf, 1973; Cuprak *et al.,* 1977). Nonadrenal cells that round up during mitosis (Wessels *et al.,* 1971) or following exposure to cAMP (Bell *et al.,* 1978) exhibit changes in the microfilaments of the cytoskeleton. The Y-1 cell rounding-up during the ACTH or cAMP response may therefore also involve microfilaments.

In the study reported in this chapter, we examined the cytoskeleton of Triton X-100 extracted Y-1 mouse adrenal tumor cells to determine whether changes occur in the cytoskeleton during ACTH stimulation or cytochalasin D inhibition of steroidogenesis.

2. Materials and Methods

2.1. Cells

Y-1 mouse adrenal tumor cells obtained from the American Type Culture Collection (Rockville, Maryland) were maintained as described previously (Mrotek and Hall, 1975, 1977, 1978). The cells used in these experiments were from the 51st population doubling.

2.2. Methods

Cells were trypsinized and added to six-well plates (Costar) containing sterile 100-mesh Formvar- and carbon-coated gold grids (Ernest Fullum) 24 hr before the start of the experiment. The cells were maintained in serum-containing culture medium; this medium was removed when the experiment began. To remove both serum proteins and steroid hormones adhering to the wells (Mrotek and Carraway, 1978, unpublished observations), each well was washed twice with serum-free medium and incubated for 30 min in 2 ml serum-free medium. After the medium was discarded, the preceding wash and incubation sequence was repeated; then the incubation medium was discarded. Following a third wash, cells were incubated for 4 hr in 2 ml serum-free medium to which ACTH, cytochalasin D, or ACTH–cytochalasin D had been added. The final concentration of these chemicals in the medium was 0.5 U ACTH/ml medium and 10^{-5} M cytochalasin D.

When the 4-hr incubation was completed, the medium was withdrawn and assayed for steroids by radioimmunoassay (Mrotek and Hall, 1977) (data not shown); then the cells were washed twice and extracted by the method of Webster *et al.* (1978). This method was modified as follows: 1% Triton X-100 was used because of the thickness of the cells; the solubilizing solutions did not contain polyethylene glycol or GTP, but 1 mM Ethyleneglycol-bis (β–amino-ethyl ether)N,N'-tetraacetic acid and 1 mM MgSO$_4$ were added instead. Cells were examined with a JEOL 1000 electron microscope (University of Colorado) and by scanning electron microscopy.

3. Results and Discussion

Control cells exhibited either a flattened spindle-shaped morphology (not shown) or a flattened polymorphic shape similar to that seen in Fig. 1. Similar morphologies have been observed by others (Yasumura *et al.,* 1966; Masui and Garren, 1971; Temple and Wolf, 1973; Cuprak *et al.,* 1977). Figure 2 is a high-voltage electron micrograph of the cytoskeleton in the nuclear area of a control cell; micrographs of other areas (not shown) were also studied. The cytoskeleton consisted of thin and intermediate filaments and microtubules; these were associated in three different patterns:

1. Cablelike bundles of microtubules and intermediate filaments that also contained a few thin filaments radiated from the nuclear area toward

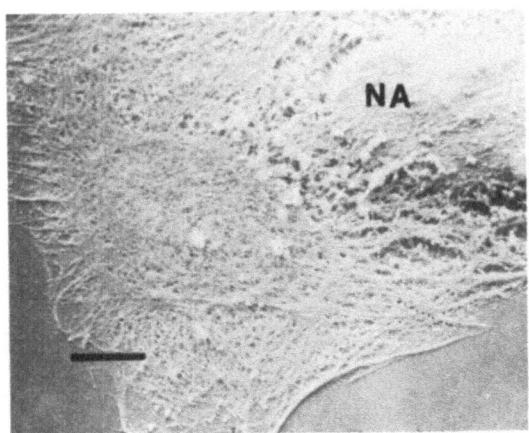

Figure 1. Scanning electron micrograph of a control Y-1 mouse adrenal tumor cell. (NA) Nuclear area. ×2400.

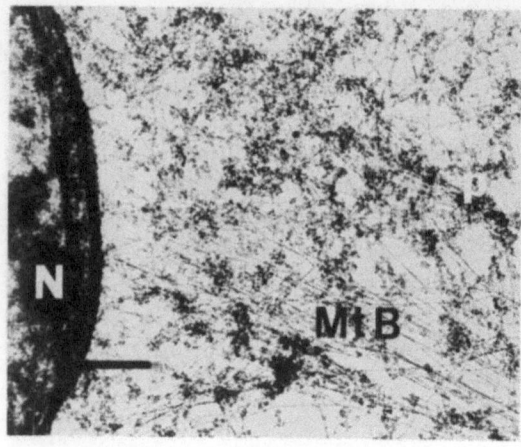

Figure 2. Transmission electron micrograph of a nuclear area of a control adrenal cell. (N) Nucleus; (p) polyribosomes; (MtB) microtubule and intermediate filament cable, which contains some thin filaments. ×13,500).

tips of pseudopodia at the cell margin. Similar cablelike arrangements occur in other cells (Pollack and Rifkin, 1975).

2. Between the radiating cables, a random latticework was observed that consisted of mostly thin filaments. Connected to this latticework were numerous granules resembling polyribosomes; these granules are similar to those first reported by Lenk *et al.* (1977).

3. Associated with the periphery of the cell were bundles of thin filaments resembling those described in cultured fibroblasts (Goldman and Knipe, 1972).

The nuclei of control cells exhibited an uncondensed, relatively translucent, appearance.

When ACTH-treated cells round up, they exhibit a stellate appearance due to filopodialike strands radiating from the top of the central spherical, nuclear area (Fig. 3). This morphological response to stimulation has been observed by others (Yasumura *et al.*, 1966; Masui and Garren, 1971; Temple

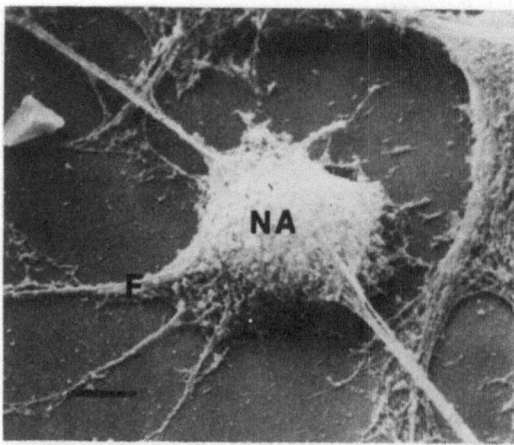

Figure 3. Scanning electron micrograph of an ACTH-treated adrenal cell. (F) Filopodia. ×2400.

and Wolf 1973; Cuprak *et al.*, 1977). Transmission electron micrographs from the nuclear area (Fig. 4) and filopodial regions (not shown) show that microtubules, together with a few intermediate filaments, were the predominant cytoskeletal component of the filopodial strands. Clark and Shay (1980a) also reported that the major cytoskeletal components of filopodia were microtubules. A large number of thin filaments and granules resembling polyribosomes surround the nucleus in ACTH-treated cells. We believe the filaments and ribosomes are derived from the filamentous latticework that connected the cables in control cells. In addition, in ACTH-treated cells, the bundles of thin filaments on the periphery of the control cells diminish. Peripheral thin-filament bundles in fibroblasts also diminish when these cells round up (Goldman and Knipe, 1972). The nuclei of ACTH-treated cells were slightly smaller and more electron-dense than control nuclei.

The morphology and cytoskeletal arrangement of cells treated with either cytochalasin D or ACTH–cytochalasin D were similar. A scanning electron micrograph (Fig. 5) illustrates cytochalasin-treated cells. The stellate morphology observed when cells received either cytochalasin treatment differed in several ways from the morphology of cells incubated with ACTH-containing medium:

1. The nuclear area was elevated above the filopodia following incubation with cytochalasin D; several investigators have used this cytochalasin-related phenomenon as a means of enucleating cells (Chen and Auersperg, 1974; Veomett *et al.*, 1974; Clark and Shay, 1980).
2. Wider filopodial strands were observed in cytochalasin-treated cells.
3. Compared to the ACTH-treated cells, the fibers in the cables of the filopodial strands were more loosely associated when the microfilament inhibitor was included in the incubation medium.
4. The entire cytoskeleton of cytochalasin-treated cells appeared to be composed of coarser fibers.

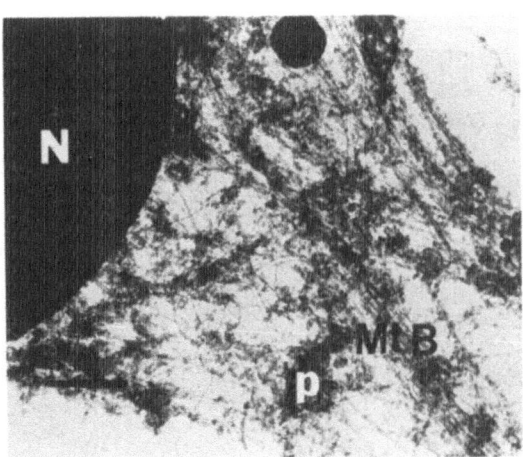

Figure 4. Transmission electron micrograph of an ACTH-treated adrenal cell. Note the condensation of the nucleus and the accumulation of polyribosomes and thin filaments in the perinuclear area. ×13,500.

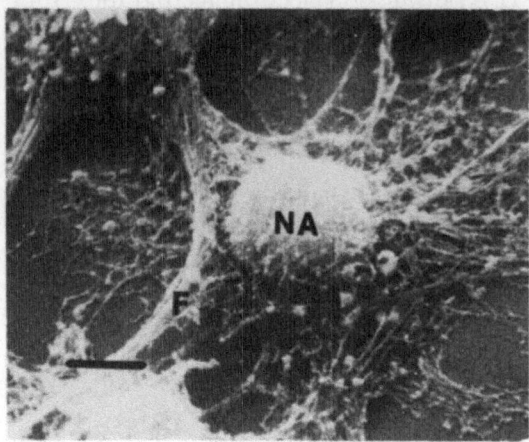

Figure 5. Scanning electron micrograph of a cytochalasin-D-treated adrenal cell. Note the broader filopodia, the coarser cytoskeleton, and the elevated nuclear area. ×2400.

Figure 6 is a transmission electron micrograph of a cytoskeleton from a cell incubated with ACTH–cytochalasin; like cytoskeletons from cells treated with cytochalasin alone (not shown), these Triton-solubilized cells contained few thin filaments and polyribosomes. As with ACTH alone, a large number of microtubules, together with a few intermediate filaments, form the core of the filopodial strands. Juxtanuclear condensations of intermediate filaments were observed after either of the cytochalasin treatments; while the condensation of the nuclear area was intermediate between that of control and ACTH-treated cells.

This study demonstrates that the Y-1 mouse adrenal tumor cell contains a microfilamentous and microtubular cytoskeleton. Changes in thin-filament distribution were observed in the ACTH-stimulated cell, and increased steroidogenesis accompanied this morphological change (data not shown). On the other hand, cells incubated with the microfilament inhibitor cyto-

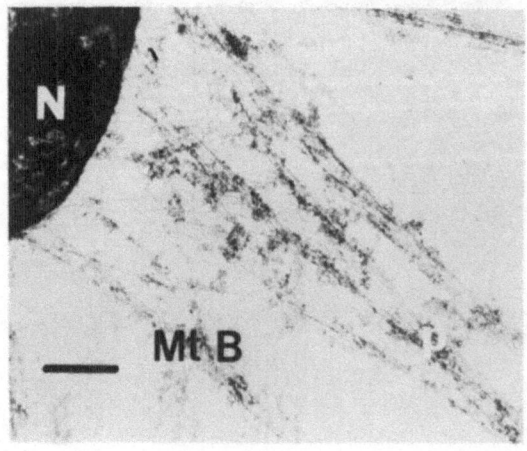

Figure 6. Transmission electron micrograph of an ACTH–cytochalasin-D-treated adrenal cell. Note the reduction in the number of thin filaments and polyribosomes; the nucleus is less condensed than in Fig. 4. ×13,500.

chalasin D (MacLean *et al.*, 1978; Lin *et al.*, 1980) (also see Chapter 2) alone, or in combination with ACTH, contained fewer thin filaments. In addition, ACTH-stimulated steroid production was reduced by at least two thirds in the presence of cytochalasin D (data not shown). A similar result was obtained when ACTH or cAMP was added to Y-1 cells with various concentrations of cytochalasin B (10^{-8} to 10^{-4} M) (Mrotek and Hall, 1975).

In the present study, the involvement of thin filaments in steroidogenesis was suggested by the observations that cytochalasin D reduced ACTH-stimulated steroid production, and that treatment with ACTH, or cytochalasin D, caused changes in thin filaments. We have previously reported that microfilament inhibitors prevented the ACTH-stimulated accumulation of cholesterol by adrenal-cell mitochondria but have no effect on steroid secretion (Mrotek and Hall, 1977). Since cytochalasin D prevents the formation of microfilaments (Lin *et al.*, 1980) the observations from this study, together with our previous observations, suggest that ACTH-stimulated cholesterol accumulation by the adrenal-cell mitochondria may depend on the ability of ACTH to cause formation of microfilaments.

ACKNOWLEDGMENTS. The authors gratefully acknowledge the use of the JEOL 1000, located at the University of Colorado at Boulder. We are indebted to Karen Mrotek for her editing and typing of this manuscript. Mark Gonzales, Bea Pena, and Joanne Youngblood provided excellent assistance. Portions of this work were presented during the 1980 FASEB meeting, Anaheim, California. This work was supported in part by North Texas State University Faculty Research Grant 34700 and NIA Grant AG01055-02A1.

References

Bell, P. B., Miller, M. M., Carraway, K. L., and Revel, J. P., 1978, SEM-revealed changes in the distribution of the Triton-insoluble cytoskeleton on Chinese hamster ovary cells induced by dibutyryl cyclic AMP, *Scanning Electron Microsc.* **2**:899.

Chen, L. M., and Auersperg, N., 1974, Response to ACTH by cultured adrenocortical tumor cells following enucleation with cytochalasin B, *J. Cell Biol.* **63**[2 (Part 2)]:57a (abstract).

Clark, M. A., and Shay, J. W., 1980a, The role of tubulin in steroidogenesis, *J. Cell Biol.* **87**[2 (Part 2)]:168a (abstract).

Clark, M. A., and Shay, J. W., 1980b, A method for producing mitochondrial chimeras, *J. Cell Biol.* **87**[2 (Part 2)]:293a (abstract).

Crivello, J. F., and Jefcoate, C. R., 1978, Mechanism of corticotropin action on rat adrenal cells. I. The effects of inhibitors of protein synthesis and of microfilament formation on corticosterone synthesis, *Biochim. Biophys. Acta* **542**:31.

Cuprak, L. J., Lammi, C. J., and Bayer, R. C., 1977, Scanning electron microscopy of induced cell rounding of mouse adrenal tumor cells in culture, *Tissue Cell* **9**:667.

Goldman, R. D., Knipe, D. M., 1972, Functions of cytoplasmic fibers in nonmuscle cell motility, *Cold Spring Harbor Symp. Quant. Biol.* **37**:523–524.

Hall, P. F., and Koritz, S. B., 1964, Influence of calcium ions and freezing upon the conversion of cholesterol-^{3}H to corticosterone-^{3}H by homogenates of rat adrenal, *Endocrinology* **75**:135.

Hall, P. F., and Koritz, S. B., 1965, Influence of interstitial cell-stimulating hormone on the conversion of cholesterol to progesterone by bovine corpus luteum, *Biochemistry* **4**:1037.

Karaboyas, G. C., and Koritz, S. B., 1965, Identity of the site of action of 3',5'-adenosine monophosphate and adrenocorticotropic hormone in beef adrenal cortex slices, *Biochemistry* **4:**462.

Kowal, J., 1970, ACTH and the metabolism of adrenal cell cultures, *Recent Prog. Horm. Res.* **26:**623.

Lenk, R., Ransom, L., Kaufmann, Y., and Penman, S., 1977, A cytoskeletal structure with associated polyribosomes obtained from HeLa cells, *Cell* **10:**67.

Lin, D. C., Tobin, K. D., Grumet, M., and Lin, S., 1980, Cytochalasins inhibit nuclei-induced actin polymerization by blocking filament elongation, *J. Cell Biol.* **84:**455.

MacLean, S., Griffith, L. M., and Pollard, T. D., 1978, A direct effect of cytochalasin-B upon actin filaments, *J. Cell Biol.* **79**[2 (Part 2)]**:**267a (abstract).

Mrotek, J. J., and Hall, P. F., 1975, The influence of cytochalasin B on the response of adrenal tumor cells to ACTH and cyclic AMP, *Biochem. Biophys. Res. Commun.* **64:**891.

Mrotek, J. J., and Hall, P. F., 1977, Response of adrenal tumor cells to adrenocorticotropin: Site of inhibition by cytochalasin B, *Biochemistry* **16:**3180.

Mrotek, J. J., and Hall, P. F., 1978, The action of ACTH on adrenal tumor cells is not inhibited by anti-tubular agents, *Gen. Pharmacol.* **9:**269.

Masui, H., and Garren, L. D., 1971, Inhibition of replication in functional mouse adrenal tumor cells by adrenocorticotropic hormone mediated by adenosine 3',5'-cyclic monophosphate, *Proc. Natl. Acad. Sci. U.S.A.* **68:**3206.

Pollack, R., and Rifkin, D., 1975, Actin containing cables within anchorage-dependent rat embryo cells are dissociated by plasmin and trypsin, *Cell* **6:**495.

Schulster, D., Burstein, S., and Cooke, B. A., 1976, The adrenal cortex, in *Molecular Endocrinology of Steroid Hormones*, pp. 53–57, John Wiley, New York.

Silavin, S. L., 1979, The role of microfilaments in luteal steroidogenesis, *Biol. Reprod.* **20**(Suppl. 1)**:**67a (abstract).

Tait, J. F., and Tait, S. A. S., 1979, Recent perspectives on the history of the adrenal cortex, *J. Endocrinol.* **83:**1p.

Temple, R., and Wolf, J., 1973, Stimulation of steroid secretion by antimicrotubular agents, *J. Biol. Chem.* **248:**2691.

Veomett, G., Prescott, P. M., Shay, J., and Porter, K. W., 1974, Reconstruction of mammalian cells from nuclear and cytoplasmic components separated by treatment with cytochalasin B, *Proc. Natl. Acad. Sci. U.S.A.* **71:**1999.

Webster, R. E., Osborn, M., and Weber, K., 1978, Visualization of the same PtK2 cytoskeleton by both immunofluorescence and low power electron microscopy, *Exp. Cell Res.* **117:**47.

Wessels, N. K., Spooner, B. S., Ash, J. F., Bradley, M. O., Luduena, M. A., Taylor, E. L., Wrenn, J. T., and Yamada, K. M., 1971, Microfilaments in cellular and developmental processes, *Science* **171:**135.

Yasumura, Y., Buonassisi, V., and Sato, G., 1966, Clonal analysis of differentiated function in animal cell cultures. I. Possible correlated maintenance of differentiated function and the diploid karyotype, *Cancer Res.* **26:**529.

4

The Role of Tubulin in Steroidogenesis of Mouse Adrenal Y-1 Cells and Rat Leydig CCL 43 Cells

Mike A. Clark and Jerry W. Shay

1. Introduction

The mechanisms of adrenocorticotropic hormone (ACTH)-induced steroid secretion in the adrenal cortex have not been clearly elucidated. However, it has been demonstrated that ACTH acts in the regulation of the steroidogenic pathway by controlling the conversion of cholesterol into pregnenolone. Cellular-fractionation studies have demonstrated that the enzymes necessary for this conversion are located in the mitochondria, whereas the cholesterol is stored in the cytoplasm. Kowal (1970) demonstrated that all the enzymes necessary to synthesize steroids are present prior to ACTH stimulation, and he concluded that steroidogenesis was regulated by controlling the transport of cholesterol to the mitochondria. Temple and Wolff (1973) showed that colchicine, and other drugs that depolymerize microtubules, could induce steroidogenesis independent of ACTH, and from these data they hypothesized that microtubules prevented the transport of cholesterol to the mitochondria and that the mechanism of ACTH-induced steroidogenesis involved the removal of this restriction.

The experiments reported herein utilized the Y-1 cell line as a model system for studying the role of tubulin in the regulation of steroidogenesis. This cell line was originally derived from a murine adrenal tumor (Buonassi *et al.,* 1962) and has retained the ability to respond to ACTH and secrete steroids in cell culture. This report will present data obtained using indirect

Mike A. Clark and Jerry W. Shay • Department of Cell Biology, The University of Texas Health Science Center at Dallas, Dallas, Texas 75235.

immunofluorescence, cell fractionation, and electron microscopy, and our results indicate that tubulin, not microtubules, is involved in the sequestering of cholesterol in Y-1 cells and that steroidogenesis may be regulated by dissociating the tubulin from the cholesterol-containing inclusions, thus allowing the transport of cholesterol to the mitochondria. To test the generality of this observation in other steroidogenic tissues, we studied the rat Leydig cell line CCL 43 and have found, in preliminary experiments, that the role of tubulin in steroidogenesis in these cells is essentially the same as in Y-1 cells.

2. Materials and Methods

2.1. Cell Lines

The Y-1 cell lines used in these experiments were obtained from the American Type Culture Collection (ATCC) and Dr. Bernard Schimmer. These cells were maintained in Ham's F12 medium supplemented with 15% horse serum and 2.5% fetal calf serum (FCS). The Leydig cell line (CCL 43) was also obtained from the ATCC, maintained in Dulbecco's medium plus 10% FCS. All cells were grown antibiotic-free and were found by two different assays (Tai and Quinn, 1977; Schneider *et al.*, 1974) not to be mycoplasma-contaminated.

2.2. Antibodies and Indirect Immunofluorescence

The antitubulin antibody was produced and affinity-column-purified as previously described (Fuller *et al.*, 1975). Indirect immunofluorescence was done as previously described by Lazarides and Weber (1974).

2.3. Transmission Electron Microscopy

Routine electron microscopy was done as previously described (Clark and Shay, 1979); however, it was necessary to modify this procedure in order to use antibodies for the localization of tubulin in the electron microscope. Cells were grown on glass coverslips for 48 hr, and the coverslips were then quickly rinsed in Pucks Saline G and fixed in 3% glutaraldehyde, 0.2 M sucrose, and 0.2 M cacodylate buffer for 1 hr at 37°C. The coverslips were then rinsed in phosphate-buffered saline (PBS) and incubated, cell-side-down, on a small drop ($\approx$40 μl) of 0.5 mg/ml of antitubulin antibody, 1.0 mg/ml of bovine

---

Figure 1. Y-1 cells prior to ACTH treatment stained using antitubulin antibody. Very few microtubules are observed in these control cells; rather, most of the tubulin appears to be concentrated in association with many brightly fluorescing granules.

Figure 2. Cells treated with 100 mU ACTH for 30 min prior to fixation. The cells round up and increase steroidogenesis 8- to 10-fold, and the granular form of tubulin now appears to be largely replaced by many organized microtubules seen in the cell periphery.

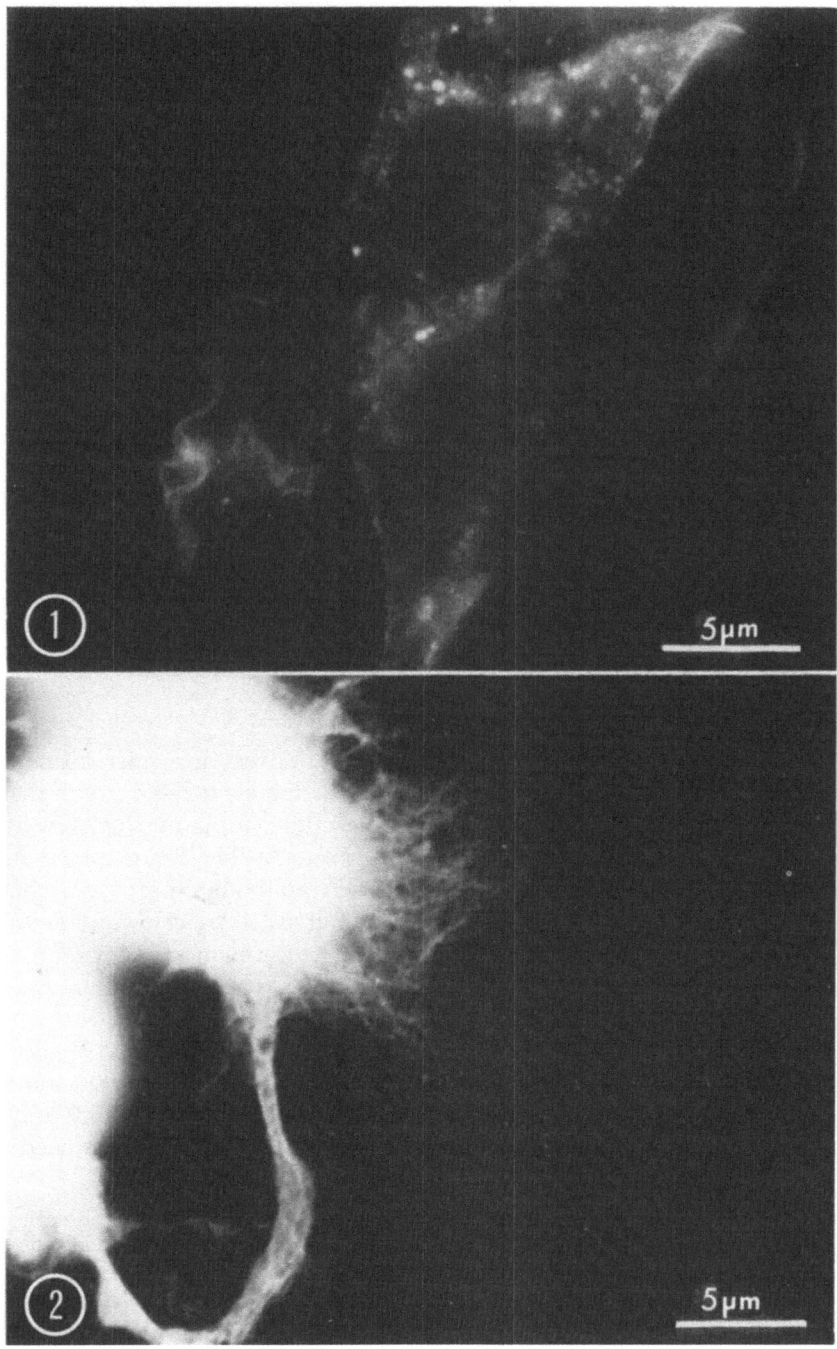

serum albumin, and 0.5% Saponin in PBS, pH 7.5, for 1 hr at 37°C. The coverslips were then rinsed in four changes of PBS, 15 min each, and placed cell-side-down on a small drop of ferritin-conjugated goat-produced anti-rabbit IgG at 0.2 mg/ml, 1.0 mg/ml of BSA, and 0.5% Saponin in PBS for 24–36 hr. The coverslips were then rinsed for 48 hr with frequent changes in PBS, pH 8.0. The specimens were then fixed again in glutaraldehyde for 1 hr, rinsed in PBS, postfixed in 1.0% OsO_4 and 0.3 M cacodylate buffer, and finally dehydrated in acetone and embedded in Epon Araldite. The glass coverslip was then removed and the monolayer of cells sectioned and observed using the JEOL 100-B electron microscope.

2.4. Cell Fractionation

The cells, grown in roller bottles, were removed using a rubber policeman and quickly homogenized in buffer, 0.1 M piperazine-N,N'-bis(2-ethane sulfonic acid), 1.0 M ethyleneglycal bis-(β amino ethyl ether)N,N'-tetraacetic acid, 0.5 M $MgCl_2$, pH 7.2, at 4°C, using a Teflon-glass homogenizer. The homogenate was then placed on a continuous 20–60% sucrose gradient in the aforementioned buffer and centrifuged for 2 hr at $8.2 \times 10^4 g$ in an SW27 Beckman rotor at 4°C.

3. Observations

The tubulin in Y-1 cells, as seen in the fluorescence microscope, was primarily localized in discrete granules, approximately 0.2–0.6 μm in diameter (Fig. 1), randomly distributed throughout the cytoplasm. Very few intact microtubules were observed in these cells prior to ACTH treatment. Identical results were obtained with other antitubulin antibodies prepared, characterized, and generously provided by Drs. Bill Brinkley, Howard Feit, Bill Garrard, and Tim Reudelhuber. This staining was obliterated if preimmune serum was substituted for antitubulin antibody or if the antitubulin antibody was absorbed with purified tubulin prior to use.

If the Y-1 cells were treated for 30 min with 100 mU ACTH prior to fixation (Fig. 2) and then stained using antitubulin antibody, the cells rounded up and increased the production of steroids approximately 10-fold, and the granular form of tubulin was largely replaced by many organized mi-

Figure 3. Tubulin visualized in a control cell using a ferritin-conjugated second antibody. Note that most of the aggregate of the dense ferritin is located over the cholesterol crystals.

Figure 4. Cells after addition of ACTH results in the appearance of organized microtubules and of coated vesicles. The cholesterol crystals do not disappear with ACTH treatment. However, they now contain electron-lucent areas.

Figure 5. Cholesterol crystals isolated from Y-1 cells prior to ACTH treatment, with associated tubulin as demonstrated using indirect immunofluorescence and antitubulin antibody. This was the only fraction isolated from the homogenate that stained intensely.

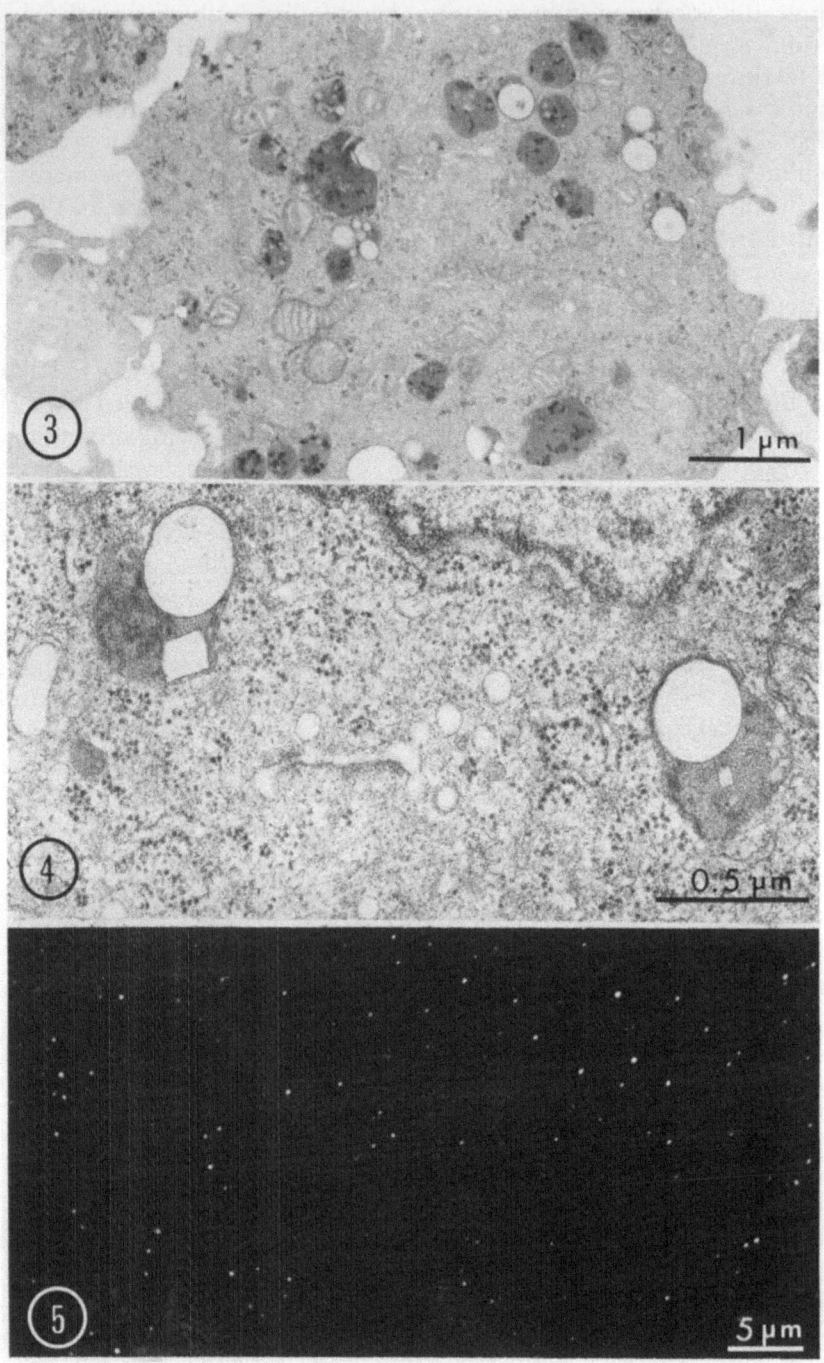

crotubules. To investigate this phenomenon on the ultrastructural level, techniques for localizing tubulin using antitubulin antibody visualized with ferritin-conjugated antibody were utilized. Prior to ACTH treatment, most of the ferritin was localized over electron-opaque structures that other investigators have previously termed cholesterol crystals (Mattson and Kowal, 1978). These structures were mostly 0.2–0.6 μm in size, and few intact microtubules were observed (Fig. 3). When examined in the electron microscope, these cholesterol crystals were not observed to disappear after ACTH treatment; however, these structures did undergo morphological alterations in that they then contained numerous electron-lucent areas. In addition, microtubules were now frequently seen in the cytoplasm of these stimulated cells (Fig. 4).

To isolate these structures, the cells were labeled with $10\,\mu$Ci [^{3}H]cholesterol for 24 hr. The cells were then homogenized and centrifuged in a sucrose gradient. This resulted in four visible bands of material being produced. The top fraction contained approximately 85% of the ^{3}H and was found in the electron microscope to be largely composed of membranes and lipid droplets. The second layer of particulate material contained approximately 6–10% of the ^{3}H label and in the electron microscope was identified as containing the cholesterol crystals. The cholesterol in these structures was composed of approximately 94% free cholesterol and approximately 6% cholesterol ester as determined by thin-layer chromatography. These structures contained approximately 8% of the total cellular protein and had the ability to hydrolyze β-glycerol phosphate. When a drop of this fracton was placed on a glass coverslip and prepared for indirect immunofluorescence using tubulin antibody, it was found to be the only fraction that stained intensely (Fig. 5).

To determine whether these observations were unique to the Y-1 cell line, the rat Leydig cell line (CLL 43) was also investigated. When these cells were stained using antitubulin antibody, many brightly fluorescent granules were observed (Fig. 6). These cells secrete steroids in response to dibutyryl cyclic AMP, and when they were treated with 5×10^{-5} M dibutyryl cyclic AMP for 30 min, the granular form of tubulin was replaced by organized microtubules (Fig. 7).

4. Discussion

The most important observation of this study was the finding that in the Y-1 cell line, most of the tubulin was located in discrete granules prior to ACTH treatment. These granules, which had previously been termed choles-

---

Figure 6. Rat Leydig cell line (CCL 43) stained using antitubulin antibody. These cells are larger than the Y-1 cells and appear to have most of their tubulin organized into granules.

Figure 7. Rat Leydig cell line (CCL 43) treated with dibutyryl cyclic AMP for 30 min prior to fixation. These cells do not round up as do the Y-1 cells. The tubulin in these cells appears to be organized into many microtubules.

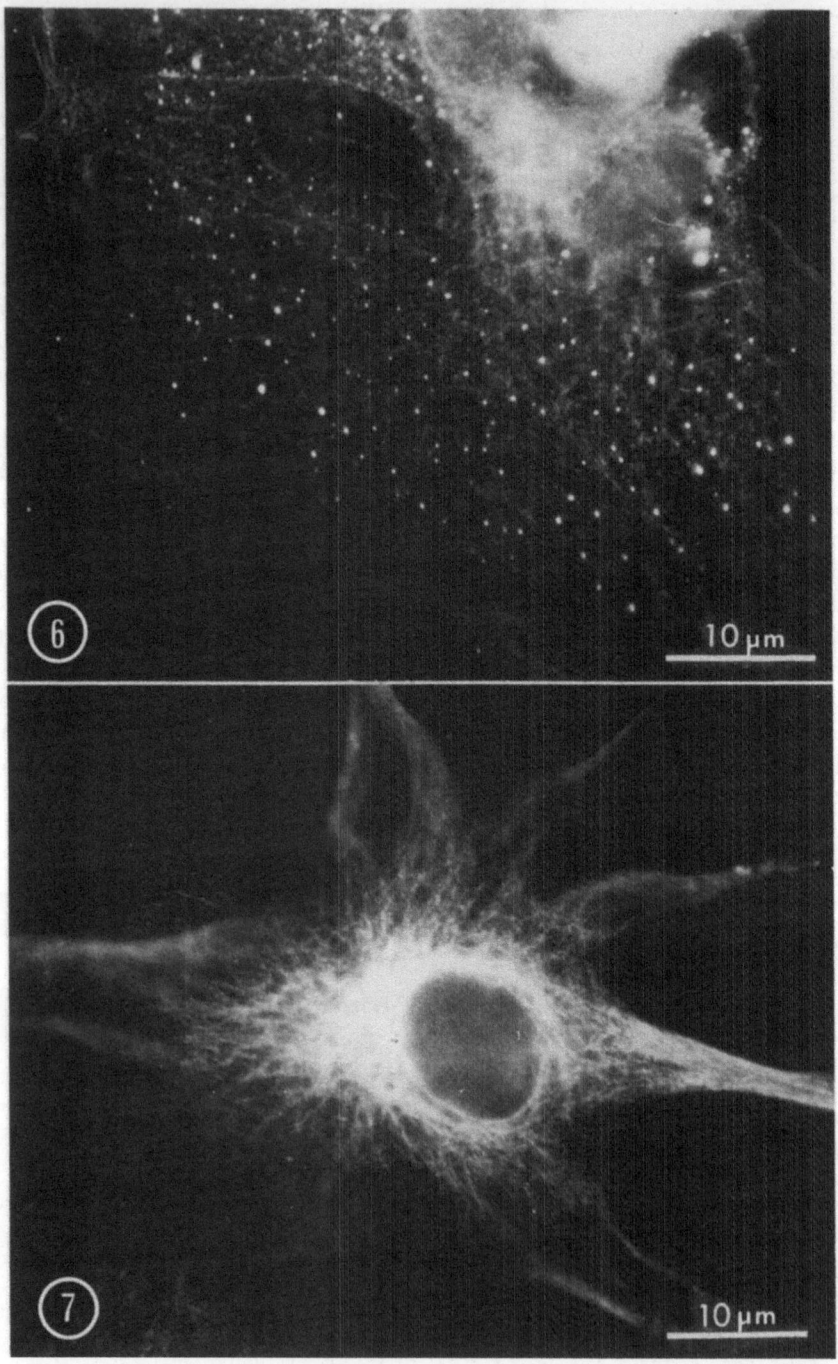

terol crystals, have been isolated and partially characterized as containing free cholesterol and having acid phosphatase activity. The discovery that tubulin was associated with these structures is both surprising and intriguing; however, since we were able to confirm this observation using antibodies prepared in other laboratories and have found a similar phenomenon in another steroidogenic cell line, we feel that these observations are correct.

We have previously reported that treatment of Y-1 cells with drugs such as colchicine, cholera toxin, and cyclic AMP disrupts the association of tubulin with the cholesterol crystals and induces steroidogenesis (Clark and Shay, 1979). These observations are consistent with those of Temple and Wolff (1973). However, the latter authors hypothesized the existence of intact microtubules, preventing the transport of cholesterol to the mitochondria. Our morphological observations indicate that it is tubulin and not microtubules that may be responsible for inhibiting the transport of cholesterol.

The possible physiological significance of this observation can be demonstrated in another series of experiments. When Y-1 cells are maintained in continuous culture for approximately 1 year, the cells lose their ability to respond to ACTH and produce steroids. In one series of experiments, we observed Y-1 cells that had been in culture for 3, 6, and 12 months. When these unstimulated, control cells were stained using antitubulin antibody, we observed, with increasing time in culture, progressively fewer cells with the granular form of tubulin and more cells containing intact microtubules. In addition, when we looked at Y-1 cells obtained from the ATCC, approximately 40% of the cells had intact microtubules; however, these cells were not able to produce steroids nearly as efficiently as the highly responsive cells obtained from Dr. Bernard Schimmer.

At present, the mechanisms by which adrenal cells regulate steroidogenesis is unknown. The studies presented in this chapter may shed some light on the controlling mechanisms involved in steroidogenesis not only of cells in culture but also of the adrenal gland *in vivo*.

ACKNOWLEDGMENTS. This work was supported by NIH Grant GM 29261 to Jerry W. Shay.

References

Buonassi, V., Sato, G., and Cohen, A. I., 1962, Hormone producing cultures of adrenal and pituitary tumor origin, *Proc. Natl. Acad. Sci. U.S.A.* **48:**1184.

Clark, M. A., and Shay, J. W., 1979, The response of whole and enucleated adrenal cortical tumor cells (Y-1 cells) to ACTH treatment, *Scanning Electron Microsc.* **3:**527.

Fuller, G. M., Brinkley, B. R., and Boughten, J. M., 1975, Immunofluorescence of mitotic spindle using monospecific antibodies against bovine brain tubulin, *Science* **187:**948.

Kowal, J., 1970, ACTH and the metabolism of adrenal cell cultures, *Recent Prog. Horm. Res.* **26:**623.

Lazarides, E., and Weber, K., 1974, Actin antibody and the specific visualization of actin filaments in non-muscle cells, *Proc. Natl. Acad. Sci. U.S.A.* **71:**2268.

Mattson, P., and Kowal, J., 1978, The ultrastructure of functional mouse adrenal cortical tumor cells *in vivo, Differentiation* **11**:75.

Schneider, E. L., Stanbridge, E. J., and Epstein, C. J., 1974, Incorporation of ^{3}H-uridine and ^{3}H-uracil into RNA, *Exp. Cell Res.* **84**:311.

Tai, Y. H., and Quinn, P. A., 1977, Rapid detection of mycoplasma contamination in tissue cultures by SEM, *Scanning Electron Microsc.* **2**:291.

Temple, R., and Wolff, J., 1973, Stimulation of steroid secretion by antimicrotubular agents, *J. Biol. Chem.* **248**:2691.

5

Use of Monoclonal Antibodies to Study Cytoskeleton

Jim Jung-Ching Lin, Keith Burridge, Stephen H. Blose,
Anne Bushnell, Sharon A. Queally, and
James R. Feramisco

The lymphocyte hybridoma technique developed by Köhler and Milstein (1975) has been widely and successfully used in the field of immunology and virology (Melchers *et al.*, 1978; Kennett *et al.*, 1980; Milstein and Lennox, 1980). However, there are few reports describing the monoclonal antibodies to cytoskeletal proteins. Monoclonal antibodies against cytoskeletal components have the potential not only for improving the immunofluorescent localization of specific proteins within cells but also for analyzing the functional sites of specific proteins and identifying the previously unidentified proteins. Furthermore, they can be used together with microinjection techniques (Feramisco, 1979; Lin and Feramisco, 1981) to investigate the physiological roles of specific proteins. In this chapter, we describe some examples to illustrate the advantages of monoclonal antibodies directed against cytoskeletal components.

We have immunized a mouse with the crude cytoskeletal preparations from chicken gizzard. After the animal was boosted, spleen cells were fused to nonsecretor myeloma cells (NS1). Cell fusion was carried out according to the method described by Kennett *et al.* (1978) with a slight modification (Lin, 1981). The resulting hybrids were then screened for the production of antibodies that reveal interesting staining patterns of chicken-embryo fibroblasts by indirect immunofluorescence (Blose, 1979). To obtain stable monoclones, the positive hybrids were cloned at least three times by the agarose cloning

Jim Jung-Ching Lin, Keith Burridge, Stephen H. Blose, Anne Bushnell, Sharon A. Queally, and James R. Feramisco • Cold Spring Harbor Laboratory, Cold Spring Harbor, New York 11724.

method (Sato *et al.*, 1972). The stable clones were grown in mass culture or in ascites tumors to produce valuable monoclonal antibodies. Using these procedures, we have been successful in producing several interesting monoclonal antibodies.

A hybridoma clone (JLA20) that secretes monoclonal antibody against actin has been isolated (Lin, 1981). Figure 1 shows the actin stress fibers (microfilaments) of gerbil fibroma cells (Fig. 1A) and human skin fibroblasts (Fig. 1B) and the I Band of rat myofibrils (Fig. 1C) stained with JLA20 antibody (1 : 4000 dilution) by indirect immunofluorescence. It should be noted that the very low noise level (or background) of immunofluorescence is probably due to the homogeneous, monospecific nature of monoclonal antibody and the very high titer (5–20 mg/ml) readily obtained from ascites fluids of hybridoma-bearing mice. The JLA20 antibody also reacts with actin from a variety of cell types including chicken-embryo fibroblasts, PtK-1 cells, BSC-1 cells, mouse 3T3 cells, and rat L8 myoblasts. The wide cross-reactivity of the antibody indicates that an antigenic determinant with identical or similar amino acid sequence is present in all actin molecules from different cell types. This is consistent with the results obtained from amino acid sequence studies showing certain homologous regions for the actins (Pollard and Weihing, 1974; Korn, 1978). In theory, one should be able to obtain monoclonal antibodies that recognize different types of actin such as α-, β-, and γ-actins. These antibodies will be most useful to study the immunofluorescent localization and function of actin isoforms.

Three hybridoma clones (JLA8, JLA10, and JLA17) produce monoclonal antibodies against chicken filamin. They can be used together with proteolytic cleavage and immunoautoradiography (Burridge, 1978) to dissect the filamin molecule. Figure 2 shows one such analysis. Purified filamin (lane 1 in Fig. 2) and its tryptic fragments (lane 2 in Fig. 2) were run on 12.5% sodium dodecyl sulfate (SDS)–polyacrylamide gels. The gel slices were cut out and reacted first with monoclonal antibodies (JLA8, JLA10, or JLA17) and then with ^{125}I-labeled rabbit anti-mouse IgG. The resulting autoradiogram (Fig. 2) clearly shows that different monoclonal antibodies recognize the different tryptic fragments. The relevant fragment can be purified and further cleaved by other proteolytic enzymes. If more independent monoclonal antibodies to filamin are obtained, one would theoretically be able to use them to probe the functional site (i.e., the sequence required for the binding of filamin to actin).

Other hybridoma clones that produce monoclonal antibodies against unidentified proteins have also been obtained from the same fusion experiment. The specific antigens recognized by these antibodies remain to be determined. They may be structural or regulatory proteins. Thus, monoclonal antibodies have the potential to identify previously unidentified proteins.

Figure 1. Fluorescence micrographs of gerbil fibroma cell (A), human skin fibroblast (B), and rat skeletal myofibril (C) stained with monoclonal antibody JLA20 against actin by indirect immunofluorescence. (1) Phase micrograph of myofibril; (2) fluorescence micrograph of myofibril; (Z) Z-line of myofibril.

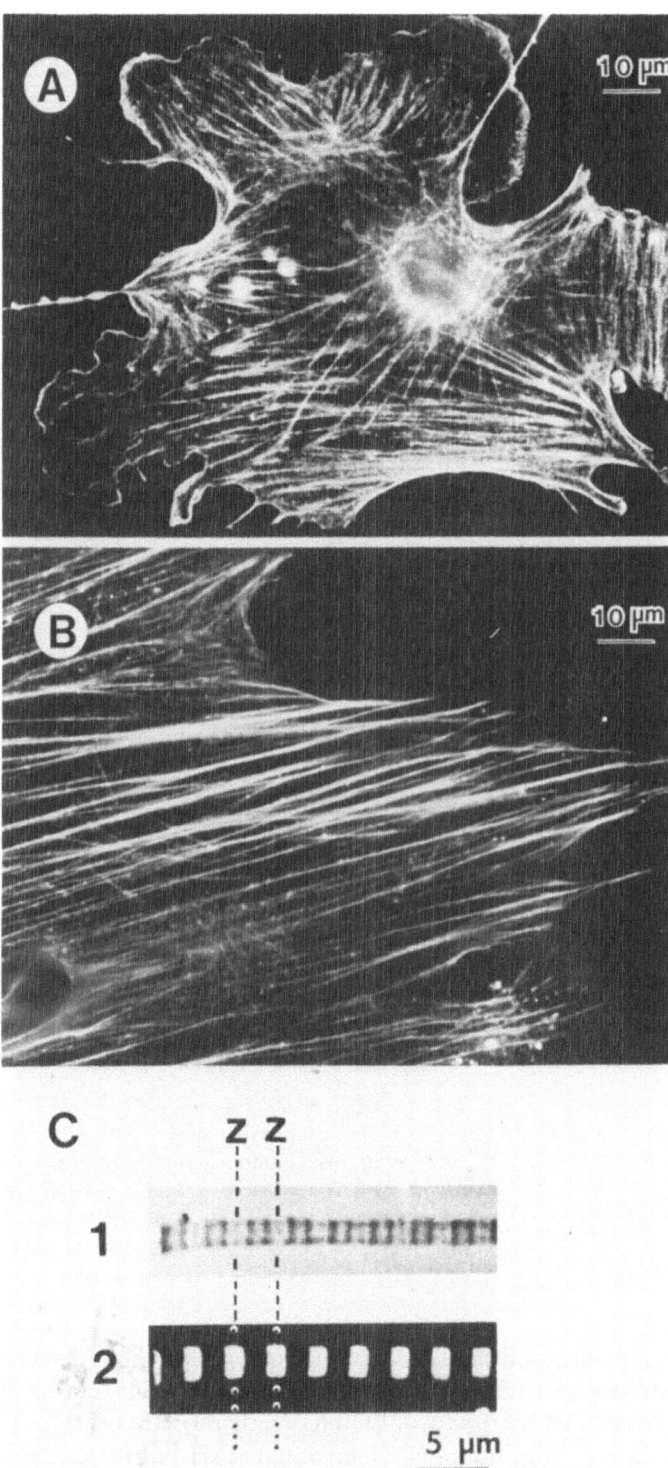

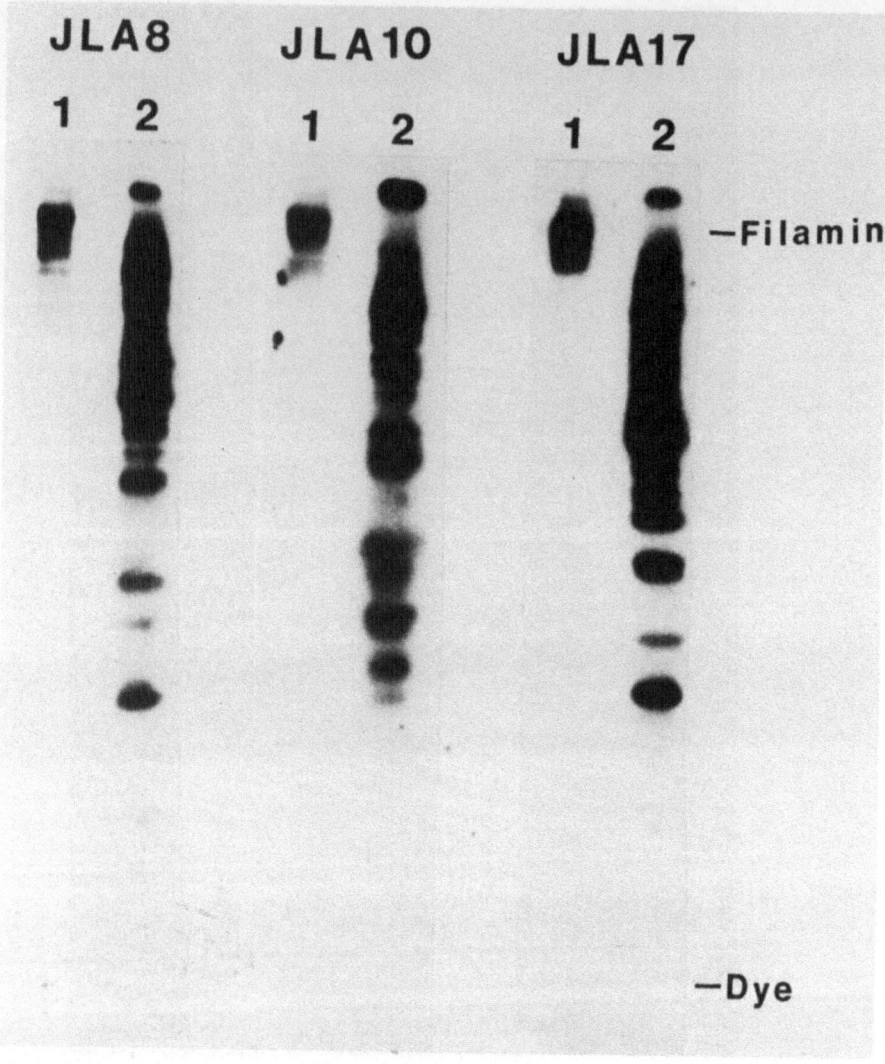

Figure 2. Analysis of filamin fragments on SDS–polyacrylamide gels with different monoclonal antibodies (JLA8, JLA10, and JLA17) by indirect immunoautoradiography. Purified filamin (lane 1) and its tryptic fragments (lane 2) were analyzed by 12.5% SDS–polyacrylamide gel electrophoresis. The gel slices were reacted first with different monoclonal antibodies and then with [125] I-labeled goat anti-mouse IgG antibodies. The radioactivity on the gel slices was detected by autoradiography on Kodak XR-1 film.

We have performed another fusion experiment using a total myofibril fraction as the antigen and screening the resulting hybrids on fibroblasts (Lin, 1981). This screening method would detect only those hybrids that produced antibodies reacting with antigenic determinants shared in common between the myofibrils and fibroblasts. The first antibody, JLB1, recognizes an antigen that is distributed in the M-line region and on either side of the Z-line of myofibrils, while the second antibody, JLB7, reacts only with an antigen located at the M-line region of myofibrils. However, both JLB1 and JLB7 antibodies gave the intermediate-filament-staining patterns on cultured cells from a variety of cell types. Immunoprecipitation of cell extracts from chicken-embryo fibroblasts with JLB1 or JLB7 antibody revealed the antigen band at a molecular weight of 210,000 or 95,000, respectively, rather than vimentin (the major subunit of fibroblast intermediate filaments). The biochemical properties of these proteins remain to be determined. Starger *et al.* (1978) have found a high-molecular-weight protein (250,000–300,000) that copurified with the major subunit (vimentin) of intermediate filaments. It is of interest to know whether or not the protein recognized by JLB1 antibody is similar to this high-molecular-weight protein reported by Starger *et al.* (1978).

Another high-molecular-weight protein (230,000) that is localized with the intermediate filaments of chicken smooth-muscle cells has also been reported by Granger and Lazarides (1980). They have found that antibodies against the 230,000-dalton protein decorate the Z-line of skeletal myofibrils by immunofluorescence. Therefore, it is unlikely that the 230,000-dalton protein is identical to one of the proteins recognized by JLB1 or JLB7 antibody, since its localization within myofibrils differs from that revealed by antibody against the 230,000-dalton protein.

Since monoclonal antibodies can be obtained in very high titers (5–20 mg/ml of specific antibody), and they are homogeneous with respect to their antigenic specificity, using the microinjection technique to introduce them into cells, one should be able to utilize them as potential perturbants to inhibit the functions of the antigen molecules in injected cells. The microinjection of JLB7 monoclonal antibody into the cultured fibroblasts causes a rapid aggregation of all the intermediate filaments to a perinuclear region. By contrast, when cells are injected with JLB1, little or no aggregation of the intermediate filaments is observed (Lin and Feramisco, 1981). Figure 3 shows the fluorescent micrographs of the JLB1-antibody-injected cell (Fig. 3A–C) and JLB7 antibody-injected cell (Fig. 3D–F) by indirect double-label immunofluorescence. Gerbil fibroma cells were microinjected with either JLB1 or JLB7 monoclonal antibody and, 3 hr after injection, were fixed, permeabilized, and counterstained with rabbit antibody against vimentin. The distribution of injected antibodies and vimentin within the cells was selectively viewed under different fluorescence optics. Some of the injected JLB1 antibodies (Fig. 3B) were located with the intermediate filaments, which are identical to those filaments revealed by the vimentin antibody (Fig. 3C), but the majority of the injected JLB1 antibodies were distributed diffusely (Fig. 3B). On the other hand, a tight aggregate near the nucleus was visualized in the cell injected with JLB7 monoclonal antibody (Fig. 3E). Furthermore, all the intermediate filaments decorated by vimentin antibody (Fig. 3F) coincided with the filamentous caps revealed by JLB7 monoclonal antibody (Fig. 3E). The distributions of actin microfilaments and microtubules within the JLB7-antibody-injected cells appeared to be undisturbed at both the fluorescent-microscopic and the

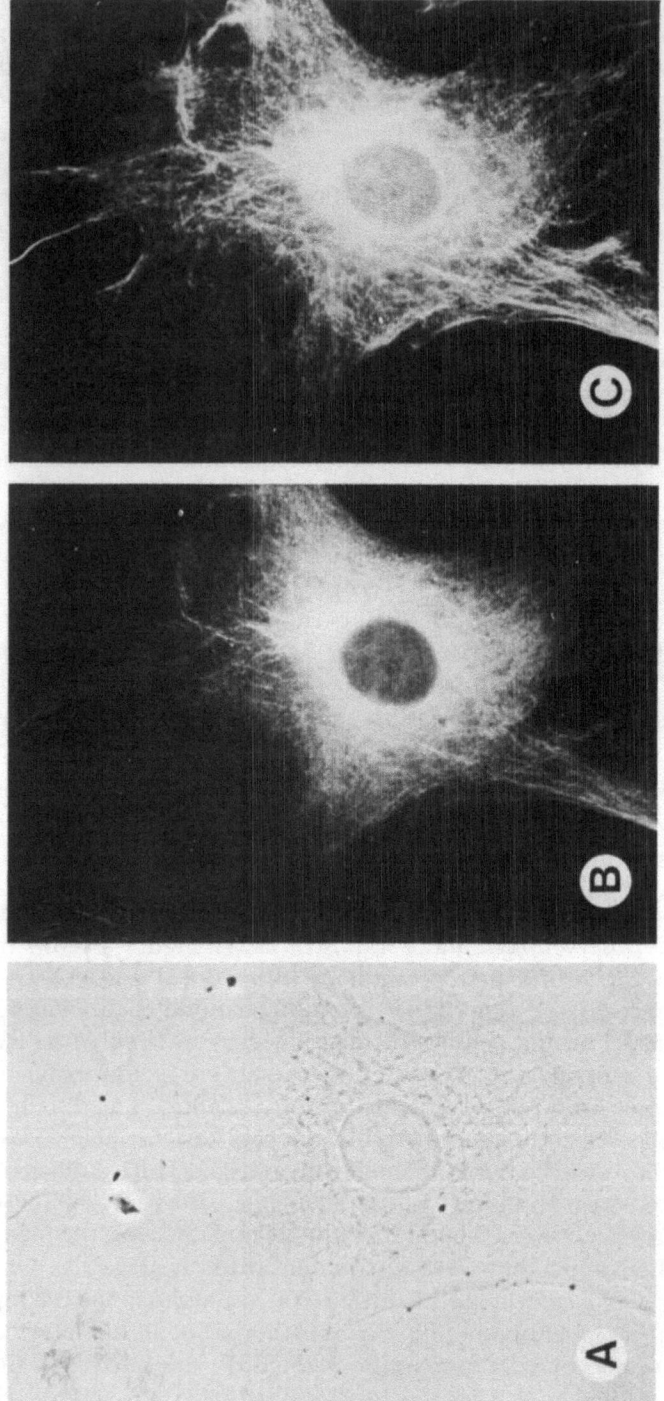

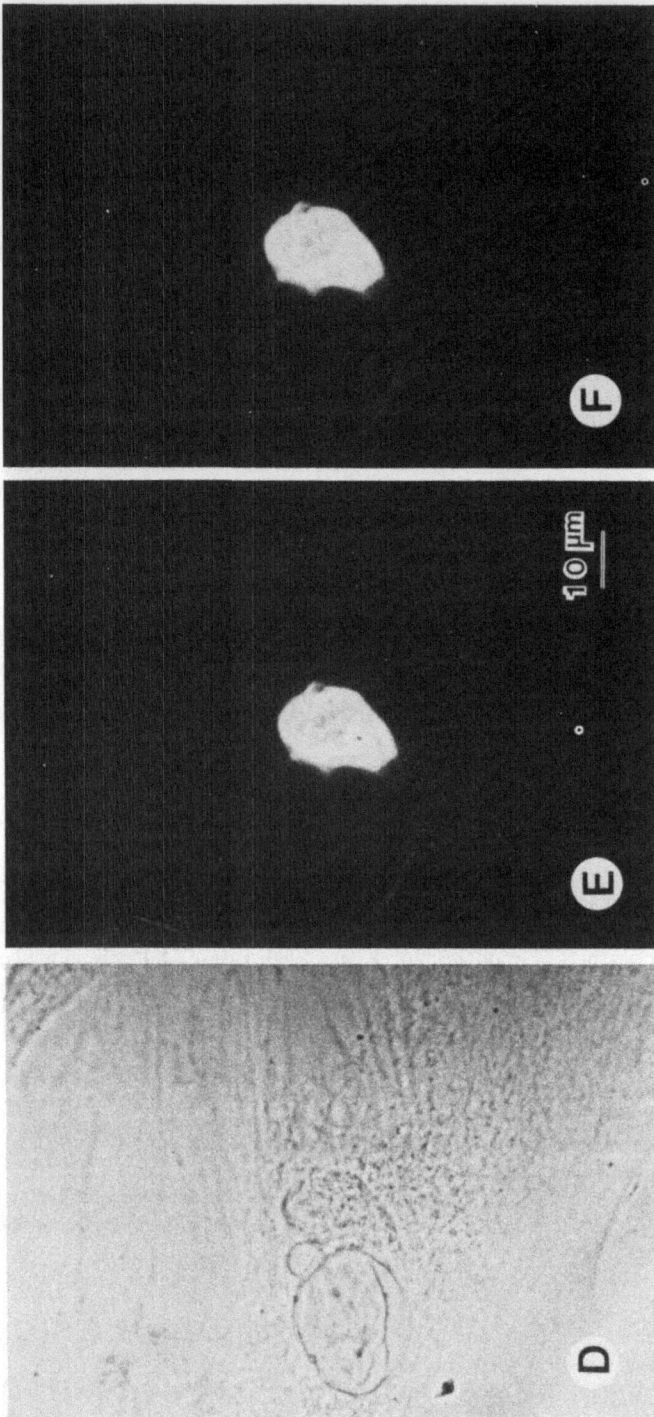

Figure 3. Indirect double-label immunofluorescence of the injected cells. Gerbil fibroma cells were microinjected with either JLB1 (A–C) or JLB7 (D–F) monoclonal antibody and, 3 hr after injection, counterstained with rabbit antibody against vimentin. The distributions of injected mouse antibody and vimentin were detected by a mixture of second antibody containing fluorescein-conjugated goat anti-mouse IgG and rhodamine-conjugated goat anti-rabbit IgG. (A,D) Phase-contrast micrographs. (B,E) Injected cells viewed selectively for fluorescein fluorescence to allow the microinjected mouse antibody to be visualized. (C,F) Same fields seen in (B) and (E), respectively, except viewed selectively for rhodamine fluorescence to allow the distribution of vimentin to be visualized.

electron-microscopic level (Lin and Feramisco, 1981). No obvious changes in cell morphology and the normal movements of intracellular granules and mitochondria were observed in the cells injected with JLB7 antibody, suggesting that the intermediate filaments may not be involved in these functions. However, the JLB7-antibody-injected cells at late telophase revealed a new class of intermediate filaments distinct from those induced to form caps by the antibody. This may suggest that this new class of filaments may play an important role in mitosis. These results demonstrate, for the first time, the usefulness of monoclonal antibodies in conjunction with microinjection to study *in vivo* functions of the antigen.

ACKNOWLEDGMENTS. We are most grateful to Dr. J. D. Watson for his enthusiastic support of this work and to Dr. G. Albrecht-Buehler for his advice. We thank Ms. Madeline Szadkowski for her careful typing of the manuscript and Mr. Ted Lukralle for his photographic skills. This work has been supported by Grant CA13106 to the Cold Spring Harbor Laboratory from the National Cancer Institute and by a Damon Runyon–Walter Winchell Cancer Fund Postdoctoral Fellowship.

References

Blose, S. H., 1979, Ten-nanometer filaments and mitosis: maintenance of structural continuity in dividing endothelial cells, *Proc. Natl. Acad. Sci. U.S.A.* **76**:3372.

Burridge, K., 1978, Direct identification of specific glycoproteins and antigens in sodium dodecyl sulfate gels, *Method. Enzymol.* **50**:54.

Feramisco, J. R., 1979, Microinjection of fluorescently labeled α-actinin into living fibroblasts, *Proc. Natl. Acad. Sci. U.S.A.* **76**:3967.

Granger, B. L., and Lazarides, E., 1980, Synemin: a new high molecular weight protein associated with desmin and vimentin filaments in muscle, *Cell* **22**:727.

Kennett, R. H., 1978, Cell fusion, *Method. Enzymol.* **58**:345.

Kennett, R. H., McKearn, T. J., and Bechtol, K. B., 1980, Part V. Monoclonal antibodies to microorganisms in *Monoclonal antibodies, Hybridomas: A new dimension in biological analysis*, pp. 317–351, Plenum Press, New York.

Köhler, G., and Milstein, C., 1975, Continuous cultures of fused cells secreting antibody of predefined specificity, *Nature* **256**:495.

Korn, E., 1978, Biochemistry of actomyosin-dependent cell motility, *Proc. Natl. Acad. Sci. U.S.A.* **75**:588.

Lin, J. J. C., 1981, Monoclonal antibodies against myofibrillar components of rat skeletal muscle decorate the intermediate filaments of cultured cells, *Proc. Natl. Acad. Sci. U.S.A.* **78**:2335.

Lin, J. J. C., and Feramisco, J. R., 1981, Disruption of the *in vivo* distribution of the intermediate filaments in fibroblasts through the microinjection of a specific monoclonal antibody, *Cell* **24**:185.

Melchers, F., Potter, M., and Wanner, N. L., 1978, Lymphocyte hybridomas, *Curr. Top. Microbiol. Immunol.* **81**:IX.

Milstein, C., and Lennox, E., 1980, The use of monoclonal antibody techniques in the study of developing cell surfaces, *Curr. Top. Dev. Biol.* **14**:1.

Pollard, T. D., and Weihing, R. R., 1974, Actin and myosin and cell movement, *CRC Crit. Rev. Biochem.* **2**:1.

Sato, K., Slesinski, R. S., and Littlefield, J. W., 1972, Chemical mutagenesis at the phosphoribosyl-transferase locus in cultured human lymphoblasts, *Proc. Natl. Acad. Sci. U.S.A.* **69:**1244.

Starger, J. M., Brown, W. E., Goldman, A. E., and Goldman, R. D., 1978, Biochemical and immunological analysis of rapidly purified 10 nm filaments from baby hamster kidney (BHK21) cells, *J. Cell Biol.* **78:**93.

6

Interaction of Calcium–Calmodulin in Microtubule Assembly in Vitro

George Perry, B. R. Brinkley, and Joseph Bryan

1. Introduction

In most eukaryotic cells, microtubules, intermediate filaments, and microfilaments constitute a fibrous network that is termed the cytoskeleton. Microtubules have been implicated in a number of cell functions including cell motility, maintenance of cell shape, organelle movement and distribution, secretion, and cell-surface modulation.

Apparently, the ability of microtubules to accomplish many of these diverse functions arises from their capacity to undergo regulated assembly-disassembly. Variations in intracellular free calcium have been suggested to be important in regulating microtubule assembly (Weisenberg, 1972). The data of Weisenberg (1972) suggested that the calcium sensitivity of microtubule assembly *in vitro* was in the micromolar range. This finding has since been supported by other laboratories (Haga *et al.,* 1974; Rosenfeld *et al.,* 1976; Salmon and Jenkins, 1977; Nishida, 1978; Nishida and Sakai, 1977). However, other investigators have found that the calcium sensitivity of microtubule polymerization *in vitro* is in the millimolar range (Olmsted and Borisy, 1973, 1975; Lee *et al.,* 1974; Fuller and Brinkley, 1975; Gaskin and Shelanski, 1975). The reasons for this difference have not been determined.

There is evidence that Ca^{2+} may be involved in the control of microtubule assembly *in vivo*. Spindle-microtubule depolymerization can be initiated in sea urchin eggs by the injection of micromolar calcium. Since calcium is rapidly sequestered in sea urchin eggs, this value is an upper limit on the calcium concentration necessary to cause depolymerization (Kiehart and Inoué, 1977).

George Perry, B. R. Brinkley, and Joseph Bryan • Department of Cell Biology, Baylor College of Medicine, Houston, Texas 77030.

Using the divalent cation ionophore A23187, Fuller and Brinkley (1975) found that increasing intracellular free calcium diminished the cytoplasmic microtubule complex. Schliwa (1976) used A23187 to depolymerize the microtubules of a heliozoan. Approximately 10 μM external calcium was required to induce depolymerization. Since neither of the ionophore studies monitored the intracellular calcium concentration, it is still not certain that normal intracellular Ca^{2+} variations, which are reported to be in the micromolar range (Baker, 1977), are sufficient to effect microtubule disassembly.

Some recent biochemical work has indicated that nontubulin protein factors may confer a higher calcium sensitivity to neurotubulin preparations. Nishida and Sakai (1977) have suggested that a heat-labile protein, which is purified away from microtubule protein by the usual cycling procedure, is responsible for the higher calcium sensitivity of crude preparations. The heat-stable protein calmodulin has also been shown to increase the calcium sensitivity of tubulin assembly (Marcum *et al.*, 1978; Nishida *et al.*, 1979; Jemiolo *et al.*, 1979). Marcum *et al.* (1978) and Jemiolo *et al.* (1979) found that a 6 : 1 calmodulin : tubulin ratio would increase the calcium sensitivity to the micromolar level. In these studies, the calcium sensitivity was defined by simple percentage inhibition at a single protein concentration, usually 1–3 mg/ml.

Several investigators (Brinkley *et al.*, 1978; Welsh *et al.*, 1978; Anderson *et al.*, 1978; Wood *et al.*, 1980) have used indirect immunofluorescence or immunoelectron microscopy (Lin *et al.*, 1979) to localize calmodulin within cells. With the use of these methods, calmodulin appears to be localized in the mitotic spindle during mitosis. One study on interphase cells (Welsh *et al.*, 1978) reports finding calmodulin associated with stress fibers, in addition to being rather uniformly distributed throughout the cytoplasm. In the spindle, the localization of calmodulin is more intense in the centrosome region and the midbody and is not totally coincident with microtubules.

2. Materials and Methods

2.1. Preparation of Microtubule Protein

Microtubule protein was obtained by a modification of the procedure of Gaskin and Shelanski (1975). Brains were removed from dogs within 30 min of death, immediately cooled on ice, and homogenized in an equal volume of ice-cold homogenization buffer (HB) with 0.2 mM GTP. HB consisted of 0.1 M piperazine-N,N'-bis(2 ethane sulfonic acid) (PIPES), 1 mM $MgSO_4$, and 1 mM ethyleneglycol bis(β-aminoethyl ether)N,N'-tetraacetic acid (EGTA), at pH 6.94. This homogenate was centrifuged at 10,000g for 30 min, and the supernatant was recentrifuged at 100,000g for 1 hr. GTP and glycerol were added to a final concentration of 1 mM and 25%, respectively. The extract was then warmed to 37°C for 1 hr. Microtubules were collected by centrifugation at 40,000g for 1 hr. The pellet was homogenized, then depolymerized in ice-cold HB with 1 mM GTP for 30 min. The tubulin was taken through three assembly cycles with glycerol. The final pellet was resuspended in HB with 1 mM GTP, depolymerized, and spun at 100,000g for 60 min, and the supernatant was frozen in liquid nitrogen and stored at $-80°C$ until used.

2.2. Preparation of Calmodulin

Calmodulin from bovine testes was prepared by the procedure of Dedman *et al.* (1977), with the exception that lypholyzation instead of ammonium sulfate was used to concentrate the protein.

2.3. Assay of Microtubule Polymerization

Microtubule polymerization was initiated by warming tubulin to 32°C in HB plus 1 mM GTP. The change in absorbance at 350 nm was used as a measure of polymer formation. The change in absorbance at 350 nm is a linear function of the protein in polymer (Bryan and Nagle, 1976).

2.4. Stellazine

Trifluoperazine (Smith, Kline and French), commonly known as stellazine, binds reversibly to calmodulin in the presence of calcium and blocks the interaction of calmodulin with various target proteins (Weiss and Levin, 1978; La Porte *et al.*, 1980). To form the calmodulin–calcium–stellazine complex, calmodulin was dialyzed overnight in the dark against 10 μM stellazine in HB with 1 mM $CaCl_2$ to yield a Ca^{2+}/EGTA ratio of 1.0. To inhibit endogenous calmodulin, stellazine was incubated with microtubule protein at 0°C for 4 hr in HB with 1 mM GTP plus 1 mM $CaCl_2$.

2.5. Other Procedures

The computer program of Potter and Gergely (1975) for the calcium–EGTA buffer system was used to determine the free calcium concentration. Protein was determined using a modification of the Lowry procedure (Hartree, 1972) with bovine serum albumin (Sigma) as standard. Sodium dodecyl sulfate (SDS)-polyacrylamide gel electrophoresis was done according to Laemmli (1970). The calmodulin affinity column was prepared as described by Dedman *et al.* (1978). Phosphodiesterase activity was assayed using the method of La Porte and Storm (1978). All chemicals were of reagent grade.

3. Results

Figure 1 shows the extent of polymer formation, plotted as the difference in absorbance at 350 nm as a function of the free calcium concentration for a

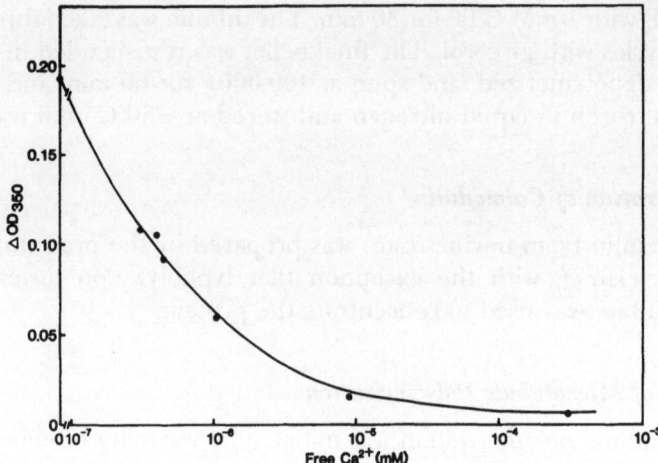

Figure 1. Calcium sensitivity of microtubule assembly. The program of Potter and Gergely (1975) for the Ca^{2+}-EGTA buffer system was used to determine free calcium concentration. The microtubule-protein concentration was 1 mg/ml.

microtubule-protein concentration of 1 mg/ml. In this experiment, there is an apparent inhibition of assembly at micromolar levels of free calcium ions.

Figure 2 shows the extent of polymer formation as a function of protein concentration. For a given calcium concentration, specified by the Ca^{2+}/EGTA ratio, the quantity of polymer formed increases as a linear function of the protein concentration. With increasing free calcium, the critical concentration

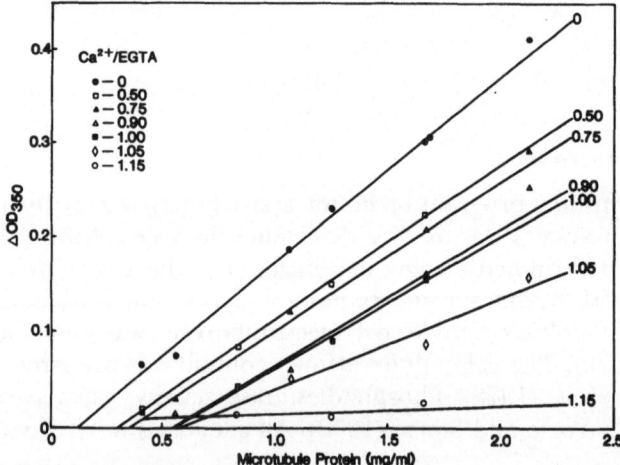

Figure 2. Microtubule assembly vs. microtubule-protein concentration in the presence of various free calcium ion concentrations. Free calcium concentration was calculated using the program of Potter and Gergely (1975) and assuming no calcium contribution from either water or reagents. (●) Ca^{2+}/EGTA = 0, $[Ca^{2+}]$ = 0; (□) Ca^{2+}/EGTA = 0.50, $[Ca^{2+}]$ = 2.1×10^{-7}M; (▲) Ca^{2+}/EGTA = 0.75, $[Ca^{2+}]$ = 6.5×10^{-7}M; (△) Ca^{2+}/EGTA = 0.90, $[Ca^{2+}]$ = 9.0×10^{-6}M; (■) Ca^{2+}/EGTA = 1.00 $[Ca^{2+}]$ = 1.1×10^{-5}M; (◇) Ca^{2+}/EGTA = 1.05, $[Ca^{2+}]$ = 6.1×10^{-5}M; (○) Ca^{2+}/EGTA = 1.15, $[Ca^{2+}]$ = 10^{-4}M.

is increased, and there is a reduction in slope. The effect of calcium on the critical concentration is shown more clearly in Fig. 3.

The use of percentage inhibition at a single protein concentration is potentially misleading with a critical-concentration phenomenon. For example, in Fig. 2, compare the percentage inhibition of 11 μM free calcium (Ca^{2+}/EGTA = 1.00) with the control (Ca^{2+}/EGTA = 0). When the protein concentration is increased from 0.5 to 1.5 mg protein/ml, the apparent percentage inhibition drops from 100 to less than 50%. To quantitatively describe the inhibition of polymerization by calcium or by other agents, several protein concentrations should be used and the effect on the critical concentration reported. Figure 3 shows that the change of calcium concentration over the physiological range increases the critical concentration 2- to 3-fold. The data suggest that at protein concentrations near the critical concentration, calcium could be an effective regulator of tubulin polymerization.

The addition of stoichiometric quantities of calmodulin to microtubule protein increased the calcium sensitivity of microtubule formation. The experiment shown in Fig. 4 gives the result with a 6 : 1 molar ratio of calmodulin to tubulin dimer; less impressive effects were seen with a lower ratio. The effect of calmodulin on the critical concentration is shown in Fig. 5. Calmodulin alone increases the critical concentration slightly. In combination with 11 μM free calcium, calmodulin increases the critical concentration to about 2.5 mg protein/ml, and there is some synergism between calcium and calmodulin at the 6 : 1 calmodulin/tubulin ratio. Determination of the endogenous calmodulin in these microtubule protein preparations by radioimmunoassay (RIA) gave values for the calmodulin/tubulin ratio of 1 : 250–7500. This low calmodulin/tubulin ratio argues that the mechanism of the endogenous calcium sensitivity might be different than that of the increased inhibition seen with exogenous calmodulin.

Four approaches have been used to study the specificity of the presumed calmodulin–tubulin interaction. The calcium-binding proteins parvalbumin

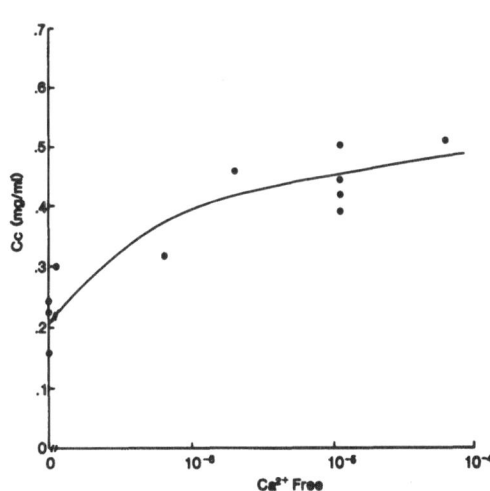

Figure 3. Effect of calcium on the critical concentration (Cc) of microtubule assembly. Derived from Fig. 2.

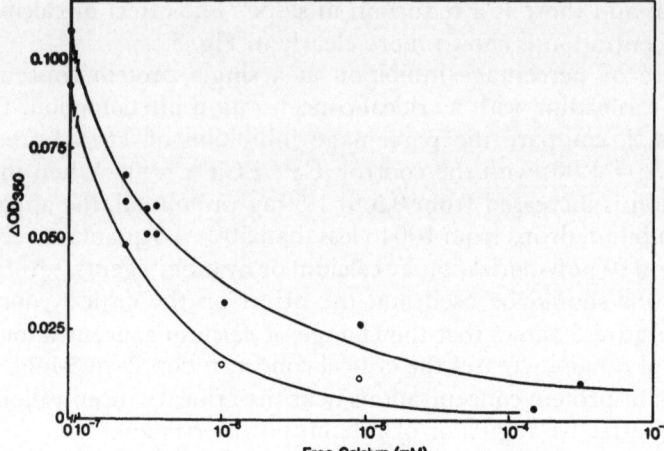

Figure 4. Effect of calmodulin on the calcium sensitivity of microtubule assembly. (●) Microtubule protein; (○) microtubule protein with exogenous calmodulin at a 6 : 1 calmodulin/tubulin molar ratio.

and troponin C, which are similiar to calmodulin, were tested. All three are small acidic proteins. Troponin C and calmodulin have similiar amino acid sequences and bind 4 moles of calcium per mole of protein; parvalbumin binds 2 moles of calcium per mole of protein. If the inhibition by calmodulin is purely a charge effect, we might expect parvalbumin to be more effective than calmodulin, since it is a more highly charged molecule. The results show that parvalbumin alone or in combination with calcium did not significantly inhibit microtubule polymerization. Troponin C, on the other hand, is more effective than calmodulin in inhibiting tubulin polymerization in the presence

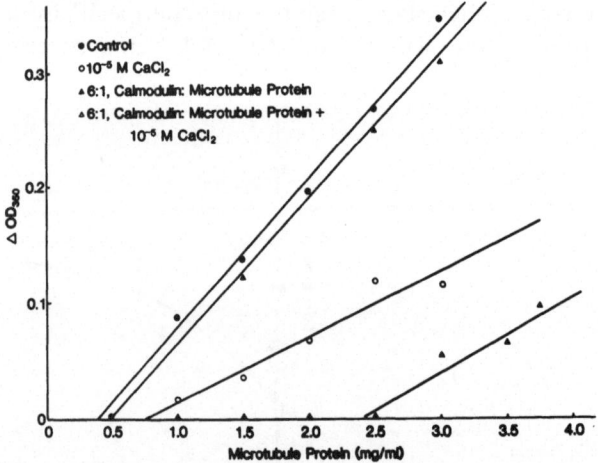

Figure 5. Microtubule assembly vs. microtubule-protein concentration in the presence of calcium and calmodulin. A 6 : 1 calmodulin/tubulin molar ratio and a free calcium concentration of 11 μM were used.

or in the absence of calcium. Since troponin C is more homologous to calmodulin than is parvalbumin, this argues that there may be some specificity to the interaction between calmodulin and microtubule protein.

A second experiment suggests that the interaction between calmodulin and microtubules is weak and confirms the RIA data. Iodinated calmodulin was added to a brain homogenate and followed through three cycles of polymerization. Table 1 shows the specific activity of the microtubule pellets after each cycle. The specific activity drops as expected for simple dilution, as we go from the first cycle, H_1P to the third, H_3P. These results clearly demonstrate that calmodulin does not interact directly with microtubule protein during the cycling procedure. A problem with interpreting this result is that the cycling procedure requires that calcium ions be sequestered.

A third approach was to determine which microtubule-protein components directly interact with calmodulin. This was done by passing microtubule proteins through a calmodulin–Sepharose column in the presence of 1 mM $CaCl_2$. The column was then washed, with 100 times the column volume, with HB with 1 mM GTP plus 1 mM $CaCl_2$, and was then washed with HB in the absence of calcium. The gel shown in Fig. 6 indicates that the only protein to be retained in a calcium-dependent fashion by the column under these conditions was a protein of the same molecular weight as tubulin. This binding appears to be weak, since less than binding capacity of the column is retained and tubulin tends to "bleed" from the column even in the presence of calcium. The reverse experiment, in which a 6 S tubulin–Sepharose column was used, indicates that significant amounts of calmodulin are not retained.

The last approach was to inactivate calmodulin with the phenothiazine derivative trifluoperazine. Figure 7 shows that stellazine was ineffective in blocking the calmodulin inhibition of microtubule assembly even though it inhibited over 90% of the phosphodiesterase-stimulating activity. Furthermore, stellazine did not affect the calcium inhibition of assembly that our preparations exhibit normally, suggesting that this inhibition is not mediated by calmodulin.

Table 1. Binding of Iodinated Calmodulin to Neurotubulin during Cycles of Assembly–Disassembly

Fraction[a]	Total protein (mg)	[^{125}I]Calmodulin specific activity (cpm/mg microtubule protein)
Supernatant[b]	322	2169
H_1P	45.6	917
H_2P	25.6	274
H_3P	15.8	32

[a] The notation of Borisy *et al.* (1975) is used to identify pellets from each "cycle" of tubulin purification.
[b] This was a 100,000*g* supernatant from a brain homogenate (see Section 2.1).

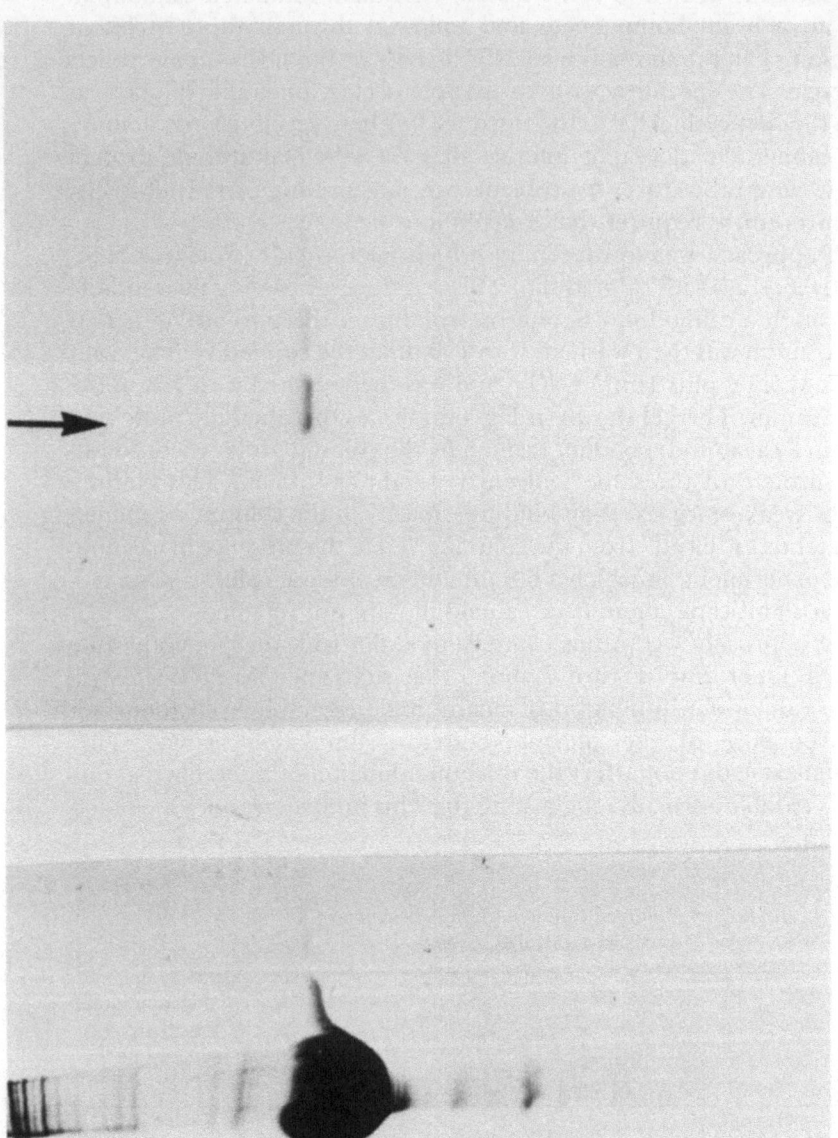

Figure 6. SDS–polyacrylamide gel of proteins retained by a calmodulin–Sepharose column. The portion of the gel to the left is the microtubule protein loaded into the column in HB with 1 mM CaCl₂ and 1 mM GTP. The arrow indicates when buffer with no calcium was run through the column. The protein eluted by removing calcium has the same molecular weight as tubulin.

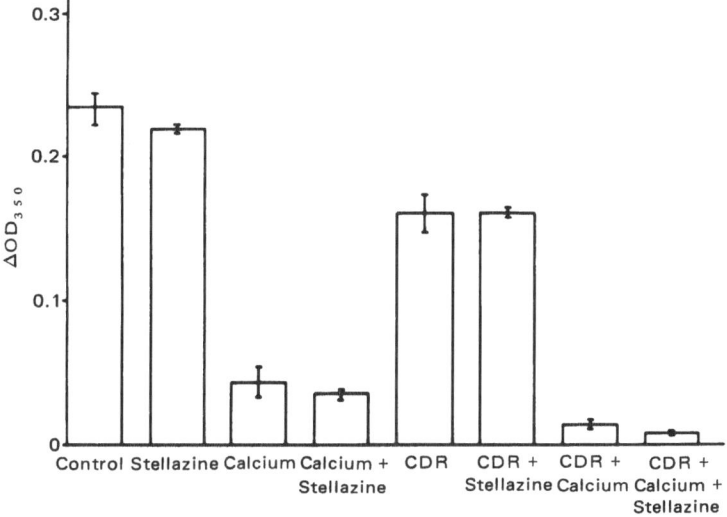

Figure 7. Effect of stellazine on the calcium and calmodulin inhibition of microtubule assembly. A 6 : 1 calmodulin (CDR)/tubulin molar ratio was used with microtubule proteins at 2.5 mg/ml. Free calcium was at 11 μM and stellazine at 10 μM. See the text for the incubation procedures. The vertical brackets indicate the standard deviation (N = 3).

4. Discussion

Previous reports on the calcium sensitivity of microtubule polymerization *in vitro* have not considered protein concentration as an important factor. For example, microtubule polymerization has been reported to be calcium-insensitive when protein concentrations of 5–10 mg/ml are used (e.g., Olmsted and Borisy, 1973, 1975), while others have reported greater sensitivity using either lower protein concentrations or poorly defined extracts (Nishida, 1978; Weisenberg, 1972). We observe that physiological free calcium levels can increase the critical concentration by a factor of about 3 in a cycled neurotubulin preparation. The importance of this increase in the critical concentration with increasing calcium in the cell would depend on the intracellular tubulin concentration. If the tubulin concentration is near the critical concentration, calcium alone might be an effective regulator of tubule formation.

We have confirmed the results of Marcum *et al.* (1978) and find that calmodulin increases the calcium sensitivity of microtubule assembly. We question the specificity of this interaction, because we cannot demonstrate tight binding between calmodulin and any microtubule components. A recent report (Jemiolo *et al.*, 1980) indicates that some microtubule-associated proteins, from microtubule protein isolated in the absence of glycerol, are retained by calmodulin–Sepharose in a calcium-dependent fashion, but these proteins can be eluted by a salt wash. One possibility for the differences in our results is that Jemiolo and co-workers isolated microtubule protein in the absence of glycerol. Glenney and Weber (1980), using a gel-overlay technique, have shown that iodinated calmodulin will bind to two brain extract components (molecular weights of 47,000 and 56,000) in a calcium-dependent fashion. It is possible that the 56,000-dalton protein is one of the tau proteins, and this technique may offer greater sensitivity than the methods reported here. There is, however, no direct evidence that either of these proteins is present in cycled tubulin.

The inability of stellazine to block the calmodulin inhibition of microtubule polymerization argues strongly that the effect on tubulin assembly is not specific and that calmodulin is not interacting with a microtubule component at the same site that it interacts with phosphodiesterase (La Porte *et al.,* 1980). Furthermore, stellazine does not affect the endogenous calcium sensitivity of our microtubule preparations, suggesting that this calcium sensitivity is not mediated by calmodulin or that the calmodulin effect is by a completely different mechanism than other known calmodulin-regulated functions.

ACKNOWLEDGMENTS. The authors wish to thank Drs. J. R. Dedman and A. R. Means for providing iodinated calmodulin, determining calmodulin by RIA, and many thoughtful discussions. We also wish to thank L. L. Wang for critical discussion of various aspects of this study. This study was supported by Research Grants NCI CA22610 and CA23022 to B. R. Brinkley, NIH Grant GM26091 to Joseph Bryan, and a Muscular Dystrophy Association postdoctoral fellowship to George Perry.

References

Anderson, B., Osborn, M., and Weber, K., 1978, Specific visualization of the distribution of the calcium dependent regulatory protein of cyclic nucleotide phosphodiesterase (modulator protein) in tissue culture cells by immunofluorescence microscopy: Mitosis and intracellular bridge, *Cytobiologie* **17**:354.

Baker, P. F., 1977, Calcium and the control of neurosecretion, *Sci. Prog. (Oxford)* **64**:95.

Borisy, G. G., Marcum, J. M., Olmsted, J. B., Murphy, D. B., and Johnson, K. A., 1975, Purification of tubulin and associated high molecular weight proteins from porcine brain and characterization of microtubule assembly *in vitro,* in: *The Biology of Cytoplasmic Microtubules* (D. Soifer, ed.), *Ann. N. Y. Acad. Sci.* **253**:107.

Brinkley, B. R., Marcum, J. M., Welsh, M. J., Dedman, J. R., and Means, A. R., 1978, Regulation of spindle microtubule assembly–disassembly: Localization and possible functional role of calcium dependent regulator protein, in: *Cell Reproduction: In Honor of Daniel Mazia* (E. R. Dirksen, D. M. Prescott, and C. F. Fox, eds.), pp. 299–314, Academic Press, New York.

Bryan, J., and Nagle, B. W., 1976, Microtubules: Inhibition of spontaneous *in vitro* assembly by non-neural cell extracts, in: *Molecular Basis of Motility* (L. Heilmeyer, J. C. Rüegg, and Th. Wieland, eds.), pp. 161–174, Springer-Verlag, Berlin.

Dedman, J. R., Potter, J. D., Jackson, R. L., Johnson, J. D., and Means, A. R., 1977, Physicochemical properties of rat testes: Ca^{2+}-dependent regulator protein of cyclic nucleotide phosphodiesterase, *J. Biol. Chem.* **252**:8415.

Dedman, J. R., Welsh, M. J., and Means, A. R., 1978, Ca^{2+}-dependent regulator, *J. Biol. Chem.* **253**:7515.

Fuller, G. M., and Brinkley, B. R., 1975, Structure and control of assembly of cytoplasmic microtubules in normal and transformed cells. *J. Supramol. Struct.* **5**:437.

Gaskin, C. R. C., and Shelanski, M. L., 1975, Biochemical studies on the *in vitro* assembly and disassembly of microtubules, in: *The Biology of Cytoplasmic Microtubules* (D. Soifer, ed.), *Ann. N.Y. Acad. Sci.* **253**:133.

Glenney, J. R., and Weber, K., 1980, Calmodulin-binding proteins of the microfilaments present in isolated brush borders and microvilli of intestinal epithelial cells, *J. Biol. Chem.* **255**:10551.

Haga, T., Abe, T., and Kurokawa, M., 1974, Polymerization and depolymerization of microtubules *in vitro* as studied by flow birefringence, *FEBS Lett.* **39**:291.

Hartree, E. F., 1972, Determination of protein: A modification of the Lowry method which gives a linear photometric response, *Anal. Biochem.* **48**:422.

Jemiolo, D. K., Keller, T. C. S., Burgess, W. H., and Rebhun, L. I., 1979, Tubulin–CDR interactions in: *The Cytoskeleton: Membranes and Movement* (J. Cordeelis, P. Satir, and K. Burridge, eds.), p. 51, Cold Spring Harbor Laboratory, New York.

Jemiolo, D. K., Burgess, W. H., Rebhun, L. I., and Kretsigner, R. H., 1980, Calmodulin interaction with cycle-purified brain tubulin components, *J. Cell Biol.* **87**:248a.

Kiehart, D. P., and Inoué, S., 1977, Local depolymerization of spindle microtubules by microinjection of calcium ions, *J. Cell Biol.* **70**:230a.

Laemmli, U. K., 1970, Cleavage of structural proteins during the assembly of the head of bacteriophage T4, *Nature (London)* **227**:680.

La Porte, D. C., and Storm, D. R., 1978, Detection of calcium-dependent regulatory protein binding components using ^{125}I-labeled calcium-dependent regulatory protein, *J. Biol. Chem.* **253**:3374.

La Porte, D. C., Wierman, B. M., and Storm, D. R., 1980, Calcium-induced exposure of a hydrophobic surface on calmodulin, *Biochemistry* **19**:3814.

Lee, Y. C., Samson, F. E., Houston, L. L., and Himes, R. H., 1974, The *in vitro* polymerization of tubulin from beef brain, *J. Neurobiol.* **5**:317.

Lin, C. T., Dedman, J. R., Welsh, M. J., Brinkley, B. R., and Means, A. R., 1979, Immunoelectron microscopic localization of calmodulin in the mitotic apparatus, *J. Cell Biol.* **83**:378a.

Marcum, J. M., Dedman, J. R., Brinkley, B. R., and Means, A. R., 1978, Control of microtubule assembly–disassembly by calcium-dependent regulator protein, *Proc. Natl. Acad. Sci. U.S.A.* **75**:3771.

Nishida, E., 1978, Effects of solution variables on the calcium sensitivity of the microtubule assembly system, *J. Biochem. (Toyko)* **84**:507.

Nishida, E., and Sakai, H., 1977, Calcium-sensitivity of the microtubule reassembly system, *J. Biochem. (Toyko)* **82**:303.

Nishida, E., Kumagai, H., Ohtsuki, I., and Sakai, H., 1979, The interactions between calcium-dependent regulator protein of cyclic nucleotide phosphodiesterase and microtubule proteins, *J. Biochem. (Toyko)* **85**:1257.

Olmsted, J. B., and Borisy, G. G., 1973, Characterization of microtubule assembly in porcine brain extracts by viscometry, *Biochemistry* **12**:4282.

Olmsted, J. B., and Borisy, G. G., 1975, Ionic and nucleotide requirements for microtubule polymerization *in vitro*, *Biochemistry* **14**:2996.

Potter, J. D., and Gergely, 1975, The calcium and magnesium binding sites on troponin and their role in the regulation of myofibrillar adenosine, *J. Biol. Chem.* **250**:4628.

Rosenfeld, A. C., Zackroff, R. V., and Weisenberg, R. C., 1976, Magnesium stimulation of calcium binding to tubulin and calcium induced depolymerization of microtubules, *FEBS Lett.* **65**:144.

Salmon, E. D., and Jenkins, R., 1977, Isolated mitotic spindles are depolymerized by μM calcium and show evidence of dynein, *J. Cell Biol.* **75**:295a.

Schliwa, M., 1976, The role of divalent cations in the regulation of microtubule assembly, *J. Cell Biol.* **70**:527.

Weisenberg, R. C., 1972, Microtubule formation *in vitro* in solutions containing low calcium concentrations, *Science* **172**:1104.

Weiss, B., and Levin, R. M., 1978, Mechanism for selectively inhibiting the activation of cyclic nucleotide phosphodiesterase and adenylate cyclase by antipsychotic agents, (W. J. George and L. J. Ignarro, eds.), *Adv. Cyclic Nucleotide Res.* **9**:285.

Welsh, M. J., Dedman, J. R., Brinkley, B. R., and Means, A. R., 1978, Calcium-dependent regulator protein: Localization in mitotic apparatus of eukaryotic cells, *Proc. Natl. Acad. Sci. U.S.A.* **75**:1867.

Wood, J. G., Wallace, R. W., Whitaker, J. N., and Cheung, W. Y., 1980, Immunocytochemical localization of calmodulin and a heat-labile calmodulin-binding protein (CaM-BP$_{80}$) in basal ganglia of mouse brain, *J. Cell Biol.* **84**:66.

7

Microtubules in Adult Mammalian Muscle

M. A. Goldstein and J. Cartwright, Jr.

Microtubules are a consistent feature of adult mammalian cardiac muscle and slow skeletal muscle. They form a network around and between the myofilament bundles (Goldstein and Entman, 1979). In longitudinal sections, the microtubules are seen in longitudinal profile over the myofibrils in grazing sections where the sarcoplasmic reticulum (SR) is seen, and between the myofibrils. In cross sections at various levels, microtubules are seen in cross-sectional view.

In cardiac muscle, the microtubules encircle the centrally placed nucleus. Changes in microtubule orientation are correlated with changes in shape of the nucleus. In longitudinal sections where the nucleus has a smooth contour, longitudinal profiles of microtubules are seen parallel to the nuclear envelope. Where irregular profiles of the nucleus are seen in longitudinal sections, the profiles of microtubules have a more random orientation. Cross sections of both cardiac and skeletal muscle through the region of the nucleus often reveal microtubules rather evenly distributed all around the nucleus (Fig. 1).

In longitudinal sections where extensive arrays of SR membrane are seen, the microtubules are most readily seen (Fig. 2). They run axially and transversely with respect to the myofibril axis. Longitudinal profiles of microtubules seem to wrap around the myofilament bundles and go across the surface of the myofilament bundle at various angles. Longitudinal profiles of microtubules can be seen running closely apposed to the outer membrane of a mitochondrion (Fig. 3). Most of the microtubules have a predominantly axial orientation, and therefore it is possible to estimate microtubule numbers by

M. A. Goldstein and J. Cartwright, Jr. • Section of Cardiovascular Sciences, Department of Medicine, Baylor College of Medicine, and The Methodist Hospital, Houston, Texas 77030.

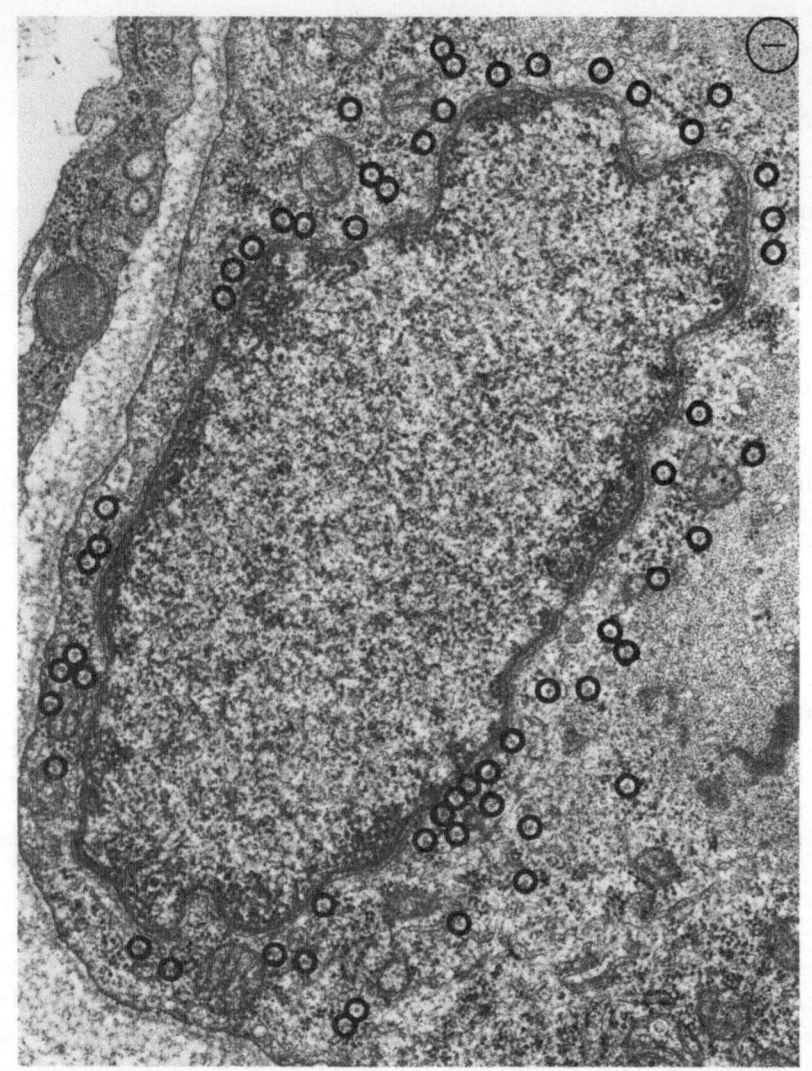

Figure 1. Cross section of soleus muscle from a young adult rat. Cross-cut microtubules (O) are distributed all around the nucleus. ×28,000.

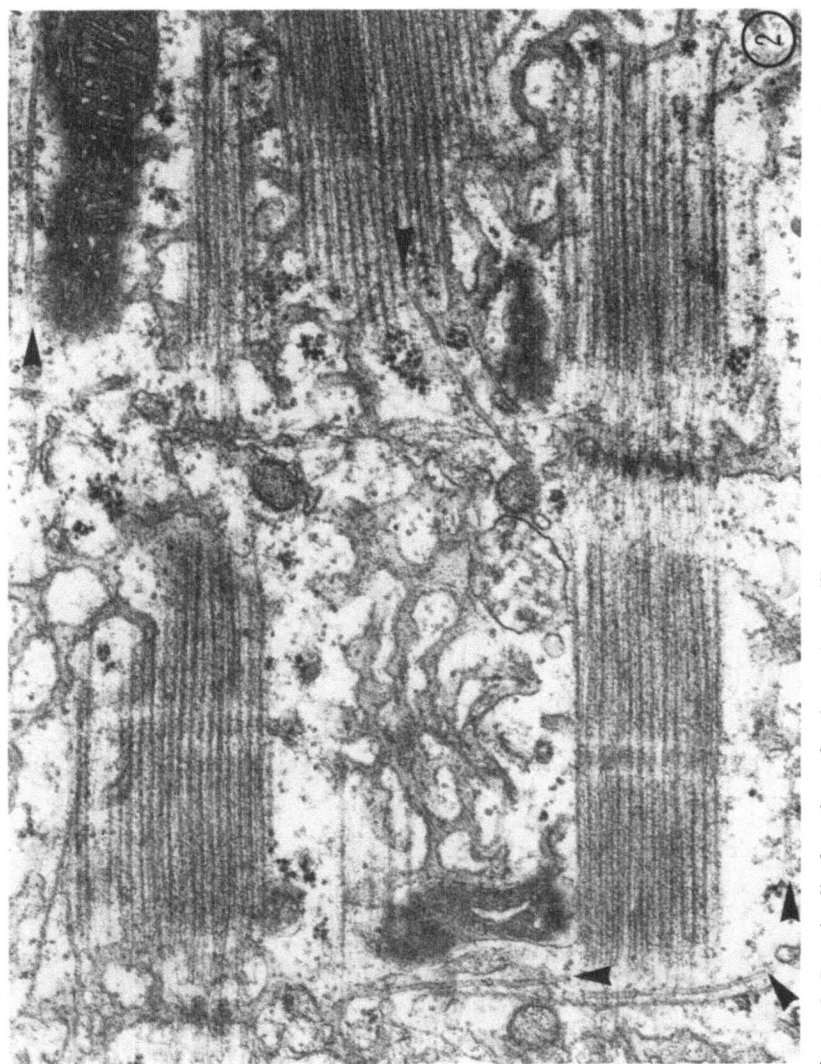

Figure 2. Longitudinal section of guinea pig papillary muscle showing microtubules (arrowheads) oriented axially and transversely to the myofibril axis. ×36,000.

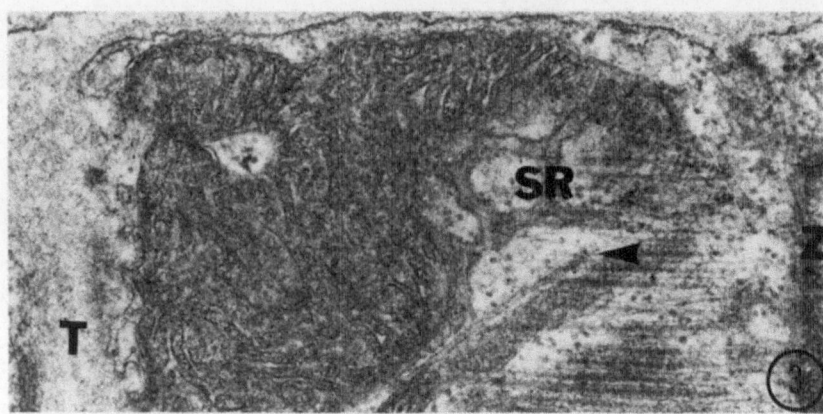

Figure 3. Longitudinal section of canine papillary muscle. Microtubule is parallel to the outer membrane of a portion of a mitochondrion. Portions of a Z-band, a T tubule, and sarcoplasmic reticulum (SR) are seen. ×42,000.

examining random cross sections of muscle. We have computed the number of microtubules per cross-sectional area of muscle in mammalian cardiac-muscle cells.

We have also examined the distribution and number of microtubules in a slow skeletal muscle—the rat soleus, which has a different cell architecture. The nuclei are located at the periphery of the cell (see Fig. 1), rather than at the center as in the case of cardiac muscle. The T–SR junctions are located at the A–I junction and not at the Z-band as in cardiac muscle (see Fig. 3). Microtubules are seen along the rim of the nucleus in longitudinal sections and encircle the nucleus in cross sections. Microtubules are found between the myofilament bundles running axially (Fig. 4) and curving across the myofila-ment bundle (Fig. 5). In some cases, the microtubules run parallel to the T–SR junction for a considerable distance and then curve over into the Z-band

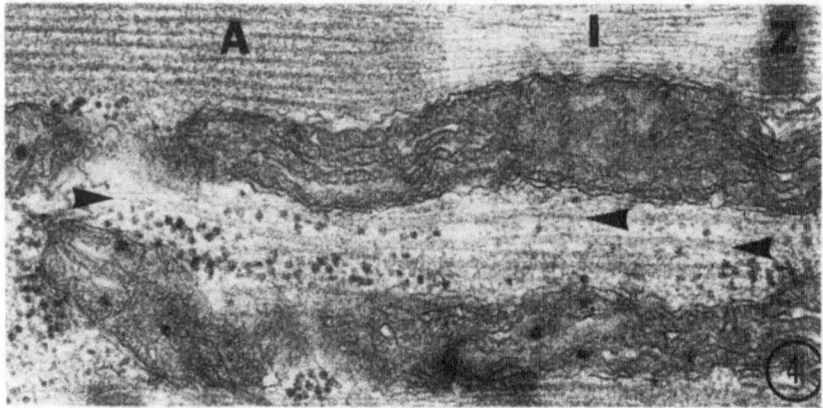

Figure 4. Longitudinal section of soleus muscle. Microtubules (arrowheads) run axially between the myofilament bundles. Portions of A-, I-, and Z-bands are seen. ×56,000.

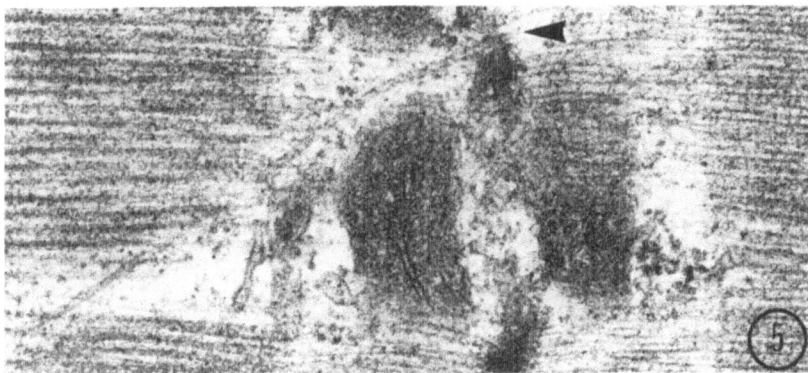

Figure 5. Longitudinal section of soleus muscle showing a microtubule bending and curving across the myofilament bundle. ×56,000.

region (Fig. 6). We did not see any consistent association between the microtubule profiles in longitudinal sections and the T–SR junctions exclusively.

Cross sections of soleus muscle reveal cross-cut profiles of microtubules. Sometimes very few microtubules are seen between the myofilament bundles, and at other times comparable areas of muscle will reveal large numbers of the microtubule profiles. The normal myofilament bundles show the distinct regions of A-, I-, and Z-bands in these cross-sectional profiles. Cartwright and Goldstein (1980) have quantitated the number of microtubules per cross-sectional area in soleus muscle. The quantitative studies substantiate this impression of the variability in the number of microtubule profiles per cross-sectional area of muscle (Fig. 7). Averages from 14 different rats of the same strain show an average value of 0.33. The range of values is from 0.0482 to 0.448 excluding a most unusual value of 6 times the average obtained in a cell where the myofilament bundle did not show the typical distinct regions of A-, I-, and Z-bands.

Next, we examined only the left soleus muscles of male sibling rats to see whether this variation in numbers of microtubules would persist in a more homogeneous population of random samples of muscle. Again, we saw considerable variation not only from animal to animal but from block to block within the same animal. We then examined sibling rats during postnatal development from 1 to 42 days after birth. The average values obtained for the animals in the 1- to 9-day group were considerably higher than those from soleus muscles of the older sibling rats (Cartwright and Goldstein, 1980). Again, there was considerable variation among siblings at the same age and variation from block to block in a single animal. In one cell, an unusually large number of microtubules was associated with an unusual distribution of myofilaments. In this muscle sample, discrete bands were not seen in the cross section of this cell, but Z-dense material was intermixed with patches of thick and thin filaments. The microtubules were distributed throughout these arrays of thick and thin filaments. However, these myofilament bundles comprised a single region in a cell where the rest of the myofilament bundles were

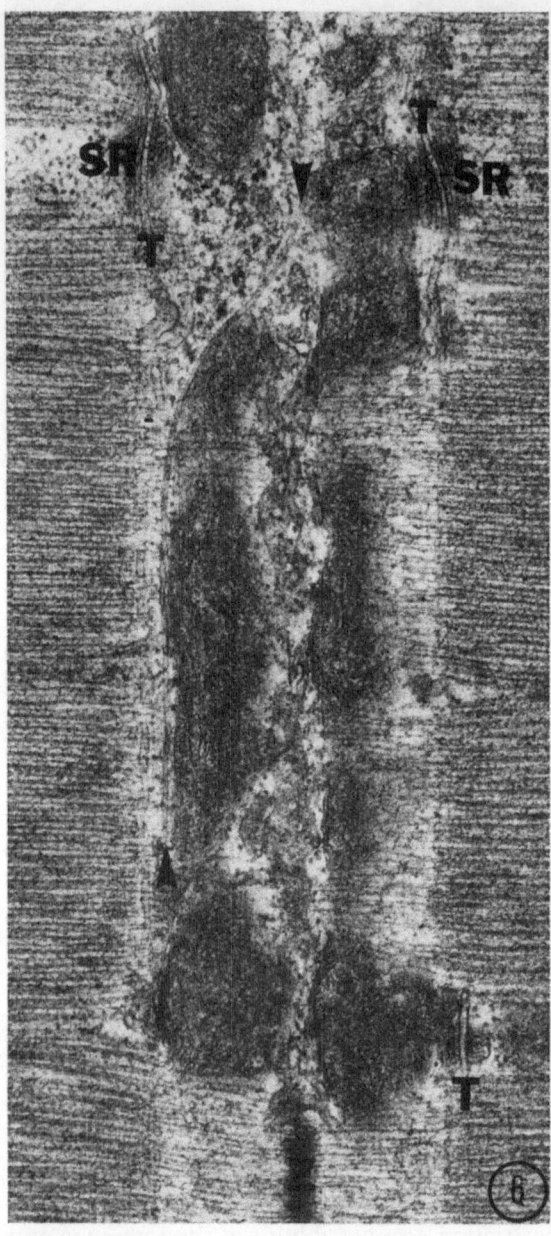

Figure 6. Longitudinal section of soleus muscle. A microtubule (arrowheads) curves from along the T–SR junction over into the Z-band region. ×42,000.

packed into the usual array typical of sarcomere structure in normal muscle. There was no consistent association between myofibril disarray and increased number of microtubules, however.

Microtubules in rat soleus muscle have been analyzed in serial electron micrographs of longitudinal and cross sections. Microtubules are helically arranged around the myofibril and have a predominantly axial orientation.

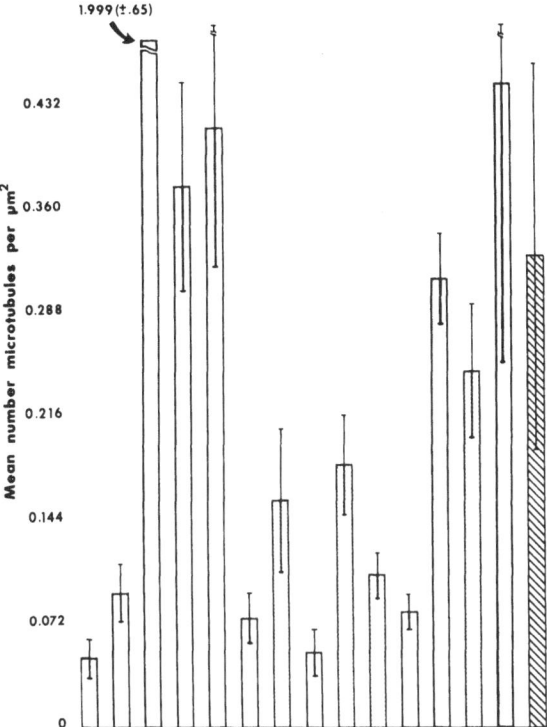

Figure 7. Number of microtubules per cross-sectional area of muscle. Each bar represents the mean number of microtubules per square-micrometer cross-sectional area of soleus muscle from a sexually mature rat. The hatched bar at right represents the population mean for the 14 rats.

The number of microtubules per unit cross-sectional area varies widely. The average number observed is 0.33 microtubule/μm^2 and is somewhat higher than the average value of 0.18 microtubule/μm^2 found in adult cardiac muscle.

Both the morphological studies and the *in vitro* studies of tubulin (Goldstein *et al.*, 1980) suggest that microtubules are quite stable, but can turn over under very specific conditions within individual cells. Large increases in microtubule number have been observed both near the nucleus and around the myofibrils in both cardiac and skeletal muscle. These findings further support the hypothesis that microtubules are an important structural feature of muscle and may perform more than a cytoskeletal function.

ACKNOWLEDGMENTS. The authors wish to thank Mr. David L. Murphy for excellent technical assistance. This research has been supported by the Muscular Dystrophy Association and the National Heart Research and Demonstration Center Grant HL 17269.

References

Cartwright, J., and Goldstein, M. A., 1980, Microtubules in rat skeletal muscle, *J. Cell Biol.* **87**:258a.

Goldstein, M. A., and Entman, M. L., 1979, Microtubules in mammalian heart muscle, *J. Cell. Biol.* **80**:183–195.

Goldstein, M. A., Bucher, J. V., Murphy, D. L., and Entman, M. L., 1980, *In vitro* features of cytoskeletal structures from adult canine myocardium, *Fed. Proc. Fed. Am. Soc. Exp. Biol.* **39**:2161.

8

Microtubules and Heart-Cell Contraction

Arthur P. Bollon, Rhonda R. Porterfield, John W. Fuseler, and Jerry W. Shay

1. Introduction

The integration of specialized cell functions with cytoskeletal structure is a subject of active interest. Clearly, a correlation between the specialized function of muscle contractility and cytoskeletal elements such as microtubules, microfilaments, intermediate filaments, and the microtrabecular lattice should improve our understanding of heart-cell contraction, cell organization, and the relationship of both to cardiac disease.

We have been studying the relationship of heart-cell contraction and cytoskeletal elements using primary heart-cell cultures of fetal rat hearts and have observed that contraction of myocytes in culture is inhibited by dibutyryl cyclic AMP and that such inhibition is reversed by colchicine but not lumicolchicine (Nath *et al.*, 1978; Nath and Bollon, 1978). This chapter will present correlations between the biochemical perturbations of myocyte contraction and the morphological alterations of myocytes as seen by scanning and transmission electron microscopy and immunological techniques utilizing indirect immunofluorescence.

2. Materials and Methods

The myocyte cultures were prepared as previously described (Bollon *et al.*, 1977). The ventricular muscles of 19-day-old embryos of Sprague–Dawley

Arthur P. Bollon • Department of Molecular Genetics, Wadley Institutes of Molecular Medicine, Dallas, Texas 75235. ***Rhonda R. Porterfield, John W. Fuseler, and Jerry W. Shay*** • Department of Cell Biology, The University of Texas Health Science Center at Dallas, Dallas, Texas 75235.

rats were minced and disaggregated by repeated incubation with 0.25% pancreatin solution (GIBCO, Grand Island, New York) at 37°C. The resultant suspension of cells was suspended in CMRL-1066 (GIBCO) medium supplemented with 50 μg/ml insulin (Pfizer or Eli Lilly), 10% horse serum, 3% fetal calf serum, 100 μg/ml penicillin, 100 U/ml streptomycin, and 0.25 μg/ml Fungizone (all from GIBCO). The cell suspension was seeded in 75-cm^2 tissue-culture flasks (Falcon) at various concentrations depending on the experimental design. For analysis of synchronous contracting cell cultures, cells were seeded in the 75-cm^2 flasks at a concentration of 2–6 × 10^6 cells/ml and incubated at 37°C in a 5% CO_2 atmosphere. Following a 90-min incubation, the culture was tapped to loosen unattached myocytes while the more adherent fibroblasts and endothelial cells remained in the 75-cm^2 flasks. The cell suspension, thus enriched with the myocytes, was poured onto 100 × 15 mm petri dishes, each containing about 15 glass coverslips (22-cm-square, Gold Seal). The contracting individual myocytes were observed within 6 hr, and the interconnecting myocyte network was observed between 36 and 48 hr.

For scanning electron microscopy, the myocyte cultures were grown on glass coverslips as indicated above and fixed in a 3% (wt./vol.) glutaraldehyde solution buffered with sodium cacodylate at pH 7.2 after a modification of the procedure described previously (Gershenbaum *et al.*, 1974; Porterfield *et al.*, 1978).

The procedure for transmission electron microscopy was a modification of the techniques described previously (Anderson *et al.*, 1976; Nath *et al.*, 1978). The cells were fixed in glutaraldehyde in 0.2 M sodium cacodylate and rinsed in cacodylate buffer. The cells were postfixed in OsO_4 and dehydrated before being embedded in Epon 812.

The techniques for indirect immunofluorescence are described in Chapters 4 and 9.

3. Results

3.1. Inhibition of Myocyte Contraction

Several agents were tested for their inhibitory effect on myocyte contraction. The most potent inhibitor of synchronously contracting myocytes was 2.00 mM dibutyryl cyclic AMP (db cAMP). Myocyte contraction was inhibited after 2.5 hr of incubation of cell cultures with the db cAMP (Nath *et al.*, 1978). As indicated in Table 1, db cAMP at concentrations of 0.20 and 0.02 mM also inhibited myocyte contraction, resulting in 82 and 58% inhibition, respectively. Several analogues were also tested for inhibitory action as indicated in Table 1. The most effective analogue was 6 N-b cAMP (2.00 mM), which inhibited myocyte contraction by 99% after 5.5 hr of incubation with the cells. In addition, 2.00 mM ^2O-b cAMP inhibited contraction by 60%, also after 5.5 hr of incubation with the cells (Nath and Bollon, 1978).

The inhibition of myocyte contraction by db cAMP is reversible by wash-

Table 1. Effect of Various Agents on the Rate of Myocyte Contraction (Beats/Minute)

	Time interval following administration of agent		
Agent	0 hr	3–4 hr	5–6 hr
Dibutyryl c̄AMP[a]			
2.00 mM	69	0.4	0
0.20 mM	45	8.0	7.0
0.02 mM	59	25.0	29.0
^2O-b cAMP[b]			
2.00 mM	75	40.0	25.0
^6N-b cAMP[b]			
2.00 mM	74	35.0	0

[a] For 2.00, 0.20, and 0.02 mM dbcAMP, 19, 8, and 6 cultures were used, and the beats/minute reported is the average value of 70, 35, and 30 observations for the three db cAMP concentrations used.
[b] The effect of 2.00 mM ^6N-bcAMP or ^2O-bcAMP was tested on 8 and 6 cultures, respectively.

ing the treated cells with culture medium or by the addition of 1μM colchicine or 1 μM colcemid, as can be seen in Table 2. The addition of water or 1 or 10 μM lumicolchicine did not reverse the db cAMP inhibitory effect.

3.2. Relationship of Microtubules and Myocyte Contraction

The morphology of myocytes treated with the various agents discussed above was examined by scanning and transmission electron microscopy and by the use of immunofluorescence techniques using tubulin and actin antibodies.

As previously shown (Porterfield *et al.*, 1978), the morphology of myocytes treated with db cAMP was dramatically altered. Treatment of control myocytes (Fig. 1) for 3 hr with 0.1–2.0 mM db cAMP resulted in the inhibition of contraction as well as the development of elongated processes (Fig. 2). The cultures treated with colchicine (100 μg/ml) displayed normal morphology compared to untreated control cells. Cultures initially treated with db cAMP

Table 2. Reversal of Dibutyryl cAMP Inhibition of Myocyte Contraction (Beats/Minute)

	Time interval following administration of agent				
Agent	0 hr	1.5 hr	3 hr	4 hr	5 hr
Water	50	49	45	42	40
Dibutyryl cAMP					
(2 mM)	40	20	0	0	0
Colchicine (1 μM)					
(added 3 hr after					
db cAMP indi-					
cated above)	—	—	0[a]	10[a]	30[a]

[a] Six cultures were tested for each of the time points after colchicine was added.

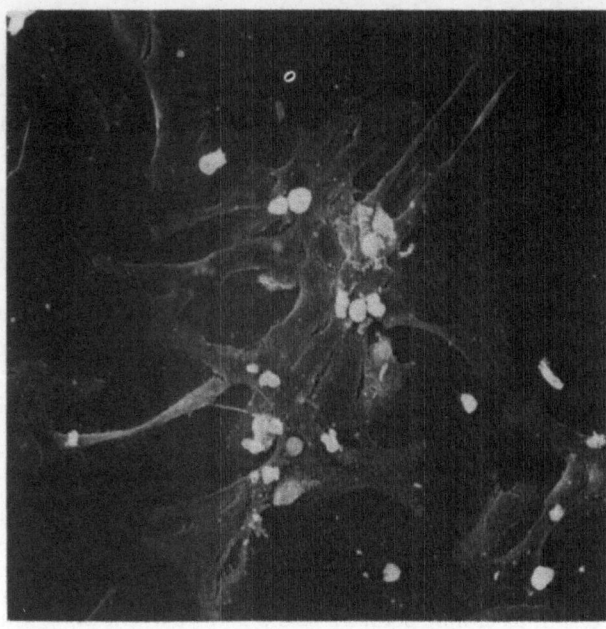

Figure 1. Scanning electron micrograph of untreated myocytes. Contracting myocyte cultures that had not been treated with any agents such as db cAMP were analyzed by scanning electron microscopy as described in Section 2.

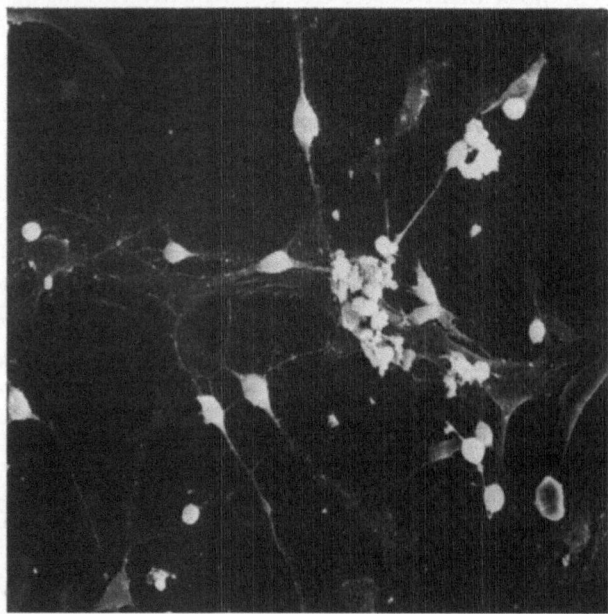

Figure 2. Scanning electron micrograph of myocytes treated with db cAMP. Myocyte cultures were treated for 3 hr with 2.0 mM db cAMP, and the cells were analyzed by scanning electron microscopy as described in Section 2. The elongated processes described in the text can be observed.

followed by treatment with colchicine also showed myocytes with relatively normal morphology.

The altered myocyte morphology as observed by scanning electron microscopy could be correlated with altered microtubule orientation by observing db cAMP-treated myocytes with transmission electron microscopy. In cultures treated with db cAMP for 3 hr, alignment of microtubules into parallel arrays was frequently observed (Fig. 3). The parallel arrays of microtubules observed in the db cAMP-treated cells were rarely observed in untreated myocytes (Figure 4). Intact microtubules were not observed in colchicine-treated myocytes.

To further analyze the arrangement of microtubules in db cAMP-treated myocytes, immunofluorescence was done using affinity-purified tubulin antibodies. As indicated in Fig. 5, use of an actin antibody permitted the identification of sarcomere structures, thus verifying that the cells being analyzed were myocytes. The use of antitubulin confirmed our previous findings obtained by transmission electron microscopy (Nath *et al.*, 1978) that the microtubules of myocytes treated with db cAMP were arranged in parallel arrays (Fig. 6) when compared with untreated myocytes (Fig. 7).

3.3. Intercellular Organization

The inhibition of myocyte contraction by db cAMP was observed by analyzing the synchronous contraction of myocyte networks. If individual cells that are usually on the periphery of such myocyte networks were examined in db cAMP-treated cultures, it was observed that their contraction was

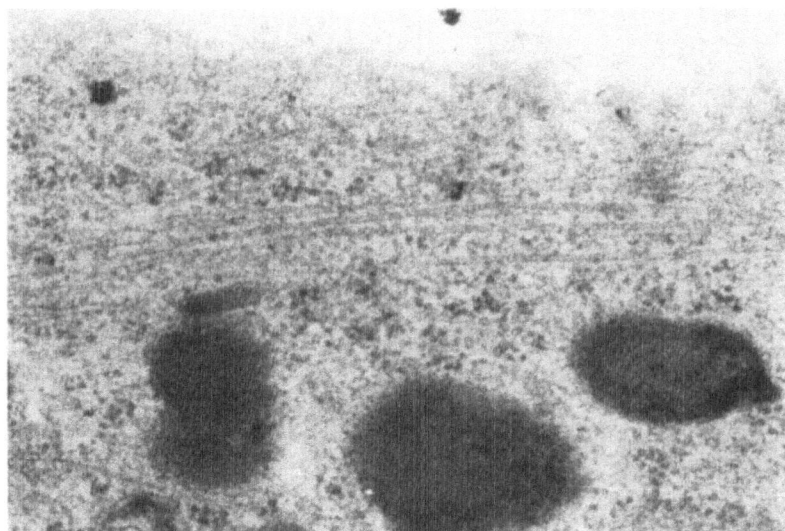

Figure 3. Transmission electron micrographs of myocytes treated with db cAMP. Myocyte cultures were treated for 3 hr with 2.0 mM db cAMP, and the cells were analyzed by transmission electron microscopy as described in Section 2. The parallel alignment of microtubules, as indicated in the text, can be detected.

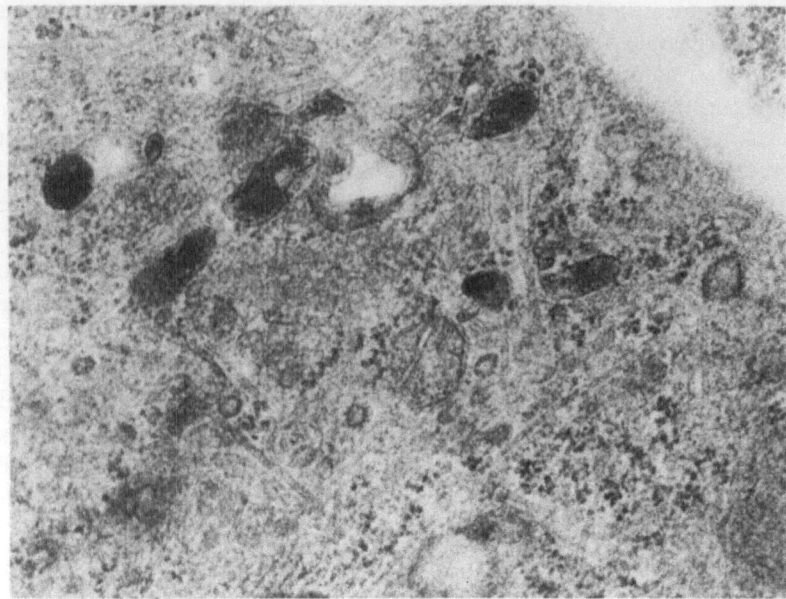

Figure 4. Transmission electron micrograph of untreated myocytes. Contracting myocyte cultures that had not been treated with any agents such as db cAMP were analyzed by transmission electron microscopy as described in Section 2. A random alignment of microtubules can be detected.

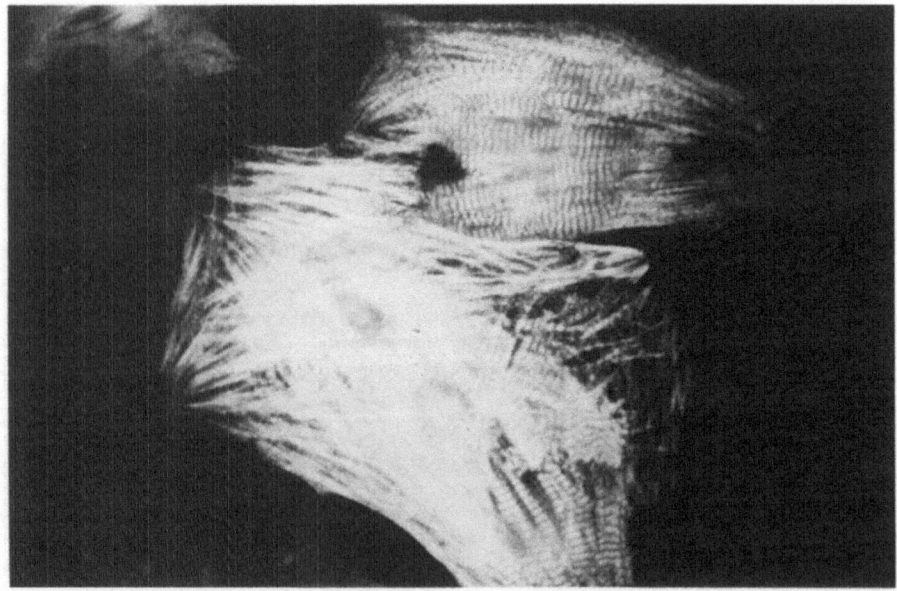

Figure 5. Immunofluorescence micrograph of myocytes using antiactin antibody. Untreated myocyte cultures were examined by indirect immunofluorescence using antiactin as indicated in Section 2. The detection of the I-band cross-striations confirms that the cells being analyzed are heart-muscle cells.

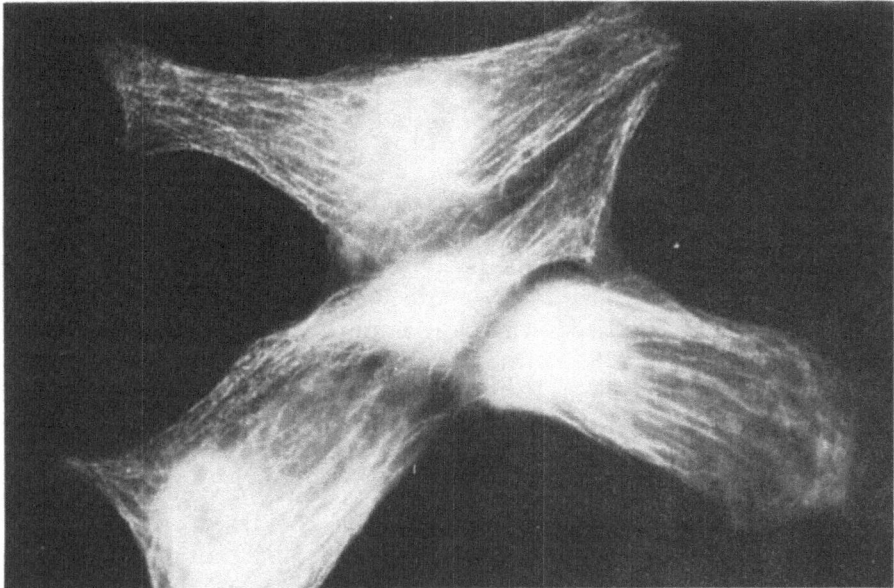

Figure 6. Immunofluorescence micrograph of db cAMP-treated myocytes using antitubulin antibody. Myocyte cultures treated with 2.0 mM db cAMP for 3 hr were analyzed by indirect immunofluorescence using antitubulin antibody as indicated in Section 2. Parallel alignment of microtubules can be detected.

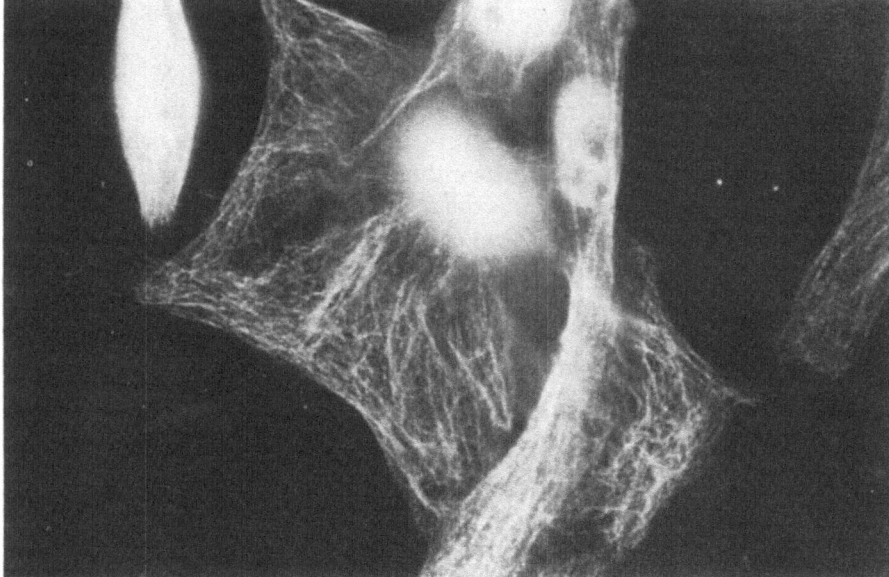

Figure 7. Immunofluorescence micrograph of untreated myocytes using antitubulin antibody. Contracting myocyte cultures that had not been treated with any agents such as db cAMP were analyzed by indirect immunofluorescence using antitubulin antibody as indicated in Section 2. A less ordered alignment of microtubules could be observed when compared to db cAMP-treated cells (Fig. 6).

not completely inhibited. Furthermore, if cultures were prepared less densely, by plating fewer cells, a variety of culture morphologies could be generated and analyzed for altered contraction. The results generated from such studies indicated that only myocyte networks were completely inhibited.

4. Discussion

Although in prior studies the purpose of treating myocytes with various agents such as db cAMP has been primarily for physiological studies (Krause *et al.*, 1972), we have utilized such drugs as physiological and morphological probes (Nath *et al.*, 1978; Nath and Bollon, 1978). The use of agents for probing cytoskeletal elements has been well established, especially in the work using db cAMP on Chinese hamster ovary cells, where the treated cells change orientation from a randomly oriented multilayer of cells into a monolayer of elongated fibroblastlike cells (Hsie and Puck, 1971). In such studies, it was demonstrated that the morphological changes could be correlated with a change in microtubule arrangement into a longitudinal configuration (Porter *et al.*, 1974). In addition, the correlation between elongation of eukaryotic cells and alignment of longitudinally oriented microtubules has been observed in a number of cell systems studied (Handel and Roth, 1971; Warren, 1974).

Hence, the use of various agents that affect myocyte morphology, myocyte contraction, and microtubule organization permit us to postulate that the organization of microtubules can influence myocyte contraction. Clearly, extensive microtubule organization is not needed for contraction, since myocytes treated with colchicine contract normally. Whether dbcAMP has a direct effect on microtubule organization or some other indirect action such as endogenous calcium availability for contraction remains to be established. It is possible that the alignment of microtubules in parallel arrays may confer a physical constraint on the myocytes that prevents them from contracting. This interpretation may be further supported by the fact that only myocyte networks and not individual cells appear to be inhibited by db cAMP. Studies presented in this chapter may improve our understanding of the relationship between myocyte contraction and cytoskeletal intercellular interactions not only in cultured myocytes arranged in networks but also in the more complex arrangements in the heart.

References

Anderson, R. G., Goldstein, J. L., and Brown, M. S., 1976, Localization of low density lysoprotein receptors on plasma membrane of normal human fibroblasts and their absence in cells from a familial hypercholesterolemia homozygote, *Proc. Natl. Acad. Sci. U.S.A.* **73**:2434.
Bollon, A. P., Nath, K., and Shay, J. W., 1977, Establishment of contracting heart muscle cell cultures, in: *Tissue Culture Association Manual*, Vol. 3 (V. Evans, V. Perry, and M. Vincent, eds.), pp. 637–640, Tissue Culture Association, Rockville, Maryland.
Gershenbaum, M. R., Shay, J. W., and Porter, K. R., 1974, The effects of cytochalasin B on

BALB-3T3 mammalian cells cultured *in vitro* as observed by scanning and high voltage electron microscopy, in: *Proceedings of the 7th Annual Scanning Electron Microscope Symposium* (O. Johari and I. Corwin, eds.), pp. 589–596, IIT Research Institute, Chicago, Illinois.

Handel, M. A., and Roth, L. E., 1971, Cell shape and morphology of the neurol tube: Implications for microtubule function, *Dev. Biol.* **25:**78.

Hsie, A. W., and Puck, T. T., 1971, Morphological transformation of Chinese hamster cells by dibutyryl adenosine cyclic 3′:5′-monophosphate and testosterone, *Proc. Natl. Acad. Sci. U.S.A.* **68:**358.

Krause, E. G., Halle, W., and Wollenberger, A., 1972, Effect of dibutyryl cyclic GMP on cultured beating rat heart cells, *Adv. Cyclic Nucleotide Res.* **1:**301.

Nath, K., and Bollon, A. P., 1978, Effect of dibutyryl cyclic AMP and analogs on the rate of contractions of myocytes in culture, *Experientia* **34:**1282.

Nath, K., Shay, J. W., and Bollon, A. P., 1978, Relationship between dibutyryl cyclic AMP and microtubule organization in contracting heart muscle cells, *Proc. Natl. Acad. Sci. U.S.A.* **75:**319.

Porter, K. R., Puck, T. T., Hsie, A. W., and Kelley, D., 1974, An electron microscope study of the effects of dibutyryl cyclic AMP on Chinese hamster ovary cells, *Cell* **2:**145.

Porterfield, R. R., Kagan, T. M., Bollon, A. P., and Shay, J. W., 1978, Scanning electron microscopic observations on normal and drug-treated primary heart cell cultures, *Scanning Electron Microsc.* **II:**465.

Warren, R. H., 1974, Microtubular organization in elongating myogenic cells, *J. Cell. Biol.* **63:**550.

9

The Association of Creatine Phosphokinase with the Mitotic Spindle

John W. Fuseler, Barry S. Eckert, Stephen J. Koons, and Jerry W. Shay

1. Introduction

The M-isozyme of creatine phosphokinase (M-CPK) has been proposed as a
phosphorylation intermediate in the respiratory control of striated muscle
(Bessman *et al.*, 1980). This isozyme has been shown previously to be a constituent of the M line of avian cardiac (Wallimann *et al.*, 1978a) and skeletal
muscle (Wallimann *et al.*, 1978a,b). Additionally, M-CPK has been demonstrated in the region of the A-band (Sharov *et al.*, 1977) and associated with
myosin (Turner *et al.*, 1973), as well as being found in abundance within
mitochondria (Jacobus and Lehninger, 1973; Sharov *et al.*, 1977). In striated
muscle, a close functional relationship has been suggested between myofibrillar CPK and ATPase in that ATP generated from creatine phosphate apparently has a preferred access to the myofibrillar ATPase system (Bessman *et al.*,
1980). M-CPK has been found in association with cytoskeletal elements in
both muscle (Fuseler and Shay, 1980) and nonmuscle cells (Eckert *et al.*,
1980), and is predominantly associated with intermediate filaments in the
cytoplasm of muscle (Fuseler *et al.*, 1981) and nonmuscle (Eckert *et al.*, 1980)
cells grown in culture. Additionally, this isozyme can associate with actin
stress fibers in mammalian cardiac cells in primary culture, but does not
associate with cytoplasmic microtubules in various cell types (Fuseler and
Shay, 1980; Fuseler *et al.*, 1981). These observations have given rise to the
hypothesis that cytoskeletal elements may function to support a CPK-

John W. Fuseler and Jerry W. Shay • Department of Cell Biology, The University of Texas
Health Science Center at Dallas, Dallas, Texas 75235. *Barry S. Eckert and Stephen J. Koons* •
Department of Anatomical and Biophysical Sciences, School of Medicine, State University of New
York at Buffalo, Buffalo, New York 14214.

catalyzed energy-generating system. By such a scheme, the distribution and organization of the elements of the cytoskeleton or an organelle could influence local levels of available ATP and hence energy in the cytoplasm.

In this chapter, we present observations that indicate that M-CPK is associated with linear, labile elements of the mammalian mitotic spindle and propose that a possible function of M-CPK in the spindle is to provide a linear energy-generation system. The ATP produced by such a system could provide for maintenance of spindle morphology, the poleward flux of spindle material, the poleward movements of the half spindles, or the directed movements of the chromosomes.

2. Material and Methods

2.1. Myocyte Isolation

The methods for isolation of embryonic myocardial cells are a modification of a previously described method (Bollon *et al.*, 1977; Nath *et al.*, 1978). This method yields an initial population of between 90 and 95% myocytes, with the remaining cells endothelial, or otherwise noncontractile. The mitotic index of such a mixed myocardial-cell population is about 1%. The myocardial cells were obtained from 18-day-old embryonic Sprague–Dawley rats. About 12–20 embryos from 2–4 female rats were used to provide approximately 10^7 cells per experiment. Immediately after its isolation, the heart was placed in adhesion-salts buffer (in g/liter): $MgSO_4 \cdot 7H_2O$, 0.2; NaCl, 6.8; KCl, 0.4; $NaH_2PO_4 \cdot H_2O$, 1.5; glucose, 1; HEPES, 4.76; pH 7.5. The auricles were then dissected away and discarded to minimize the presence of connective tissue. The ventricles were transferred to a sterile 60-mm petri dish containing adhesion-salts buffer, rinsed several times to remove blood cells, and dissected into pieces about 1 mm³. These fragments were placed in a sterile 25-ml spinner flask. To this flask was added 20 ml digestion medium (0.6 mg/ml Pancreatin; GIBCO, Grand Island, New York) and 1% nonessential amino acids and vitamins (GIBCO) in adhesion-salts buffer, and the mixture was stirred slowly in a CO_2 incubator (5% CO_2, 95% humidity, 37°C) for 30 min. The supernatant from the initial digestion was discarded, and 20 ml of fresh digestion medium was added and the mixture stirred for 60 min in the incubator. The supernatant was removed with a Pastuer pipette and placed in a 15-ml sterile conical contrifuge tube containing 1 ml Dulbecco's Modified Eagle's Medium (DMEM) for tissue culture (GIBCO) supplemented with 15% horse serum (Kansas City Biological, Lenexa, Kansas), 5% fetal calf serum (Kansas City Biological), and 1% antibiotics (GIBCO), mixed well, and placed on ice for 5 min. The cold suspension was centrifuged in an IEC clinical centrifuge for 3 min at 1000 rpm, and then 1 min at 500 rpm. The supernatant was discarded and the pellet of cells resuspended in 1 ml DMEM culture medium and placed back on ice. The procedures described above were repeated until 90–95% of the heart tissue was digested.

The myocytes were isolated from nonmyocyte (endothelial and fibroblast) cells by differential adhesion. The 1.0-ml cold ventricular-cell suspension was transferred into a 75-cm² Corning culture flask (Corning, New York) containing 10 ml prewarmed (37°C) DMEM and incubated in the CO_2 incubator for 90 min; next, the 75-cm² flask was tapped with moderate force to detach the myocytes, which were only lightly adhered to the flask surface. The nonmyocytes, which spread more rapidly and attach more strongly to the flask surface than do the myocytes, are not released by this method. The cells in suspension were counted and plated on 22 × 22 mm or 11 × 22 mm sterile, unwashed Gold Seal coverslips previously placed in a sterile 60-mm² petri dish (Corning). The cells in the drop were allowed to adhere to the coverslip for 24 hr; then the dish was flooded with 2 ml DMEM. The cells on the coverslip were allowed to grow for 5–7 days before being used for experiments.

2.2. Immunocytochemistry

The antibody against bovine-brain tubulin was prepared in rabbits by the methods described by Fuller (Fuller *et al.*, 1975), purified on immunoabsorbent columns, and found to be specific for microtubules of the spindle and cytoplasmic microtubule complex. The antibody against the isozyme of M-CPK was prepared and characterized by Dr. B. Eckert (Eckert *et al.*, 1980). This antibody (Fig. 1A), when absorbed with tubulin or intact microtubules, did not lose its capacity to label microtubules of the mitotic spindle (Fig. 1B). The anti-M-CPK, when absorbed with its antigen M-CPK, gave negative results (Eckert *et al.*, 1980) in that no CPK staining was observed in any of the

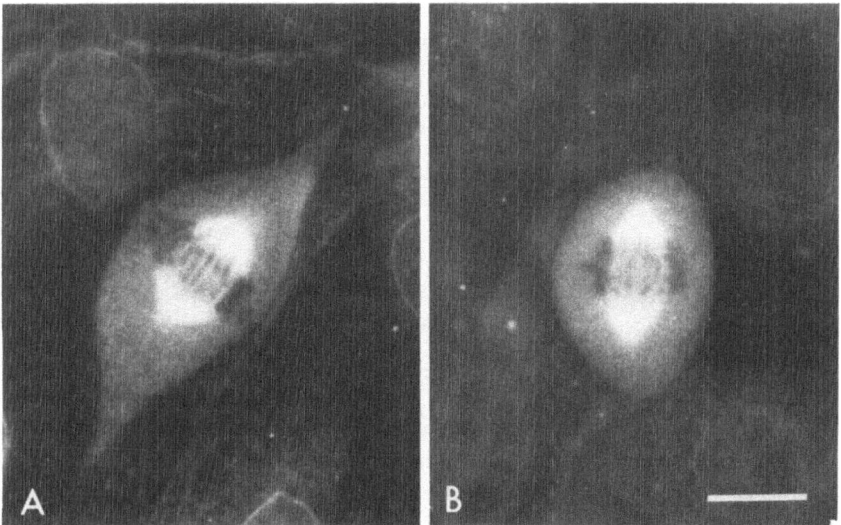

Figure 1. (A) Mitotic spindle stained with antibody against M-CPK. (B) Mitotic spindle stained with antibody against M-CPK absorbed with tubulin, showing no apparent change in morphology or spindle-fiber distribution. Scale bar: 10 μm.

cells. The cells were prepared for indirect immunofluorescence as previously described (Miller *et al.*, 1977). The antibodies were used at a protein concentration of between 0.5 and 0.1 mg/ml. The stability of the labeling with fluorescein isothiocyanate (FITC) goat anti-rabbit IgG was tested by washes with high-pH (8.0) phosphate-buffered saline (PBS).

2.3. Cold Treatment of Cells

Myocytes growing on coverslips were placed in a Coplin jar containing DMEM at 37°C. The jar containing the cells was placed on an ice–salt bath and the temperature monitored with a thermistor probe (Yellow Springs Instruments, Yellow Springs, Ohio). As the temperature dropped, coverslips were quickly removed and fixed. The fixative (3% ultrapure formalin, in PBS), also contained in a Coplin jar, was placed in the ice bath simultaneously with the jar containing the cells so that the temperature of the medium and fixative remained within ±1°C for the duration of the experiment. The cells were maintained at 0°C for upward of 2 hr, after which the Coplin jars containing cells and fixative were removed from the ice bath and placed at 37°C. As the cells warmed further, samples were taken and fixed. After all the samples had been fixed for a minimum of 20 min, the coverslips were prepared for indirect immunofluorescence.

2.4. Microscopy

The stained cells were viewed on a Leitz Orthoplan microscope (E. Leitz, Rockleigh, New Jersey) equipped for transmitted dark-field fluorescence microscopy. The excitation source was an XBO 150-W xenon lamp (Osram) filtered with a Leitz BG38 red suppression filter and a KP490 FITC excitation filter. The barrier filter was a Leitz K530. Photographic records were made with the Leitz Orthomat automatic camera system using Kodak Tri-X (ASA 400) or Kodak SO115 (ASA 50) film. Magnification was calibrated by photographing a standard stage micrometer through the optical system.

3. Results

We have previously observed by immunocytochemical methods that M-CPK is associated with the mitotic spindle of mammalian cells (Fuseler and Shay, 1980; Fuseler *et al.*, 1981). The staining of the spindle by the anti-M-CPK is specific, since the fluorescence is stable when the cells are washed with high-pH buffer. Furthermore, absorption of the anti-M-CPK (Fig. 1A) with tubulin (Fig. 1B) did not inhibit the antibody from staining the spindle, whereas absorption of anti-M-CPK with CPK completely abolished staining (Eckert *et al.*, 1980).

The changes in distribution of CPK in various stages of mitosis in cultured primary rat myocardial cells and in the L-6 established cell line are similar. These distributions of CPK in normal dividing cells and cells in which the spindles are reversibly disassembled closely follows the changes in microtubule spindle-fiber distribution. There are, however, some minor differences between microtubule distribution and CPK distribution when the spindle is reassembling from cold treatment.

3.1. Prophase

The distribution of microtubules in the prophase spindle prior to the breakdown of the nuclear membrane consists of two distinctly separated polar asters, with their large accompanying array of astral fibers. These polar asters are connected by numerous microtubular fiber bundles extending over the surface of the nucleus. The prophase-spindle asters are very large, with fibers extending well toward the margin of the cell. The prophase spindle consists entirely of thin bundles of microtubules that resemble interzonal fibers that maintain connections with both poles and cover the surface of the nucleus (Fig. 2A).

The morphology of CPK found in the prophase spindle is predominantly in a filamenteous form. The distribution of CPK in the spindle at this stage consists of diffuse thin filaments of CPK that extend over the nucleus and connect the poles. This filamentous form of CPK is localized in the asters and also is found in association with the astral fibers radiating from the poles. The CPK filaments associated with the astral fibers are much less extensive than microtubule astral fibers and do not extend very far into the cytoplasm from the pole (Fig. 2B). Filamentous CPK is not observed in the cytoplasm of the cells in prophase.

3.2. Prometaphase

The stage of prometaphase follows breakdown of the nuclear membrane and is characterized by the random orientation of the chromosomes. The distribution of microtubules in the prometaphase-spindle fibers is predominantly localized in the region between the spindle poles. Numerous thin, continous spindle fibers are observed connecting the poles. Distinct kinetochore-spindle-fiber microtubules are not clearly resolvable at this early stage (Fig. 2C).

The filamentous CPK associated with the prometaphase spindle is found predominantly at the poles. The CPK filaments of the polar asters are smaller and less extensive than those observed in prophase. In the region between the poles, the prophase spindles possess diffuse CPK staining. The few thin CPK filaments that are visible in the spindle do not appear to be associated with the chromosomes and do not represent early-forming kinetochore fibers (Fig. 2D). At this stage of mitosis, spindle microtubules and filamentous CPK possess essentially the same morphology and distribution.

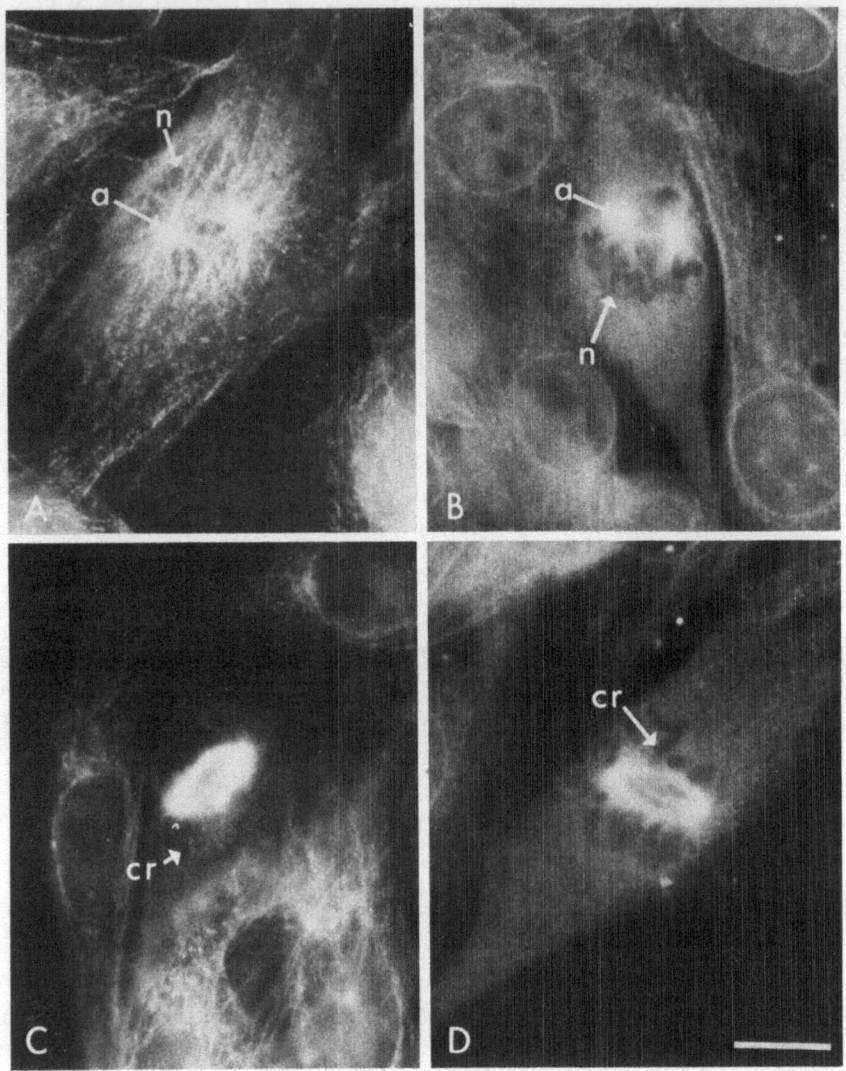

Figure 2. (A) Cell in prophase stained with tubulin antibody showing distribution of microtubules in the prophase spindle. (B) Distribution of M-CPK in the prophase spindle. (C) Distribution of microtubules in the prometaphase spindle. (D) Distribution and morphology of M-CPK associated with the prometaphase spindle. (a) Spindle asters; (n) nucleus; (cr) chromosomes. Scale bar: 10 μm.

3.3. Metaphase

During the stages of metaphase in which the daughter chromosomes become aligned in their equilibrium position on the metaphase plate, the fibers of the spindle become distinct and the asters diminish in size. The distribution of microtubules in the spindle fibers and the presence of filamentous CPK are very similar. In the daughter half-spindle, kinetochore-fiber microtubule bundles are prominent and can be easily traced from the chromosomes to the poles (Fig. 3A). The pole of the metaphase spindle stained with tubulin antibody in many cases reveals the polar centriole, which is slightly removed from the convergence point of the spindle fibers.

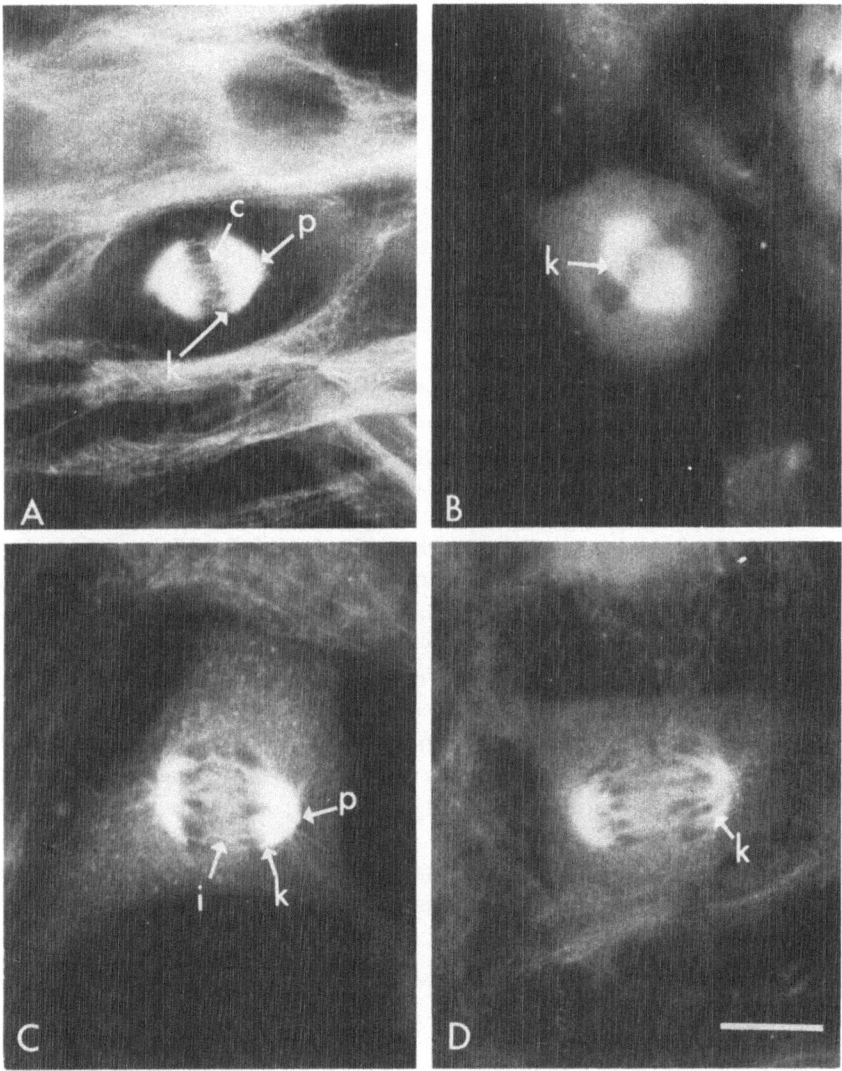

Figure 3. (A) Metaphase spindle stained with tubulin antibody showing distribution of microtubules in kinetochore and continuous spindle fibers. (B) Distribution of M-CPK in the metaphase spindle showing a close similarity to the distribution of spindle microtubules. (C) Anaphase spindle showing shorter microtubules in the region between the chromosomes and spindle and an abundance of interzonal microtubules. (D) Distribution of M-CPK in the anaphase spindle, primarily found in the daughter half-spindle, and little if any present with the interzonal fibers. (k) Kinetochore fibers; (c) continuous spindle fibers; (p) centrosome; (i) interzonal fibers. Scale bar: 10 μm.

The filamentous CPK in the metaphase spindle is more diffuse in appearance than are the microtubule bundles, and only a few distinct, individual CPK fibers are resolved. These stronger-staining CPK fibers appear to extend from the daughter chromosomes to the spindle poles and resemble the microtubule distribution of kinetochore fibers (Fig. 3B). The centriolar region of the spindle does not exhibit the presence of CPK. The highly diminished metaphase astral-fiber complexes are likewise negative for CPK during metaphase.

3.4. Anaphase

Anaphase is characterized by the segregation of the daughter chromosomes to their respective poles accompanied by progressive shortening of the spindle fibers between the chromosomes and the poles. Numerous studies (Fuseler, 1975; Salmon, 1975; Inoué and Ritter, 1975) have shown that microtubules of the anaphase half-spindle are progressively disassembled as the chromosomes move toward the poles.

The microtubules of the anaphase spindle are predominantly localized in the region between the chromosomes and poles (Fig. 3C). The kinetochore fibers are distinct during the early stages of anaphase. In the later stages of anaphase, the microtubular spindle fibers become very diffuse and stain much more weakly. During mid- to late anaphase, the asters begin a regrowth of microtubules into the cytoplasm of the daughter cells. The interzonal microtubules and microtubule bundles are well defined and quite numerous during anaphase. These interzonal microtubules extend continuously from pole to pole in most cases (Fig. 3C). There are, however, some interzonal fibers that exhibit an interruption of tubulin staining in the middle of this region where the cleavage furrow will be formed.

The distribution of CPK in anaphase is predominantly confined to the region of the daughter half-spindle between the chromosomes and the poles. The CPK in the anaphase half-spindle is diffuse, and individual CPK filaments cannot be distinguished at this stage (Fig. 3D). The interzonal region is predominantly devoid of CPK, with only occassional interzonal fibers being visible at this stage. The spindle pole and astral fibers also exhibit a paucity of CPK in mid- to late anaphase.

3.5. Telophase

From telophase to early G_1 stage, which marks the conclusion of mitosis, microtubules are found in the midbody that connects the two daughter cells. Microtubules are also localized in the cytoplasm associated with the cytoplasmic microtubule complex (CMTC) that forms from the centrosome of the spindle (Fig. 4A). At this stage, CPK is predominantly localized only in the midbody of the spindle remnant. CPK is not seen in association with the CMTC or its organization center (Fig. 4B).

The stability of the association of CPK with the mitotic spindle was further studied by subjecting the cells to cyclic temperature changes. When the

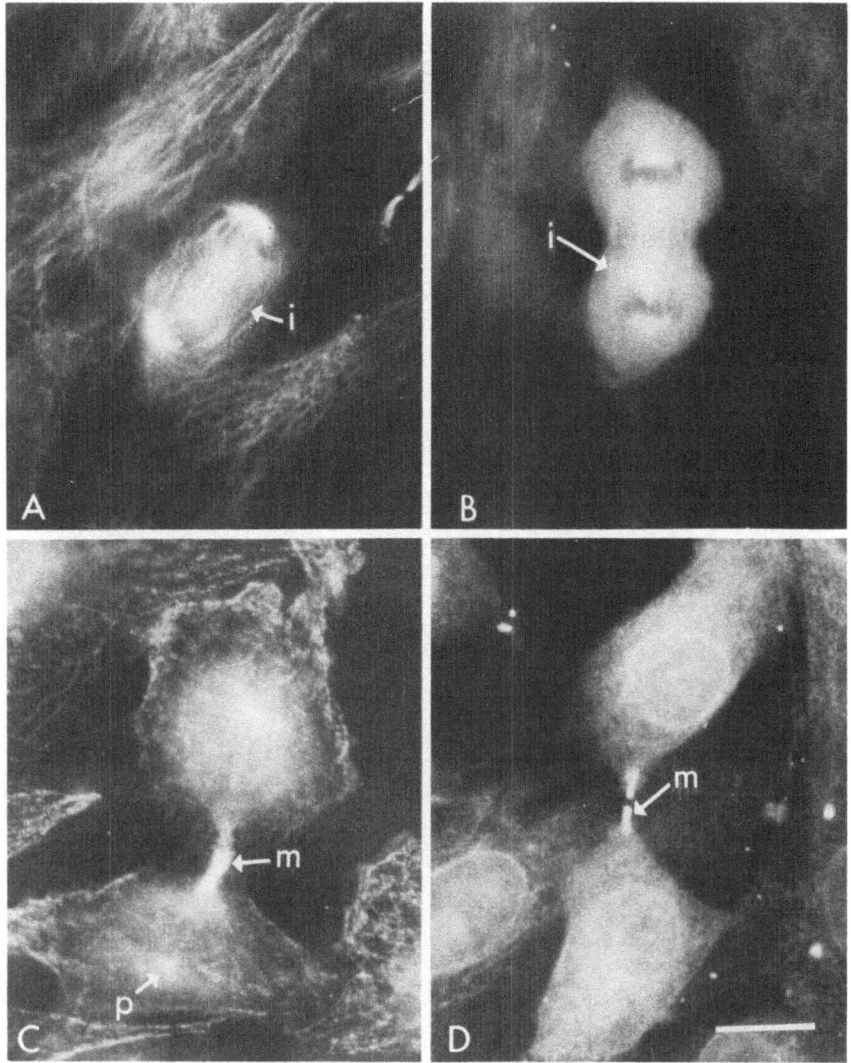

Figure 4. (A) Late anaphase to telophase. The microtubules of the centrosome begin enlarging to re-form the CMTC of the interphase cell. (B) Distribution of M-CPK. The isozyme is found only in the remaining anaphase daughter spindle and a few of the interzonal fibers, but is not associated with centrosome or cleavage furrow. (C) Late telophase. Microtubules of CMTC are well formed, and the nuclear membrane is re-formed. Microtubules are present in the midbody. (D) Distribution of M-CPK. The isozyme is found only in the midbody at late telophase. (i) Interzonal fibers; (p) centrosome; (m) midbody. Scale bar: 10 μm.

environmental temperature of the cells is lowered to around 0°C, microtubules of the spindle fibers and CMTC are induced to depolymerize (Fig. 5A–C). Conversely, when the temperature is elevated from 0°C, microtubules undergo repolymerization, and the spindle and CMTC are re-formed (Fig. 6A–C).

Slow cooling of the mammalian myocardial cells in mitosis results in de-

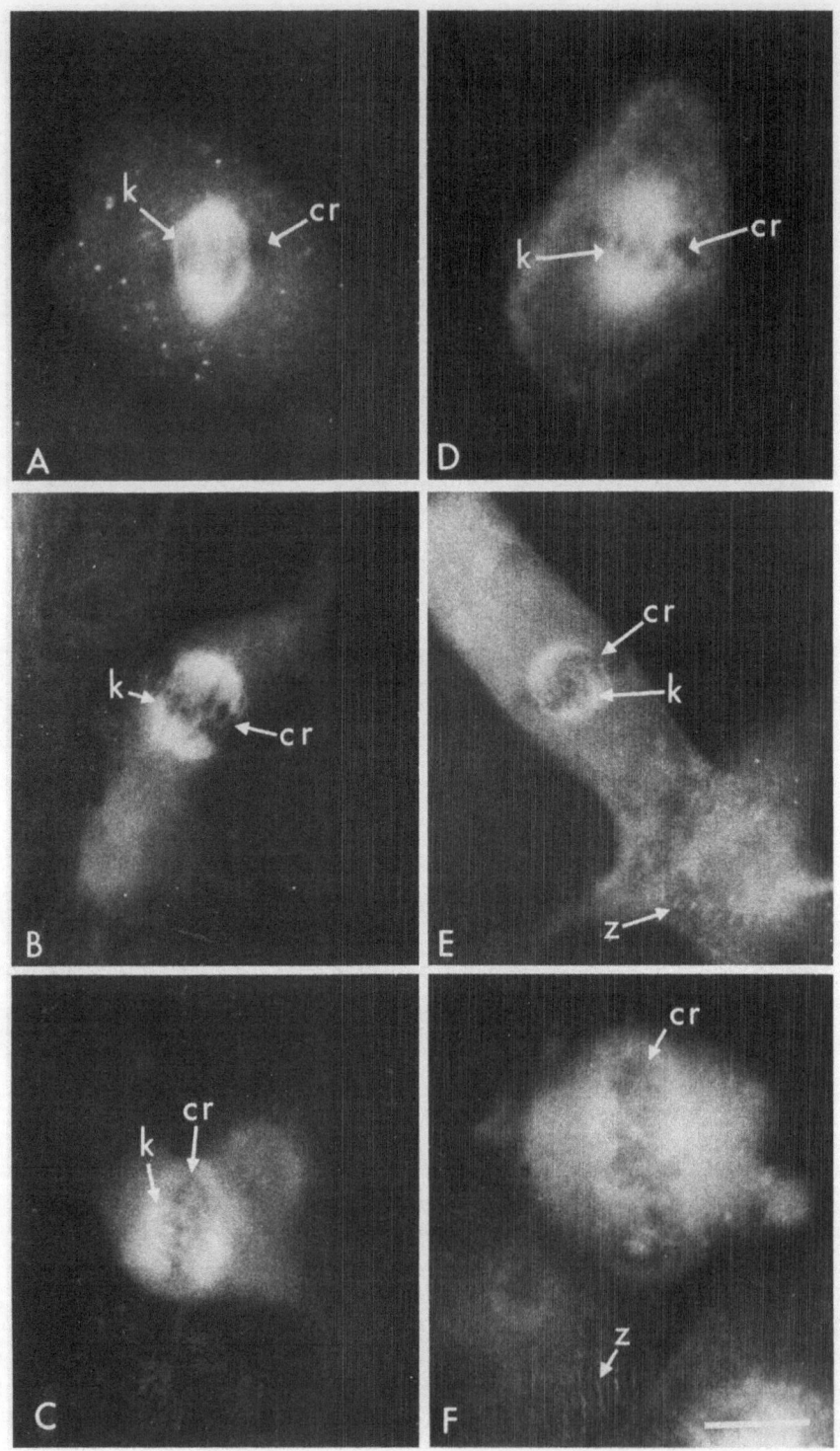

polymerization of the spindle-fiber microtubules and subsequent movement of the spindle poles toward the metaphase plate (Fig. 5B,C). Rewarming of the cells results in re-formation of the spindle by growth and enlongation of the spindle fibers, which move the spindle poles back to their original positions.

The control metaphase spindles at 37°C labeled with tubulin (Fig. 5A) and M-CPK antibody (Fig. 5D) possess a strikingly similar morphology. The metaphase tubulin spindle and CPK spindle are of the same size with respect to length and width of the mitotic apparatus. The individual kinetochore fibers are more distinct in the tubulin spindle, whereas such fibers in the CPK-labeled spindle are somewhat more diffuse in appearance. As the tubulin spindle shortens in response to depression of the temperature, its overall fluorescence becomes weaker and more diffuse. The kinetochore fibers, however, become more distinct as the background spindle fluorescence is lowered due to the loss of the more labile continous spindle fibers (Fig. 5B). At the same equilibrium temperatures (Fig. 5B), the CPK spindle is likewise shortened by about the same order of magnitude as observed for the tubulin spindle. At lower temperatures, the CPK spindle is generally more disorganized than the tubulin spindle. The CPK remains in close association with the kinetochore fibers as they persist during this treatment, but are much weaker in fluorescence than the control cells.

At temperatures between 0 and +4°C, only a remnant of the tubulin spindle remains. The poles are almost completely adjacent to the metaphase plate. The remaining daughter half-spindles possess weak, diffuse fluorescence with occasional short kinetochore-fiber remnants adjacent to individual chromosomes (Fig. 5C). The mechanical and structural integrity of the mitotic spindle is lost at these temperatures as the metaphase plate becomes disorganized and chromosomes become somewhat scattered (Figs. 5C and 6A). The presence of CPK associated with the spindle is highly diminished at these low temperatures (Figs. 5E,F and 6D). Only weak, diffuse fluorescence marks the presence of CPK immediately adjacent to the metaphase plate region (Fig. 5E). The filamentous form of CPK observed in the spindle at higher temperatures is almost completely abolished at 4°C (Fig. 5F). However, the presence of CPK in Z- and H-lines of the myofibrils (Fig. 5E,F) does not appear to be affected by the depression of temperature to 0°C for time periods of up to 2 hr.

Figure 5. Effect of temperature depression on the distribution and morphology of microtubules and M-CPK in the mitotic spindle. (A) 37°C: Distribution of metaphase-spindle microtubules. (B) 10°C: The metaphase microtubules become shorter, and kinetochore fibers are most distinct. (C) 4°C: Microtubules of the spindle are almost completely depolymerized, and the poles have moved in close to the metaphase plate. Short kinetochore fibers are still visible. (D) 37°C: The distribution of M-CPK in the metaphase spindle is very similar to the distribution of microtubules at this temperature. (E) 10°C: The M-CPK spindle shortens and becomes weaker at lower temperatures. Weak kinetochore-fiber staining is visible. M-CPK in Z- and M-lines is not affected by low temperature. (F) 4°C: The M-CPK in the spindle is all but completely lost with depolymerization of the majority of the spindle microtubules. (k) Kinetochore fibers; (cr) chromosomes; (z) Z-lines. Scale bar: 10 μm.

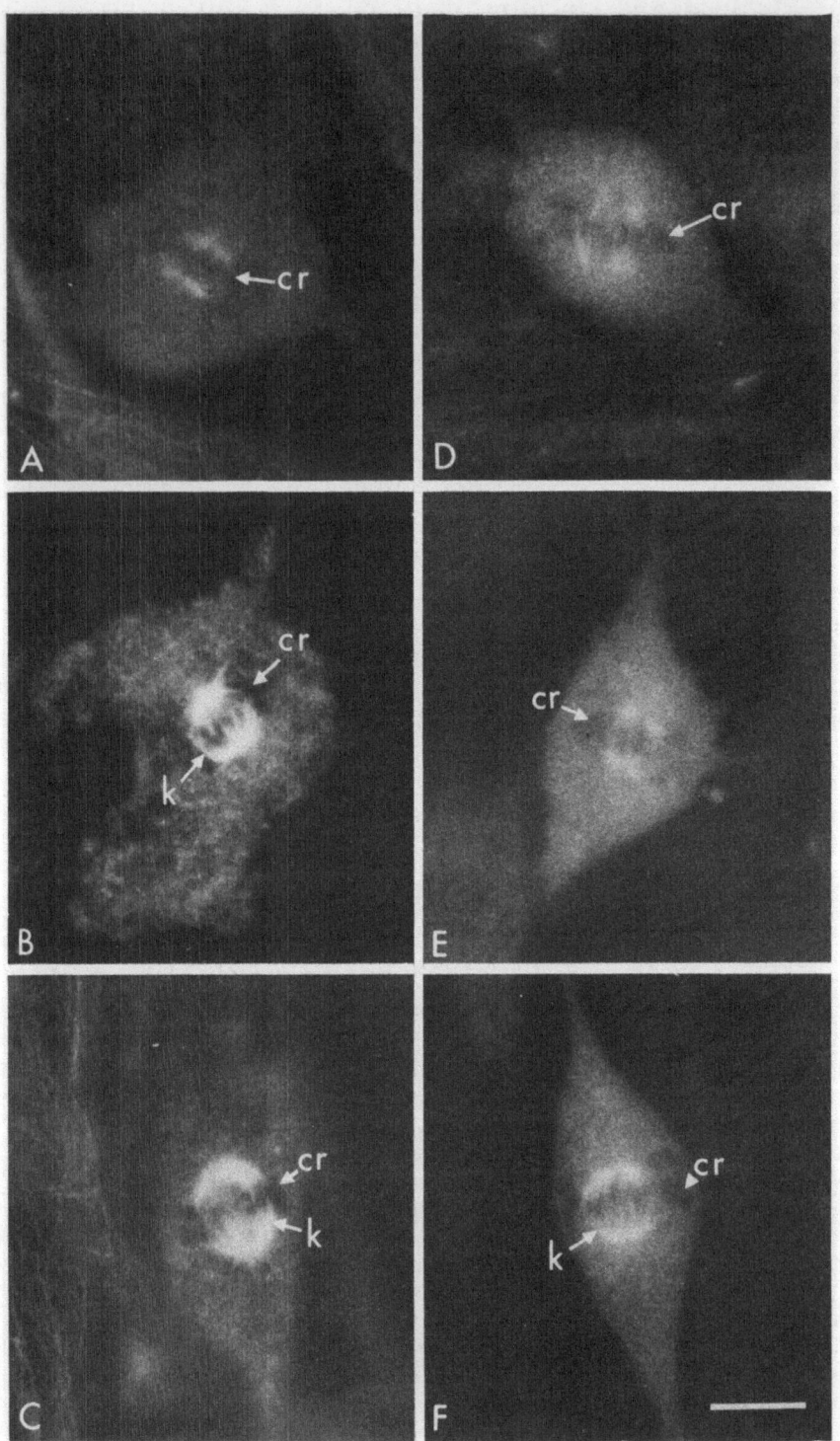

The microtubules of the dividing rat myocardial cells are not completely depolymerized by exposure to 0°C for 2 hr or more (Fig. 6A). Microtubules are continually observed as areas of very weak fluorescence adjacent to the metaphase chromosomes. A few weakly fluorescent remnants of the kinetochore fibers remain associated with individual chromosomes. The CPK that remains associated with the spindle at this temperature (Fig. 6D) is diffuse and in an area adjacent to the metaphase plate.

When the cells are slowly rewarmed, the spindle microtubules repolymerize and the spindle begins to elongate (Fig. 6B,C). At 10°C, the spindle-fiber microtubules re-form to the extent of pushing the poles about halfway back to their normal metaphase position seen at 37°C (Fig. 6B). The fluorescence of spindle fibers increases and the kinetochore microtubules become quite distinct. The distribution of CPK associated with the re-forming spindle at this equilibrium temperature resembles the tubulin spindle in morphology, but the linear recovery appears somewhat slower. The CPK spindle elongates less rapidly and CPK does not rapidly reassociate with the kinetochore fibers (Fig. 6E). The overall fluorescence of the CPK spindle remains weaker and less organized than that of the tubulin spindle. As the temperature is further elevated, the mitotic spindle continues to re-form until it resembles the control spindle in morphology. At 20–25°C, the tubulin spindle appears about the same as an untreated control spindle with respect to its size, overall morphology, and appearance of the kinetochore fibers (Fig. 6C). The CPK spindle at this same equilibrium temperature is usually shorter in pole-to-pole distance, but possesses the same width as the control spindles (Fig. 6F). The overall CPK spindle fluorescence is weaker than that of the control, and CPK has just begun to associate with kinetochore-spindle fibers. Maintainance of the treated spindles at 37°C produces complete recovery of the distribution and morphology of the tubulin and CPK spindle, which are indistinguishable from those of the untreated control cells.

4. Discussion

The M-isozyme of CPK is observed in association with the spindle fibers of the mammalian mitotic apparatus. The specificity of this antibody for the

---

Figure 6. Effect of temperature elevation on the recovery of microtubules and M-CPK in the partially depolymerized metaphase spindle. (A) 1°C: After 2 hr at this temperature, only a few microtubules of the metaphase spindle are present adjacent to the chromosomes. (B) 10°C: Repolymerization of the spindle microtubules results in pushing away of the spindle poles from the metaphase plate and re-formation of the kinetochore fibers. (C) 20°C: The distribution of microtubules at this temperature closely resembles that in the control spindle (Fig. 5A). (D) 1°C: M-CPK in the spindle after 2-hr treatment at this temperature is barely visible and is almost completely dissociated from the spindle. (E) 10°C: Recovery of M-CPK in the re-forming spindle is slower than the repolymerization of spindle microtubules. (F) 20°C: The M-CPK spindle becomes more organized but still lags behind the tubulin spindle in morphology. M-CPK associated with the kinetochore fibers is apparent at this stage of spindle recovery. (cr) Chromosomes; (k) kinetochore fibers. Scale Bar: 10 μm.

presence of M-CPK in the mitotic spindle is supported by absorption and double-immunodiffusion control experiments. Absorption of the M-CPK antibody with tubulin does not inhibit its labeling of spindle fibers in mitotic cells, the Z- or H-lines of cardiac myocytes, or intermediate and actin stress fibers of nonmuscle cells in interphase. Absorption of the antibody with its antigen, M-CPK, results in no staining in any of the cells (Eckert *et al.*, 1980). Double immunodiffusion shows no reaction between the M-CPK antibody and tubulin, actin, or desmin (Eckert *et al.*, 1980). A single precipitin line is formed between M-CPK and the M-CPK antibody. The indirect staining of the mitotic spindle with the M-CPK antibody is stable, being unaffected by repeated washes with PBS at high pH. These observations indicate that M-CPK is a real component of the mitotic spindle. This isozyme is found in the spindle at all mitotic stages from prophase to telophase. The association of CPK with the spindle is stable, since natural and experimentally induced alterations in spindle-microtubule morphology produce parallel changes in the distribution of the isozyme.

The predominant localization of CPK is in the half-spindle region between the pole of the spindle and the metaphase plate. In this region of the spindle, CPK shows distinct linear orientation that is similar to the distribution of the kinetochore fibers. CPK is more diffusely associated with continuous or interzonal spindle fibers. The spindle poles or asters stain with the M-CPK antibody during late prophase and during the dissolution of the nuclear membrane. The presence of CPK in the aster diminishes as the spindle matures toward metaphase. CPK does not reappear in polar regions of the spindle when the asters begin enlarging in late anaphase or in telophase when the nuclear membrane is re-forming.

These observations would suggest and are consistent with the hypothesis that CPK is associated with those regions of the spindle that may require a quantity of energy in the form of ATP to perform a specific function. The CPK associated with the prophase aster could provide energy for the disolution of nuclear membrane at the end of prophase. Such an energy source in this polar region of the spindle could also provide energy for the initiation of spindle assembly and maintenance of spindle morphology prior to the integration of the chromosomes into the organelle. The CPK associated with spindle fibers in the region between the chromosomes and the poles may serve to form linear energy generators that could be involved in the poleward flux of material or movement of chromosomes. Such a linear energy-generation system could provide the motive force for the constant poleward movement of cellular material that is to be excluded from the spindle. This system could also serve to deliver energy for the maintenance of the poleward flux of spindle-fiber birefringence (Forer, 1965, 1966; Inoué and Sato, 1967) or in maintenance of the spatial geometry and configuration of the spindle. Since the energy required for anaphase chromosome movement has been calculated as being small (Inoué and Ritter, 1975; Nicklas, 1965, 1971; Taylor, 1965), this CPK system might also provide energy for this particular function. This energy source is, however, present and properly oriented if energy in the form of ATP is required by the mitotic apparatus for the movement of chromosomes.

The disassembly and reassembly of the mitotic apparatus by changes in temperature induce a reversible disappearance and reappearance of the spindle-associated CPK. The disappearance of CPK when the spindle is depolymerized follows very closely the distribution and morphology of spindle microtubules at the same equilibrium temperature. The recovery of CPK when the spindle fibers are undergoing re-formation closely resembles the distribution of the repolymerizing microtubules. There is, however, a difference in the recovery of spindle microtubules and spindle-associated CPK. This difference is seen at recovery temperatures below 25°C and is manifest by the CPK spindle's being shorter in pole–pole length than its microtubule-spindle counterpart. This difference is not observed in spindles at temperatures above 25°C. The shortened morphology of the CPK spindle at suboptimal temperatures could be accounted for if the isozyme were preferentially reassembled at the kinetochore and then subsequently moved poleward in the wake of the re-forming spindle microtubules.

These data suggest that CPK is associated with a labile element of mitotic spindle and is predominantly localized in regions of the spindle that may require a localized output of energy or a linear orientation of energy production. That there are morphological similarities between tubulin and CPK distribution in the spindle, and that microtubules are the most predominant linear elements in the mitotic apparatus, would suggest that CPK could be directly associated with the spindle microtubules. The absorption of CPK antibody with tubulin has revealed no cross-reactivity, indicating that tubulin itself does not possess an antigenic site for CPK, or CPK activity. The current observations of the association of CPK with cytoplasmic fibrous proteins and its behavior in the spindle would suggest that CPK is not intercalated into the individual microtubules, but is associated with a bundle of microtubules, such as seen in the kinetochore fibers. These observations, however, do not rule out the possibility that CPK could be interacting with the microtubule-associated proteins or tau proteins that are present in the spindle and are involved in microtubule polymerization–depolymerization kinetics. Since CPK has not been observed in association with microtubules of the CMTC, this isozyme may well be mobilized to those regions of the spindle that require a specific energy source.

Neither do these observations rule out the possibility that CPK may be associated with some other labile, linear element found in the spindle. The most likely such alternate would be actin filaments that have been proposed (Ishikawa *et al.*, 1969; Gawadi, 1971, 1974; Behnke *et al.*, 1971; Sanger, 1977) to be present in the spindle between the poles and the metaphase plate. Thin, weakly fluoresent actin filaments have been observed in spindles isolated from sea urchin embryos and mammalian cells in culture utilizing various antibodies against vertebrate and invertebrate actin (Salmon and Fuseler, 1981, unpublished observations) and fluorescently labeled S-1 protein (Sanger, 1977). These thin actin fibers that extend from the individual chromosomes to the spindle pole may provide the stable framework for the linear orientation of CPK within the spindle. Such an arrangement would be similar to the proposal that CPK can associate with actin stress fibers (Fuseler *et al.*, 1981) and intermediate filaments (Eckert *et al.*, 1980; Fuseler *et al.*, 1981) in the cytoplasm of interphase cells to locally control ATP levels.

However, it is interesting that CPK in the dividing mammalian cell is not associated with actin fibers involved in formation of the cleavage furrow and is present only weakly in the interzonal region of the spindle between the separating daughter chromosomes. Apparently, events of cellular cleavage do not require the CPK energy-generation system. The precise contribution of M-CPK to the complicated mechanism of mitosis and its association with fibrous labile elements of the spindle must await further investigations.

ACKNOWLEDGMENTS. This work was supported in part by American Heart Grant 80-757 awarded to John W. Fuseler. We would like to express our thanks to Ms. Gay Lorkowski for her technical assistance.

References

Behnke, O., Forer, A., and Emmersen, J., 1971, Actin in sperm tails and meiotic spindles, *Nature (London)* **234**:408.

Bessman, S. P., Yang, W. C. T., Geiger, P. J., and Erickson-Viitanen, S., 1980, Intimate coupling of creatine phosphokinase and myofibrillar adenosinetriphosphate, *Biochem. Biophys. Res. Commun.* **96**:1414.

Bollon, A. P., Nath, K., and Shay, J. W., 1977, Establishment of contracting heart muscle cell cultures, in: *Tissue Culture Association Manual*, Vol. 3 (V. Evans, V. Perry, and M. Vincent, eds.), pp. 637–640, Tissue Culture Association, Rockville, Maryland.

Eckert, B. S., Koons, S. J., Schantz, A. W., and Zobel, C. R., 1980, Association of creatine phosphokinase with the cytoskeleton of cultured mammalian cells, *J. Cell Biol.* **86**:1.

Forer, A., 1965, Local reduction in spindle fiber birefringonce in living *Nephrotoma suturalis* (Loew) spermatocytes induced by ultraviolet microbeam irradiation, *J. Cell Biol. (Mitosis Suppl.)* **25**:95.

Forer, A., 1966, Characterization of the mitotic traction system, and evidence that birefringent spindle fibers neither produce nor transmit force for chromosome movement, *Chromosoma* **19**:44.

Fuller, G. M., Brinkley, B. R., and Boughter, J. M., 1975, Immunofluorescence of mitotic spindle by using monospecific antibody against bovine brain tubulin, *Science* **187**:948.

Fuseler, J. W., 1975, Temperature dependence of anaphase chromosome velocity and microtubule depolymerization rate, *J. Cell Biol.* **89**:737.

Fuseler, J. W., and Shay, J. W., 1980, Interrelationship of desmin and creatine phosphokinase in developing cardiac cells, *J. Cell Biol.* **87**:183a.

Fuseler, J. W., Shay, J. W., and Feit, H., 1981, The role of intermediate (10-nm) filaments in the development and integration of the myofibrillar contractile apparatus in the embryonic mammalian heart, in: *Cell and Muscle Motility*, Vol. 1 (R. M. Dowben and J. W. Shay, eds.), pp. 205–259, Plenum Press, New York.

Gawadi, N., 1971, Actin in the mitotic spindle, *Nature (London)* **234**:410.

Gawadi, N., 1974. Characterization and distribution of microfilaments in dividing locust testis cells, *Cytobios* **10**:17.

Inoué, S., and Sato, H., 1967, Cell motility by labile association of molecules, *J. Gen. Physiol.* **50:**259.

Inoué, S., and Ritter, H., Jr., 1975, Dynamics of mitotic spindle organization and function, in: *Molecules and Cell Movements* (S. Inoué and R. E. Stephens, eds.), pp. 3–30, Raven Press, New York.

Ishikawa, H., Bischoff, R., and Holtzer, H., 1969, Formation of arrowhead complexes with heavy meromyosin in a variety of cell types, *J. Cell Biol.* **43:**312.

Jacobus, W. E., and Lehninger, A. L., 1973, Creatine kinase of rat heart mitochodria, *J. Biol. Chem.* **248:**4803.

Miller, C. L., Fuseler, J. W., and Brinkley, B. R., 1977, Cytoplasmic microtubules in transformed mouse × nontransformed human cell hybrids: Correlation with *in vitro* growth, *Cell* **12:**319.

Nath, K., Shay, J. W., and Bollon, A. P., 1978, Relationship between dibutyryl cyclic AMP and microtubule organization in contracting heart muscle cells, *Proc. Natl. Acad. Sci. U.S.A.* **75:**319.

Nicklas, R. B., 1965, Chromosome velocity during mitosis as a function of chromosome size and position, *J. Cell Biol. (Mitosis Suppl.)* **25:**119.

Nicklas, R. B. 1971, Mitosis, in: *Advances in Cell Biology*, Vol. 2 (D. M. Prescott, L. Goldstein, and E. McConkey, eds.), pp. 225–297, Appleton-Century-Crofts, New York.

Salmon, E. D., 1975, Spindle microtubules: Thermodynamics of *in vitro* assembly and role in chromosome movement, *Ann. N. Y. Acad. Sci.* **253:**383.

Sanger, J. W., 1977, Nontubulin molecules in the spindle, in: *Mitosis Facts and Questions* (M. Little, N. Paweletz, C. Petzelt, H. Ponstingl, D. Schroeter, and H. P. Zimmerman, eds.), pp. 98–120, Springer-Verlag, New York.

Sharov, V. G., Saks, V. A., Smirnov, U. S., and Chazov, E. I., 1977, An electron microscopic histochemical investigation of the localization of creatine phosphokinase in heart cells, *Biochim. Biophys. Acta* **468:**495.

Taylor, E. W., 1965, Brownian and saltatory movements of cytoplasmic granules and the movement of anaphase chromosomes, in: *Proceedings of the Fourth International Congress on Rheology (1963)*, Part 4, *Symposium on Biorheology* (A. L. Copley, ed.), pp. 175–191, Interscience, New York.

Turner, D. C., Walliman, T., and Eppenberger, H. M., 1973, A protein that binds specifically to the M-line of skeletal muscle is identified as the muscle form of creatine kinase, *Proc. Natl. Acad. Sci. U.S.A.* **70:**702.

Wallimann, T. H., Kuhn, J., Pelloni, G., Turner, D. C., and Eppenberger, H. M., 1978a, Localization of creatine kinase isoenzyme in myofibrils. II. Chicken heart muscle, *J. Cell Biol.* **75:**318.

Wallimann, T., Pelloni, G., Turner, D. C., and Eppenberger, H. M., 1978b, Monovalent antibodies against MM-creatine kinase remove the M-line from myofibrils, *Proc. Natl. Acad. Sci. U.S.A.* **75:**4296.

10

Cytoskeletal Defects in Avian Muscular Dystrophy

Jerry W. Shay, John W. Fuseler, Ursula Neudeck, Gay Lorkowski, Marguerite Stauver, and Howard Feit

1. Introduction

Since there is at present no effective form of chemotherapy available to humans affected with the major types of muscular dystrophy, hereditary and experimentally induced animal models have been extensively studied to gain information about beneficial effects of various pharmacological agents and to study nerve–muscle interactions with respect to structure and function. None of the animal models, however, is directly comparable to any of the human dystrophies, and analogies drawn between them can be very misleading. However, the genetically dystrophic chicken provides a convenient model for testing drugs that may be of benefit even though the primary lesions in these animals are unknown and their relevance to human muscular dystrophies is unclear. Despite these drawbacks, the dystrophic-chicken model is an especially promising system for the study of genetically inherited myopathies because the progressive developmental onset of the disease closely mimics the human form of Duchenne muscular dystrophy in the areas of histological signs of muscle degeneration, muscle-enzyme leakage into blood, and progressive loss of functional ability. Because of the large number of similarities that the avian model of muscular dystrophy shares with Duchenne muscular dystrophy, chemotherapeutic trials in chickens can assess quantitative and

Jerry W. Shay, John W. Fuseler, Gay Lorkowski, and Marguerite Stauver • Department of Cell Biology, The University of Texas Health Science Center at Dallas, Dallas, Texas 75235. *Ursula Neudeck and Howard Feit* • Department of Neurology, The University of Texas Health Science Center at Dallas, Dallas, Texas 75235.

qualitative parameters before clinical trials are initiated. In this chapter, both morphological and biochemical data will be presented to document a cyto-skeletal defect in avian muscular dystrophy. Isaxonine, a recently synthesized drug, retards the developmental pathological processes in the dystrophic chicken. It will be shown that by daily injection of this chemotherapeutic agent, the cytoskeletal defect in avian muscular dystrophy is corrected.

2. Materials and Methods

Genetically homozygous dystrophic chickens, Line 413, and their corresponding normal controls, Line 412, were purchased from the Department of Avian Sciences, University of California, Davis. At various days *ex ovo,* small pieces (<1 mm³) of cardiac or skeletal muscle were placed on glass coverslips and allowed to attach firmly (Figs. 1 and 2) before the addition of 2 ml CMRL-1066 culture medium (GIBCO, Grand Island, New York) containing 15% horse serum and 5% fetal bovine serum (Kansas City Biologicals, Lenexa, Kansas). After 5–7 days of incubation, there grew out from these muscle explants a monolayer of cells (Figs. 3 and 4) that morphologically appeared to be primarily fibroblasts. These cells, when stained with an actin antiserum, contained numerous actin stress fibers, which substantiated that these were, in fact, fibroblasts, since epithelial cells contain few actin stress fibers and muscle fibers contain repeating actin striations (see Chapter 8).

The coverslips with attached cells were rinsed in phosphate-buffered saline (PBS), fixed in 3% ultrapure formaldehyde–PBS for 20 min, and then extracted in acetone at −20°C for 7 min. The cells were then incubated in one of three different tubulin antisera: affinity-purified rabbit antitubulin serum provided by Dr. Bill Brinkley, which has been previously characterized (Brinkley *et al.,* 1980); a non-affinity-purified rabbit antitubulin serum provided by Dr. Joe Connolly, which has been previously characterized (Connolly *et al.,* 1978; Connolly and Kalnins, 1978); and affinity-purified rabbit antitubulin serum, which we prepared. In all experiments, the normal and dystrophic cells were incubated with the antiserum for 45–60 min at 37°C and 100% humidity, followed by several rinses in PBS, and then incubated in fluorescein-conjugated goat anti-rabbit IgG (Meloy), diluted 1 : 10 with PBS, for 30 min. The cells were then rinsed several times in PBS and finally mounted on glass slides in a drop of glycerol–PBS (9 : 1 by volume at pH 9.5–10.0) as previously described (Miller *et al.,* 1977). The slides were coded and then scored in the fluorescence microscope for full or diminished complexes of microtubules by an observer unaware of the nature of each slide.

Normal and dystrophic chicken brains were processed for the tubulin-assembly assay using a modification of the procedure of Asnes and Wilson (1979). Normal and dystrophic chickens at ages 10, 20, 40, 65, 80, and 90 days *ex ovo* were sacrificed by decapitation, and the brains were quickly removed and weighed. The brains were homogenized in cold phosphate–glutamate buffer (20 mM sodium phosphate and 100 mM sodium glutamate, pH 6.90,

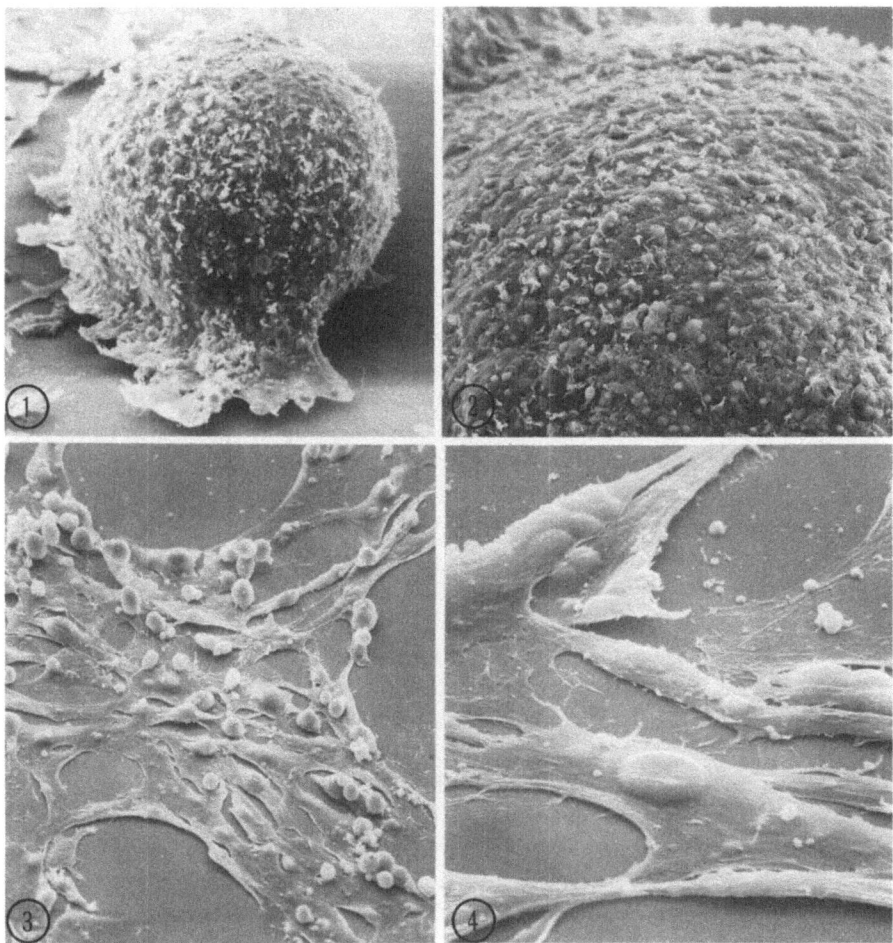

Figures 1–4. Scanning electron micrographs. Avian cardiac explant in culture 2 days (Fig. 1). Note the rounded individual cells on the surface of this explant (Fig. 2). After 5–7 days in culture, numerous cells migrate away from the primary explant (Fig. 3), most of which appear to be fibroblasts (Fig. 4).

1.5 ml/g wet weight) using a Dounce homogenizer. The homogenate was centrifuged at $39,000g$ for 40 min at 4°C, and the supernatant was removed. Aliquots for protein determination and protein electrophoresis were removed, and the volume of the remaining supernatant was measured; 1/9 volume of a 10-fold concentrated solution of polymerization cofactors [25 mM GTP, 10 mM ethyleneglycol bis-(β-aminoethyl ether)N,N'-tetraacetic acid, and 5 mM $MgCl_2$ in phosphate–glutamate buffer at pH 6.75] was added to the supernatants. The pH of each supernatant was then adjusted to 6.85 at 4°C with NaOH. The supernatants were placed in cold cuvettes in a temperature-controlled spectrophotometer (Gilford model 2400-2 equipped with Thermo-programmer model 2527), and polymerization was initiated by increasing the temperature of the cuvettes to 30°C. The polymerization was monitored by the increase in absorbance at a wavelength of 380 nm. After polymerization, an aliquot was removed and centrifuged at 39,000g for 30 min at 30–35°C. Aliquots of the assembled microtubules (designated T_1) were taken for electron microscopy and for protein electrophoresis. In some experiments, a second cycle of polymerization was performed. T_1 was resuspended in 14% of the initial volume of the supernatant in phosphate–glutamate buffer at pH 6.75. The resuspended microtubules were depolymerized at 0°C for 30 min and centrifuged at 39,000g for 40 min. A second round of polymerization was obtained by incubation at 30°C after polymerization cofactors were added.

The earliest detectable clinical sign of disease in chickens with hereditary muscular dystrophy is the inability of the dystrophic bird to right itself from the supine position. The normal (412) and dystrophic (413) chickens were subjected to a test of righting ability (Flip test) every 7 days, which involved placing each chicken in the supine position on a flat surface. The chicken was then allowed to rise to a standing position, which, if accomplished, constituted a success. The test was concluded if the bird succeeded five times or if it failed to right itself in 15 sec or less (Hudecki *et al.*, 1979).

Starting on day 3 *ex ovo*, dystrophic chickens received intraperitoneal injections of a recently synthesized neurotrophic drug, Isaxonine (N-isopropyl-amino-2-pyrimidine orthophosphate), which was provided by Biomeasure, Inc. (Hopkinton, Massachusetts). This drug was evaluated for beneficial effects because it has been reported (Tarrade and Hugelin, 1978) to significantly increase the number, length, and arborization of neurites on fetal mice dorsal-root-ganglia cultures (10^{-7} M, 48 hr). In another report (Hugelin *et al.*, 1979), it was demonstrated in the rat that after sciatic-nerve lesion by freezing, the length of the most rapidly regenerating fibers was signficantly increased by injection of Isaxonine. In the studies reported herein, dystrophic chickens received daily injections of Isaxonine at 50 mg/kg weight. At weekly intervals, saline- and Isaxonine-injected dystrophic animals as well as normal control chickens were subjected to the standardized Flip test. Unnecessary handling of the birds between tests was avoided, and all animals were tested in a random fashion, with results being compared to wing-band number following the test. In addition to the Flip test, at the time of sacrifice for tissue-culture and tubulin-assembly assays, pectoralis muscle was prepared for histological and ultrastructural examination. Muscle for histological studies was mounted on cork disks and rapidly frozen by immersion in Freon that was cooled by liquid nitrogen. Transverse serial sections (8 μm thick) were cut on a cryostat (American Optical) at −20°C, mounted on coverslips, and processed with various stains. For electron microscopy, a strip of pectoralis muscle (2.5 cm) was removed and tied to an applicator stick with a slight amount of tension. The muscle samples were fixed in 2% ultrapure glutaraldehyde (Ladd) in 0.1 M cacodylate buffer (pH 7.2–7.4) for 1 hr at room temperature. The samples were then stored at 4°C until processed by routine methods for electron microscopy. Sections were examined in a JEOL 100B electron micro-scope at 60 and 80 kV.

In the tubulin-assembly assays involving exogenous Isaxonine, the drug was dissolved (50 mg/ml) in phosphate–glutamate buffer, and the pH was readjusted to 6.85 at 20°C by the addition of NaOH. The pH of the final mixture of drug, polymerization cofactors, and protein was rechecked and adjusted to 6.85 at 4°C. Careful control of pH is necessary, since small dif-ferences in pH can produce significant changes in the rate of microtubule polymerization. Protein concentration was determined by the method of Lowry *et al.* (1951), and one-dimensional protein electrophoresis was per-formed by the techniques of Laemmli (1970).

3. Results

Figure 5 illustrates a representative cell derived from a 14-day *ex ovo* normal (412) chicken cardiac explant stained with a tubulin antiserum. In these experiments, 74% of the cells contained full cytoplasmic microtubule complexes (CMTCs). A cell with full CMTCs was defined by morphology in the fluorescence microscope as a cell containing numerous microtubules that radiate out to the plasma membrane. Figure 6 illustrates an identically treated cell representative of 85% of the cells from a 14-day *ex ovo* dystrophic chicken cardiac explant. In these experiments, only 15% of the cells contained full CMTCs, while the majority of the cells contained greatly diminished com-plexes. It is important to mention several points about these observations. First, these results were obtained with three different tubulin antisera, each prepared by a different laboratory, so it is unlikely that these diminished complexes are due to an antibody artifact. Second, these observations appear to be developmental, since explants derived from dystrophic chick embryos did not display diminished CMTCs. In addition, another line of dystrophic chicken (304) also displayed diminished CMTCs, but at a later developmental time point (20 days *ex ovo*). Third, there appears to be a correlation between other characteristics of the dystrophic phenotype and microtubule organiza-tion. Dystrophic chickens with normal righting ability and serum enzyme levels did not express diminished CMTCs.

The diminished CMTC, in dystrophic explants did not correlate with the shape of the cells, since both normal and dystrophic cells appeared to have similar morphologies. In addition, when dystrophic fibroblasts divided, their mitotic spindles appeared normal, suggesting that the decrease of micro-tubules in interphase cells is a regulatory abnormality. Interestingly, when primary cultures of dystrophic tissue were established, the diminished CMTCs were not observed. The significance of this observation is unclear, but it may indicate either that the intact explants produce a factor that produces the diminished CMTC, or that dissociating enzymes may selectively destroy most of the dystrophic cells, while precursor myogenic cells survive these

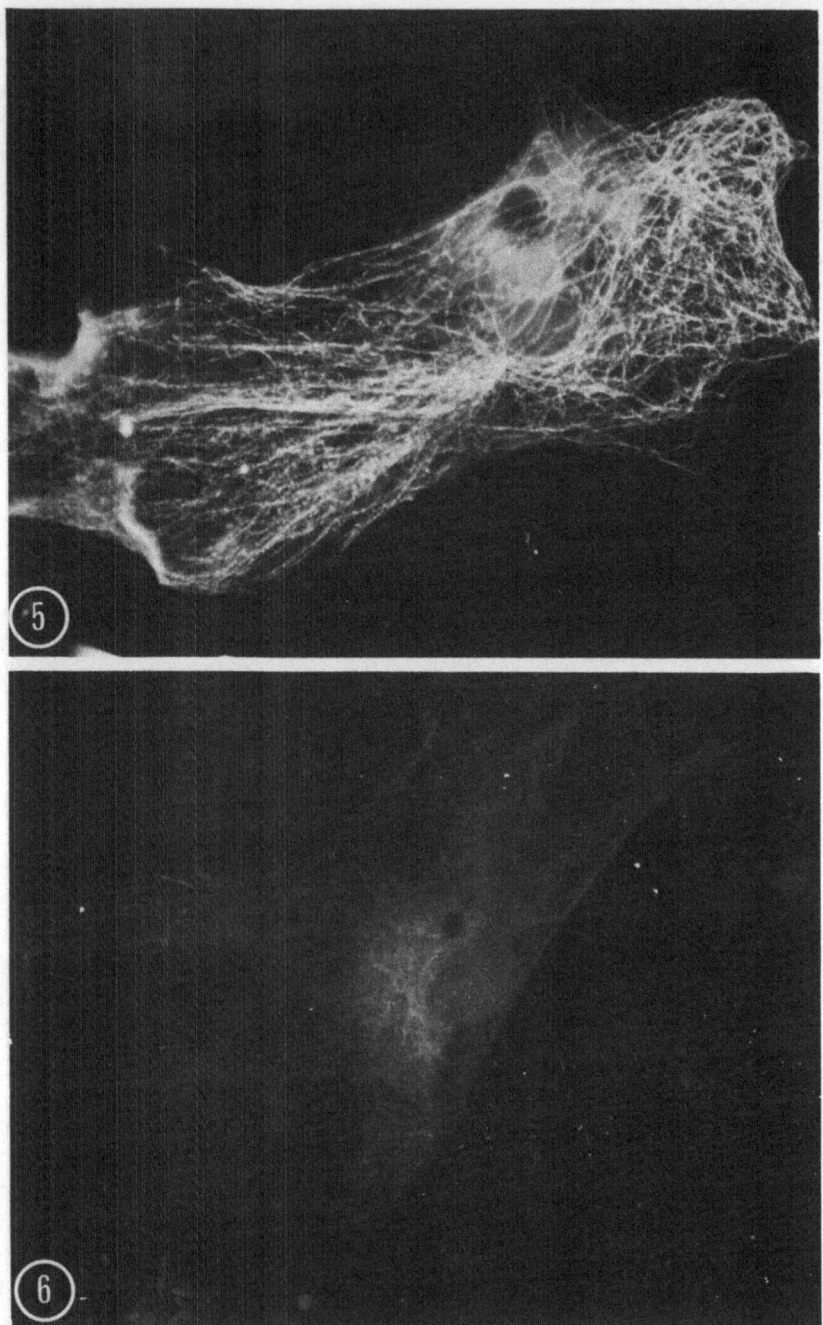

conditions. Since there is increasing evidence of membrane abnormalities in many of the human muscular dystrophies, the sensitivity of dystrophic cells to dissociating enzymes, such as trypsin, may help explain why passaged dystrophic cells appear normal (i.e., it is possible that the techniques of primary cell culture select for the healthiest cells).

To establish a clearer understanding of these morphological observations, a biochemical assay was employed. Experiments were designed to monitor the ability of normal and dystrophic tubulin to assemble into microtubules *in vitro*. A turbidimetric assay was employed, and Fig. 7 illustrates a comparison of microtubule polymerization in extracts of chicken brain from normal and dystrophic 80-day-old animals. A 50% reduction in both the rate and the extent of polymerization of microtubules was observed, even though the protein concentration of both extracts was the same (12.2 mg/ml). These results have been repeated twice with 90-day-old animals, and three times with 80-day-old animals, and have been observed once in 65-day-old animals, but these differences have not been observed in younger animals (10, 20, and 40 days), again suggesting a developmental aspect of this defect.

Protein electrophoresis of the supernatants of brain extracts from normal and dystrophic 80-day-old chickens revealed a dramatic difference in the region of the 68,000-dalton tau protein, a factor that is associated with the promotion of microtubule assembly (Fig. 8). In the gels obtained from the high-speed supernatant (S in Fig. 8), the 68,000-dalton protein is clearly resolved into a closely spaced doublet with the higher-molecular-weight band markedly reduced in the extracts of dystrophic brain. Figure 9 illustrates a densitometric tracing of the stained portion of the 68,000-dalton area previously described in Fig. 8.

Densitometry reveals that this 68,000-dalton protein is greatly reduced but not absent, and there appears to be a slight enhancement of the lower-molecular-weight protein. Because there is some 68,000-dalton tau protein present, it is possible to obtain microtubule polymerization from the dystrophic extract. If polymerized microtubules obtained from dystrophic brain extracts are subjected to a second round of temperature-dependent microtubule polymerization, the component of the 68,000-dalton region with the higher molecular weight is selectively incorporated into the microtubule fraction (Fig. 8, T_1). This gel indicates that it is possible to select for microtubules that are competent to polymerize and that these microtubules that did polymerize from brains of dystrophic animals contained the 68,000-dalton protein. In the presence of the protein, the rate of assembly of microtubules from dystrophic animals was almost equal at the second round of polymerization (Fig. 10). These results are consistent with our morphological observations that most but not all dystrophic cells contained diminished microtubules

Figures 5 and 6. Fibroblast cells that grew out from a 14-day *ex ovo* normal (Fig. 5) and dystrophic (Fig. 6) avian cardiac explant, stained with a tubulin antiserum, and examined in a fluorescence microscope. Most normal cells in such cultures contain full CMTCs, while most dystrophic cells contain diminished CMTCs.

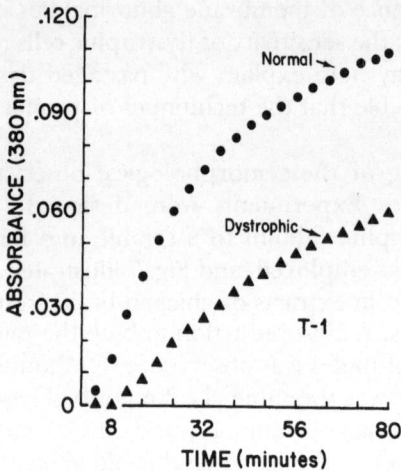

Figure 7. Comparison of relative turbidity during tubulin polymerization from 80-day *ex ovo* normal and dystrophic brain extracts. In this experiment, protein concentrations of normal and dystrophic extracts were equal, yet the dystrophic extract had an approximate 2-fold decrease in turbidity at 80 min, indicating a 2-fold decrease in microtubule assembly.

and that cells could be selected for full CMTCs by enzyme dissociations (i.e., trypsin).

Experiments were then initiated to evaluate the chemotherapeutic effects of the drug Isaxonine, not only using the indices of disease progression such as functional ability, serum enzyme levels, and muscle morphology, but also to determine the effects, if any, on the cytoskeletal defects we had observed.

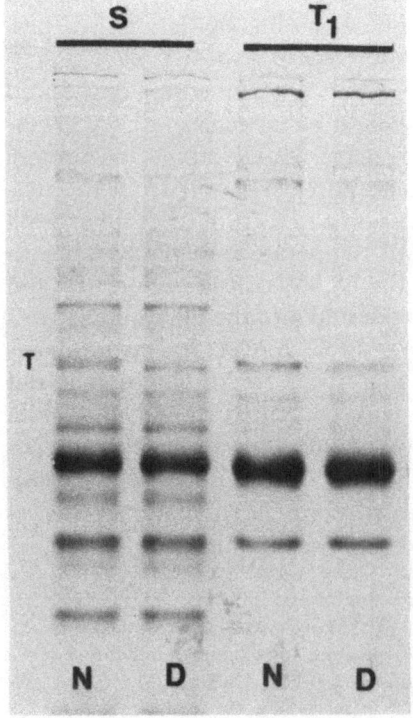

Figure 8. One-dimensional sodium dodecyl sulfate (SDS)–polyacrylamide (7½%) gel electrophoresis of the initial brain extract (S) and microtubules after one (T_1) cycle of assembly from normal (N) and dystrophic (D) chickens. The position of the 68,000-dalton tau protein is indicated.

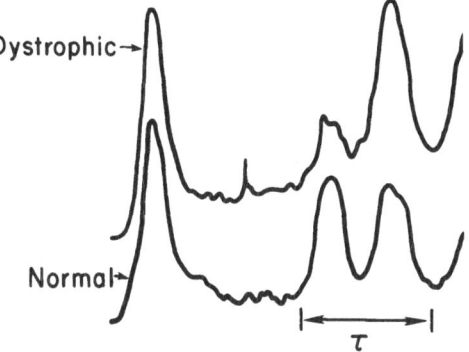

Figure 9. Densitometric tracing of SDS–polyacrylamide gels of brain extracts of normal and dystrophic chickens. The 68,000-dalton region (τ) was resolved using 5% acrylamide and allowing the tubulin to run to the bottom of the gel.

Beginning 3 days *ex ovo,* dystrophic chickens were given daily intraperitoneal injections of Isaxonine (50 mg/kg weight) and evaluated for functional ability using the standardized Flip test as illustrated in Fig. 11. The difference in mean Flip number at 14 and 20 days between saline-injected dystrophic chickens and Isaxonine-injected dystrophic chickens was not statistically significant. However, after 30 days, there was a statistically significant difference between these groups at beyond $P = 0.05$. (These results were based on 40 saline-injected animals and 17 Isaxonine-injected animals.) The differences at 40, 50, and 60 days were also statistically significant at the $P = 0.01$, $P = 0.02$, and $P = 0.02$ level, respectively.

Muscle histology and ultrastructure of pectoralis muscle were also compared among normal, dystrophic, and Isaxonine-treated dystrophic chickens at various stages. Illustrated in Fig. 12 is a micrograph of a histogical section representative of 40-day normal (412) muscle. Figure 13 is a representative micrograph of 40-day dystrophic (413) muscle revealing numerous degenerating fibers. Figure 14 illustrates a representative micrograph of pectoralis muscle obtained from a dystrophic chicken that had received Isaxonine treatment for 40 days. Clearly, the overall muscle histology in Fig. 14 resem-

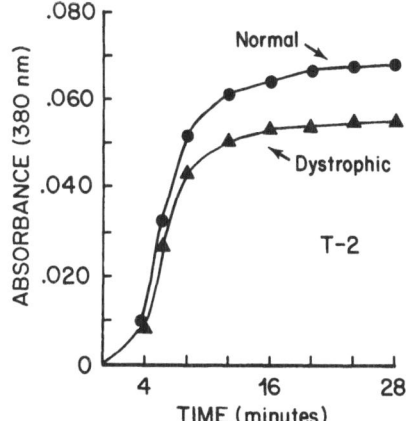

Figure 10. Microtubule polymerization as monitored by change in absorbance in the second cycle of assembly of microtubules from normal and dystrophic animals. The protein concentration was 6.3 mg/ml.

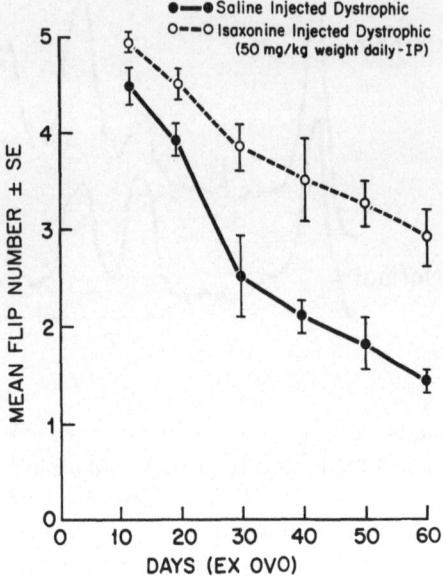

Figure 11. Effect of Isaxonine on the mean Flip number of dystrophic line 413 at various times *ex ovo*. Details of the dosage schedule are given in the text.

bles the normal histology (Fig. 12) more closely than it does the dystrophic histology (Fig. 13). Even though there are occasional degenerating fibers, these results confirm the general improvement noted in the functional test (see Fig. 11). Figure 15 is an ultramicrograph of pectoralis muscle of a normal (412) chicken, while Fig. 16 is representative of a muscle from a saline-injected dystrophic chicken. There is evidence of lipid infiltration, hypertrophied sarcoplasmic reticulum, and muscle-fiber degeneration in Fig. 16. Muscle obtained from a dystrophic animal receiving Isaxonine treatment (Fig. 17) still contained slightly hypertrophied sarcoplasmic reticulum, but in general more closely resembled the normal muscle ultrastructure. Plasma creatine phosphokinase levels in Isaxonine-treated dystrophic animals were reduced by approximately 20% compared with saline-injected dystrophic animals, but this was not statistically significant. It is important to mention that at 50 mg/kg weight, there was no detectable toxicity as evidenced by weight loss or by any detectable pathological changes. From these data, we conclude that Isaxonine has chemotherapeutic value and that it produces significant benefical effects with little or no concomitant toxic effects.

To determine whether Isaxonine had any effect on microtubules, experiments were conducted by incubating 40-day dystrophic explants with 0.5 mg/ml Isaxonine in cell culture. Figure 18 represents cells from a 40-day-old dystrophic cardiac explant stained with a tubulin antiserum and shows greatly diminished CMTCs. In Figs. 19 and 20, cells were incubated in Isaxonine for

Figures 12-14. Muscle histology of normal pectoralis muscle (Fig. 12), dystrophic pectoralis muscle (Fig. 13), muscle from a dystrophic animal receiving Isaxonine treatment (Fig. 14). Note that the muscle from the dystrophic animal receiving Isaxonine resembles that of the saline-injected control (Fig. 12) more closely than that of the saline-injected dystrophic animal (Fig. 13).

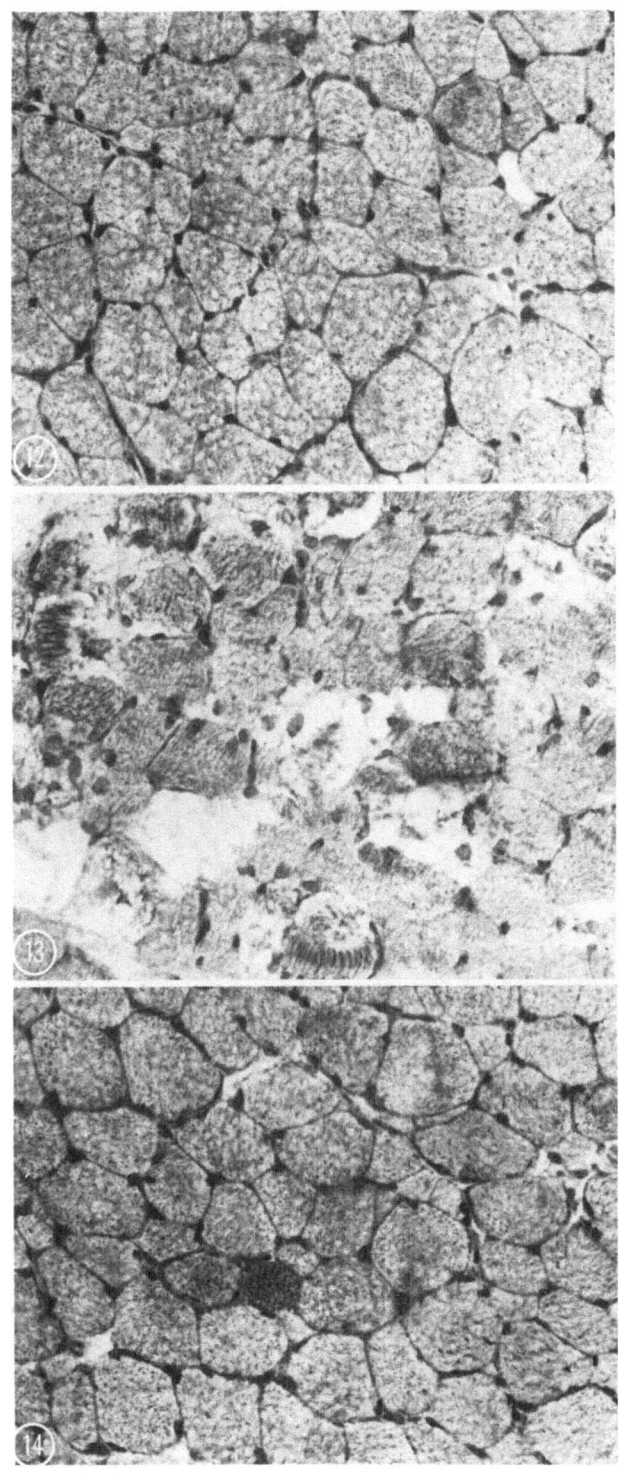

12 and 48 hr, respectively, and foci of assembling microtubules in the juxtanuclear position are observed in many of the cells. If fresh Isaxonine was added to the dystrophic cultures every other day for 7 days, approximately 56% of the cells contained full CMTCs (Fig. 21) as compared to 11% for the dystrophic controls (Fig. 18), 74% for the normal control, and 85% for the normal controls incubated in Isaxonine. These experiments indicate that Isaxonine promotes microtubule assembly in dystrophic cells by a mechanism that is still unclear.

Addition of Isaxonine (0.5 mg/ml) to the normal and dystrophic brain extracts, surprisingly, did not have any effect on microtubule polymerization (Fig. 22). However, pretreatment of dystrophic chickens by daily intraperitoneal injection of Isaxonine corrected the microtubule-polymerization defect (Fig. 23). These biochemical data suggest that the mechanism of action of Isaxonine probably involves a metabolite of the drug.

Drs. P. J. and E. A. Barnard (1980) originated the idea of testing Isaxonine on the dystrophic chicken. Their work indicates that the effect of Isaxonine on the dystrophic chicken is beneficial, and this has been confirmed both by our laboratory and M. Hudecki (personal communication). However, in an unpublished preliminary study by Barry Wilson and Richard Entrikin, beneficial effects were not observed.

4. Discussion

The data in this chapter document a developmental defect in microtubule polymerization in brain extracts of dystrophic chickens and extend the previous immunofluorescence observations of diminished cytoplasmic microtubules in cells obtained by explant-culture technique of dystrophic tissue (Shay and Fuseler, 1979; Shay *et al.*, 1980). In addition, this chapter presents data that a new drug, Isaxonine, not only has chemotherapeutic value but also corrects the cytoskeletal defect when dystrophic chickens are pretreated with it.

In addition to our observations, it has recently been reported (Thakar *et al.*, 1980) that there is a poor organization of microtubules in myotubes established from neonatal dystrophic Syrian hamster skeletal muscle. There are other reports that describe human disease with diminished microtubules in peripheral blood or fibroblasts, or both, including Duchenne muscular dystrophy (Shay *et al.*, 1981), Alzheimer's disease (Andria-Waltenbaugh and

---

Figures 15–17. Muscle ultrastructure of normal pectoralis muscle (Fig. 15), dystrophic pectoralis muscle (Fig. 16), and muscle from a dystrophic animal receiving Isaxonine treatment (Fig. 17). Note that the muscle from the saline-injected dystrophic animal (Fig. 16) contains lipid inclusions as well as hypertrophied sarcoplasmic reticulum. The muscle from the Isaxonine-injected dystrophic animal (Fig. 17) contains slightly hypertrophied sarcoplasmic reticulum, but in general the ultrastructure appears more like that in the control (Fig. 15) than that in the dystrophic animal (Fig. 16).

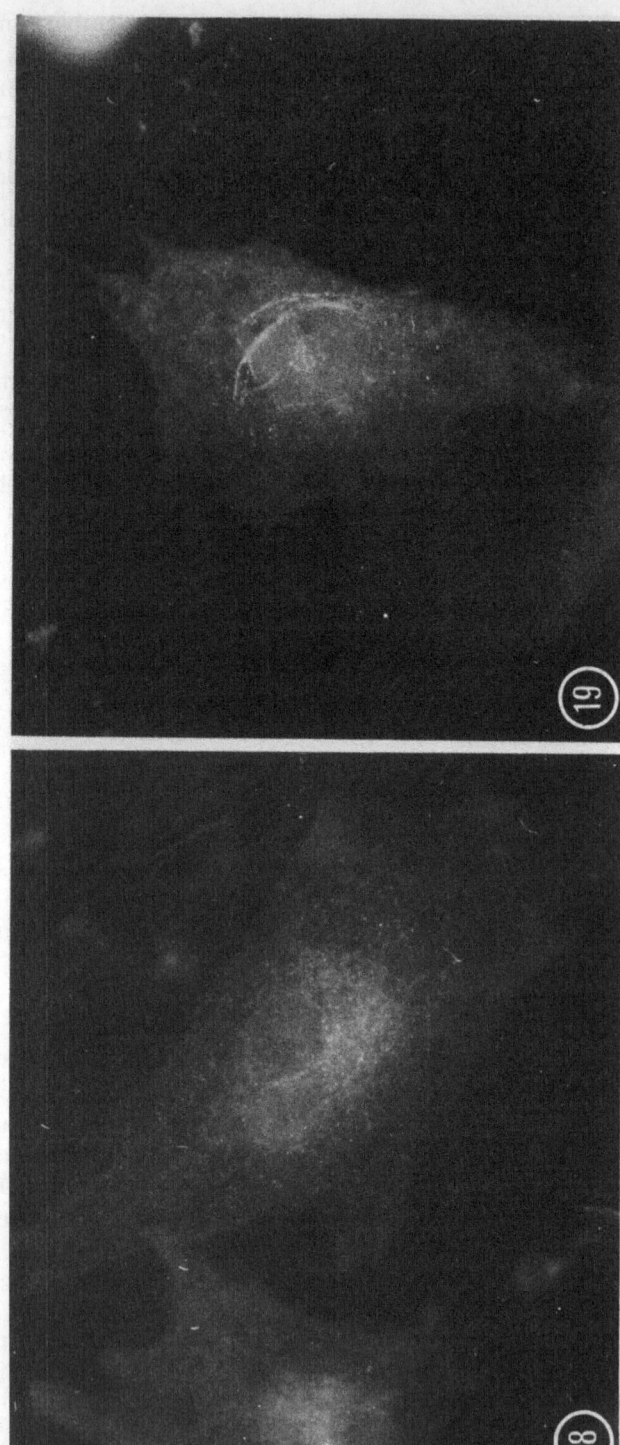

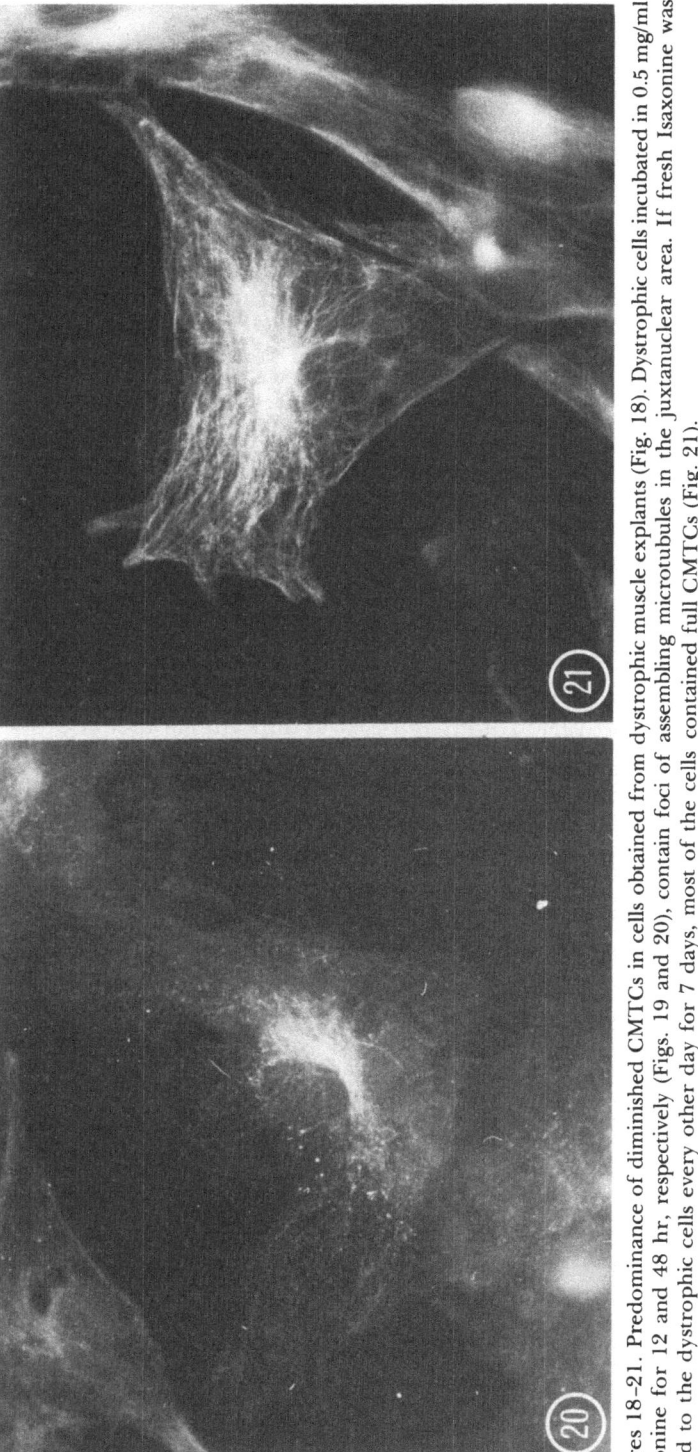

Figures 18–21. Predominance of diminished CMTCs in cells obtained from dystrophic muscle explants (Fig. 18). Dystrophic cells incubated in 0.5 mg/ml Isaxonine for 12 and 48 hr, respectively (Figs. 19 and 20), contain foci of assembling microtubules in the juxtanuclear area. If fresh Isaxonine was added to the dystrophic cells every other day for 7 days, most of the cells contained full CMTCs (Fig. 21).

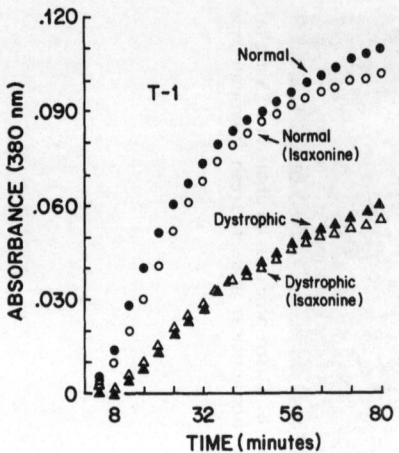

Figure 22. Microtubule polymerization as monitored by change in absorbance in extracts from brains of normal and dystrophic chickens in the presence and absence of Isaxonine (0.5 mg/ml). The protein concentration was 12.2 mg/ml for all samples.

Puck, 1977), Chediak–Higashi syndrome (Oliver, 1976), and Lesch–Nyhan syndrome (Schneider *et al.,* 1979). Almost without exception, each of these reports has been controversial in that other investigators have not been able to confirm or extend the generality of the initial observations (White and Clawson, 1979; Harper *et al.,* 1979; Connolly *et al.,* 1979). It is unlikely that all these reports on diminished or poorly organized microtubular systems are incorrect, but it does suggest a need for more careful investigations on microtubule organization in various human diseases and animal models, especially since the primary biochemical defect has not been determined in most of these diseases.

Even though, at present, we still do not fully understand the significance of the reduced 68,000-dalton protein in dystrophic chickens, the following working hypothesis is consistent with the available literature and is a testable hypothesis in a variety of systems. The microtubule-polymerization defect reported herein provides an interesting clue to the possible etiology of avian muscular dystrophy and perhaps to certain human muscular dystrophies. The altered rate of polymerization of microtubules was detected only in the

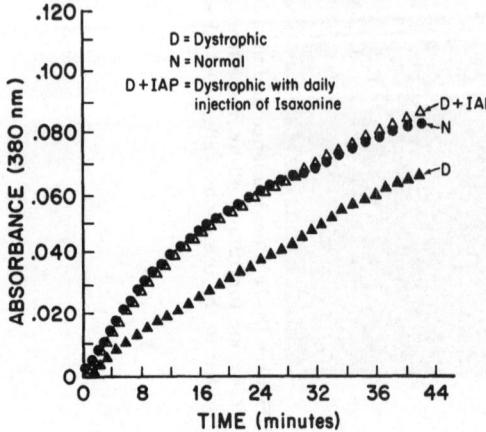

Figure 23. Microtubule polymerization as monitored by change in absorbance in extracts from normal (N), dystrophic (D), and dystrophic chickens injected daily with 50 mg/kg Isaxonine (D + IAP). The protein concentration was 8.1 mg/ml for all samples.

initial supernatants from brain extracts. The formation of microtubules in these supernatants takes place in a complex reaction mixture containing many different proteins and other factors that could either stimulate or inhibit microtubule formation. In particular, the high-molecular-weight microtubule-associated proteins and a group of 55,000- to 70,000-dalton tau proteins have been studied extensively for their effects on microtubule assembly (Weingarten *et al.*, 1979; Lockwood, 1978, Berkowitz *et al.*, 1977; Runge *et al.*, 1979; Wang *et al.*, 1980; Sandoval and Weber, 1980). We have observed that brain extracts from dystrophic animals were markedly diminished in a protein with a molecular weight of 68,000–70,000 and suggest that the decreased rate of polymerization of microtubules is a consequence of the decreased amount of this protein. We propose that intermediate filaments may be the connection between these observations and the obvious muscle degeneration that occurs in muscular dystrophy.

Huxley and Hanson (1954) proposed that the mechanochemical basis for muscle contraction was the sliding of thick and thin filaments past one another. It was appreciated by Huxley and Hanson that this model was not completely satisfactory because the muscle would tear itself apart unless some other noncontractile elements held the sarcomere together, and they proposed the existence of a hypothetical "S" filament (Hanson and Huxley, 1955). The existence of these S filaments was inferred from the observation that when myosin was extracted from muscle, the periodicity and integrity of the Z-band were maintained. Direct evidence for a cytoskeletal framework in muscle that holds the Z-bands together has come only recently (Price and Sanger, 1979). This cytoskeletal framework is the 10-nm, or intermediate, filaments that are present in virtually all cells (Fuseler *et al.*, 1981). Price and Sanger (1979) examined myosin-extracted myofibrils and were able to visualize in the electron microscope a system of 10-nm filaments existing as a parallel array extending between adjacent Z-bands.

The proposed relationship of intermediate filaments to the decreased 68,000-dalton protein in dystrophic chickens is that neuronal intermediate filaments consist in part of a 68,000-dalton component and that antibodies directed against the 68,000-dalton protein cross-react with skeletal muscle by staining the Z-band region (Gard *et al.*, 1979; Wang *et al.*, 1980). It has been shown (Berkowitz *et al.*, 1977; Runge *et al.*, 1979) that the 68,000-dalton protein that copurifies with brain microtubules during *in vitro* assembly is closely related to the 68,000-dalton component of the "neurofilament triplet" found in highly purified preparations of neurofilaments (Berkowitz *et al.*, 1977; Runge *et al.*, 1979). The evidence for the identification of the tau protein in microtubules as a protein also associated with intermediate filaments has been further documented by the recent report of Wang *et al.* (1980). These investigators showed that a 68,000-dalton protein with the same isoelectric point could be identified in diverse cytoskeletal preparations from rat spinal-cord neurofilaments, bovine microtubules, baby hamster kidney (BHK-21) cells, mouse 3T3 cells, chicken fibroblasts, and chicken muscle cells. One-dimensional peptide maps of the 68,000-dalton protein purified from chicken myofibrils, bovine-brain microtubules, and neurofilaments were identical. Thus, a protein that may be associated with both microtubules and intermediate filaments is found in many cell types including skeletal muscle (Geiger and Singer, 1980). However, the role of this 68,000-dalton protein as a promoter of microtubule polymerization remains unclear. Runge *et al.* (1979) reported that intact neurofilaments did not stimulate microtubule assembly, whereas preparations that contained the 68,000-dalton neurofilament protein in a disassembled state did stimulate microtubule assembly. Our observation that loss of one of the 68,000-dalton proteins in the supernatants from brains of dystrophic chickens was associated with a decreased rate of microtubule assembly is further evidence that this protein has a role in the promotion of microtubule assembly.

From the information presented above, it is becoming clear that the intermediate filaments of muscle may have a direct and important role in the transmission of the force of contraction. The intermediate filaments that connect Z-bands would prevent the disruption of sarcomeres from overextension to the point where thick and thin filaments no longer overlap. The intermediate filaments of the Z-band would serve as anchorage for the thin filaments and would also interconnect with other Z-bands so as to transmit force.

Therefore, one could propose that the possible pathogenesis in muscular dystrophy is that: (1) There is an abnormality in muscle cells, and many other cells, that results in or from a defect in intermediate-filament proteins. This defect could be at the level of genetic expression of the protein, proteolysis of the protein, or elsewhere. (2) The lack of intermediate filaments results in sarcomeres that are mechanically unstable because of the loss of the cytoskeletal backbone necessary to hold the Z-bands together or to anchor the thin filaments into the Z-bands. This inherent mechanical instability of dystrophic sarcomeres would result in muscle that is weak but without any defect in energy production or utilization or in the interaction of thick and thin filaments. (3) The postulated cytoskeletal defect would result in a progressive degeneration of muscle as the muscle tears itself apart. Clearly, additional studies of the 68,000-dalton protein will be useful in clarifying the relationship between proteins that stimulate microtubule assembly and proteins that are associated with intermediate filaments in brain and muscle.

ACKNOWLEDGMENTS. We would like to thank Mr. Bob Domke for his help in the preparation and characterization of the tubulin antisera, Ms. Sybil Cox for her technical work with the animals, and Ms. Leigh Thomas for her help with the CPK analysis. The support of Biomeasure, Inc., The Muscular Dystrophy Association, The American Heart Association, and NIH grant KO4 HL 00422 is appreciated.

References

Andia-Waltenbaugh, A. M., and Puck, T. T., 1977, Alzheimer's disease: Futher evidence of a microtubule defect, *J. Cell Biol.* **75**:279a.

Asnes, C. F., and Wilson, L., 1979, Isolation of bovine brain microtubule protein without glycerol: Polymerization kinetics change during purification cycles, *Anal. Biochem.* **98**:64.

Barnard, P. J., and Barnard, E. A., 1980, Chemotherapy in the genetically dystrophic chicken with a neurotrophic drug and with serotonin antagonists, in: *Proceedings of an International Symposium on Muscular Dystrophy Research* (C. Angelina, G. A. Danieli, and D. Fontanari, eds.), pp. 242–249, Elsevier/North Holland, Amsterdam.

Berkowitz, S. A., Katagiri, J., Binder, H. K., and Williams, R. C., 1977, Separation and characterization of microtubule proteins from calf brain, *Biochemistry* **16**:5610.

Brinkley, B. R., Fistel, S. H., and Marcum, J. M., 1980, Microtubules in cultured cells; indirect immunofluorescent staining with tubulin antibody, *Int. Rev. Cytol.* **63**:59.

Connolly, J. A., and Kalnins, V. I., 1978, Visualization of centrioles and basal bodies by fluorescent staining with nonimmune rabbit sera, *J. Cell Biol.* **79**:526.

Connolly, J. A., Kalnins, V. I., Cleveland, D. W., and Kirschner, M. W., 1978, Intracellular localization of the high molecular weight microtubule accessory protein by indirect immunofluorescence, *J. Cell Biol.* **76**:781.

Connolly, J. A., Kalnins, V. I., and Barber, B. H., 1979, Microtubule organization in fibroblasts from dystrophic chickens and persons with Duchenne muscular dystrophy, *Nature (London)* **282**:511.

Fuseler, J. W., Shay, J. W., and Feit, H., 1981, The role of intermediate (10-nm) filaments in the development and integration of the myofibrillar contractile apparatus in the embryonic mammalian heart, in *Cell and Muscle Motility*, Vol. 1 (R. Dowben and J. W. Shay, eds.), pp. 205–259, Plenum Press, New York.

Gard, D. L., Bell, P. B., and Lazarides, E., 1979, Co-existence of desmin and the fibroblast intermediate filament subunit in muscle and nonmuscle cells: Identification and comparative peptide analysis, *Proc. Natl. Acad. Sci. U.S.A.* **76**:3894.

Geiger, B., and Singer, S. S., 1980, Association of microtubules and intermediate filaments in chicken gizzard cells as detected by double immunofluorescence, *Proc. Natl. Acad. Sci. U.S.A.* **77**:4769.

Hanson, J., and Huxley, H. E., 1955, The structural basis of contraction in striated muscle, *Symp. Soc. Exp. Biol.* **9**:228.

Harper, C. G., Buck, D., Gonates, N. K., Guilbert, B., and Avrameas, S., 1979, Skin fibroblast microtubular network in Alzheimer's disease, *Ann. Neurol.* **6**:548.

Hudecki, M. S., Pollena, C. M., Bhargava, A. K., Hudecki, R. S., and Heffner, R. R., 1979, Delayed functional disability in dystrophic chickens receiving chemotherapy, *Muscle Nerve* **2**:57.

Hugelin, A., Legrain, Y., and Bondoux-Johan, M., 1979, Nerve growth promoting action of Isaxonine in rats, *Experientia* **35**:626.

Huxley, H. E., and Hanson, J., 1954, Changes in the cross-striation of muscle during contraction and stretch and their structural interpretation, *Nature (London)* **173**:973.

Laemmli, U. K., 1970, Cleavage of structural proteins during the assembly of the head of bacteriophage T4, *Nature (London)* **227**:680.

Lockwood, A. H., 1978, Tubulin assembly protein: Immunochemical and immunofluorescent studies on its function and distribution in microtubules and cultured cells, *Cell* **13**:613.

Lowry, O. H., Rosebrough, N. J., Farr, A. L., and Randall, R. J., 1951, Protein measurement with the Folin phenol reagent, *J. Biol. Chem.* **193**:265.

Miller, C. L., Fuseler, J. W., and Brinkley, B. R., 1977, Cytoplasmic microtubules in transformed mouse × nontransformed human cell hybrids: Correlation with *in vitro* growth, *Cell* **12**:319.

Oliver, J. M., 1976, Impaired microtubule function correctable by cyclic GMP and cholinergic antagonists in the Chediak–Higashi syndrome, *Am. J. Pathol.* **85**:395.

Price, M., and Sanger, J. W., 1979, Intermediate filaments connect Z-discs in adult chicken muscle, *J. Exp. Zool.* **280**:263.

Runge, M. S., Detrich, H. W., and Williams, R. C., 1979, Identification of the major 68,000-dalton protein of microtubule preparations as a 10-nm filament protein and its effect on microtubule assembly *in vitro*, *Biochemistry* **18**:1689.

Sandoval, I. V., and Weber, K., 1980, Different tubulin polymers are produced by microtubule associated proteins MAP$_2$ and Tau in the presence of guanosine 5-(α,β-methylene) triphosphate, *J. Biol. Chem.* **255**:8952.

Schneider, W., Morgenstern, E., and Reimers, H. J., 1979, Disassembly of microtubules in the Lesch–Nyhan syndrome, *Klin. Wochenschr.* **57**:181.

Shay, J. W., and Fuseler, J. W., 1979, Diminished microtubules in fibroblast cells derived from inherited dystrophic muscle explant, *Nature (London)* **278**:178.

Shay, J. W., Feit, H., and Neudeck, U., 1980, Microtubules and avian muscular dystrophy, in: *Microtubules and Microtubule Inhibitors* (M. de Brabander and J. De Mey, eds.), pp. 545–553, Elsevier/North Holland, Amsterdam.

Shay, J. W., Thomas, L. E., and Fuseler, J. W., 1981, Altered mononuclear cell spreading and microtubules in Duchenne muscular dystrophy patients and carriers, in: *Diseases of the Motor Unit* (D. L. Schotland, ed.) (in press).

Tarrade, T., and Hugelin, A., 1978, Nerve growth promoting action of Isaxonine *in vitro*, in: "Abstracts of the Second European Neuroscience Meeting," *Neurosci. Lett. Suppl.* **1**:545.

Thakar, J. H., Thede, A., and Strickland, K. P., 1980, On the ultrastructure of primary cultures of normal and dystrophic hamster tongue muscle, *Muscle Nerve* **3**:340.

Wang, C., Asai, D. J., and Lazarides, E., 1980, The 68,000-dalton neurofilament-associated polypeptide is a component of nonneuronal cells and of skeletal myofibrils, *Proc. Natl. Acad. Sci. U.S.A.* **77**:1541.

Weingarten, M. D., Lockwood, A. H., Hwo, S.-Y., and Kirschner, M. W., 1979, A protein factor essential for microtubule assembly, *Proc. Natl. Acad. Sci. U.S.A.* **72**:1858.

White, J. G., and Clawson, C. C., 1979, The Chediak–Higashi syndrome: Microtubules in monocytes and lymphocytes, *Am. J. Hematol.* **7**:349.

11

The Structure of Vertebrate Skeletal-Muscle Myosin Filaments

Frank A. Pepe

1. Introduction

Recently, a number of observations have been made that are highly significant with respect to our understanding of the structure of the vertebrate skeletal-muscle myosin filament. It is the purpose of this chapter to review these findings and to see how they are related to existing detailed models for the myosin filament (Pepe, 1966a, 1967a; Squire, 1973).

The structural characteristics of myosin filaments that are clearly visible in negatively stained preparations of separated filaments (Huxley, 1963) are: (1) a length of 1600 nm, (2) a diameter of about 16 nm, (3) the presence of tapered ends, and (4) a rough surface along the filaments except for a bare zone region in the middle of the filament for which a length of 149 nm has recently been reported (Craig, 1977). From X-ray diffraction studies (Huxley and Brown, 1967), it was found that myosin cross-bridges occur along the myosin filament at intervals of 14 nm and that the helical arrangement of myosin cross-bridges on the surface of the filament has an axial repeat of 43 nm. From the X-ray data, it is not possible to determine whether the helical arrangement is two-stranded, three-stranded, or four-stranded.

A considerable amount of biochemical work has been done to pin down the myosin content of the vertebrate skeletal-myosin filament. The myosin content would help to decide among the possibilities for strandedness of the helical arrangement. Values for myosin content have centered around three or four myosin molecules per 14-nm interval along the length of the filament. The value of three myosin molecules per 14-nm interval has been obtained by

Frank A. Pepe • Department of Anatomy, School of Medicine, University of Pennsylvania, Philadelphia, Pennsylvania 19104.

quantitative sodium dodecyl sulfate (SDS) gel electrophoresis (Tregear and Squire, 1973; Potter, 1974), nucleotide-binding studies (Marston and Tregear, 1972), hydrodynamic studies (Emes and Rowe, 1978), and mass measurements in scanning transmission electron microscopy (Lamvik, 1978). The value of four myosin molecules per 14-nm interval has been obtained by quantitative SDS gel electrophoresis (Morimoto and Harrington, 1974; Pepe and Drucker, 1979), nucleotide-binding studies (Maruyama and Weber, 1972; Weber *et al.*, 1969), hydrodynamic studies (Katsura and Noda, 1973), and a particle-counting technique (Morimoto and Harrington, 1974). Pepe and Drucker (1979) have shown that the source of this discrepancy, for data obtained from quantitative SDS–polyacrylamide gel electrophoresis, appears to be related to whether or not actin (to which the myosin content was standardized) was properly resolved from two closely neighboring protein bands. When actin was properly resolved, a value of four myosin molecules per 14-nm interval was obtained for the myosin content of vertebrate skeletal-myosin filaments. The quantitative data must be good enough to clearly distinguish between this difference of only 25% between values of three and four. Because of the relatively large uncertainties in all the quantitative procedures used, it is not likely that a definitive value will come from these approaches. It is more likely that the correct quantitative results will eventually be accepted by the force of structural information, rather than by the results of quantitative analytical techniques.

The shape and size of the filament, its myosin content, and the arrangement of myosin cross-bridges on the surface of the filament must be related to the structure of the shaft or backbone of the filament in some way. Information about the structure of the backbone can come from a variety of sources including transverse sections studied in electron microscopy, studies of the assembly and disassembly of the filaments, specific antibody staining, and X-ray diffraction.

2. Subfilament Spacing

The presence of subfilaments in the shaft of myosin filaments has been suggested for a long time from observations of subunits in transverse sections of the filament. Subunits with diameters in the range of 2.5–3 or 4–4.5 nm have been observed for invertebrate muscle-myosin filaments (Baccetti, 1965, 1966; Gilev, 1966a–c). In vertebrate skeletal-muscle myosin filaments, Pepe and Drucker (1972) found subunits with a diameter of about 3 nm and spaced about 3.7 nm apart. Further studies of transverse sections of the vertebrate skeletal-myosin filaments using optical diffraction of the electron micrographs have substantiated the spacing of about 4 nm (Pepe and Dowben, 1977; Pepe, 1979; Pepe *et al.*, 1981). A variety of fixation procedures were used, patterned after the multistage fixation procedure introduced by Reedy (1976). The fixation procedure that provided the best preservation of subfilament order as judged by the optical-diffraction pattern obtained from the image (Ashton and Pepe, 1981) gave the distribution of spacings shown in Fig. 1, which has a mean of 3.5 nm. Since the spacings measured from the optical-diffraction patterns represent the spacing between lattice planes, the center-to-center spacing of the subfilaments is about 4 nm.

Subfilament spacings of about 4 nm have also been observed by X-ray diffraction in vertebrate skeletal-myosin filaments (Millman, 1979) as well as in the myosin filaments of crustacean muscles (Wray, 1979a,b).

Since the α-helical myosin rod has a diameter of 2 nm, a 4-nm spacing between subfilaments suggests that there is more than one myosin molecule in a transverse section of the subfilament. If the subfilaments were constructed of the myosin dimers found to be present in myosin solutions (Harrington and Burke, 1972; Burke and Harrington, 1972), then one might expect there to be more than one tail in a transverse section of a subfilament, and the 4-nm spacing becomes reasonable.

3. Parallel Subfilaments

Evidence from electron microscopy suggests strongly that the subfilaments in vertebrate skeletal-muscle myosin filaments are parallel to the long axis of the myosin filaments. One source of evidence for parallel subfilaments

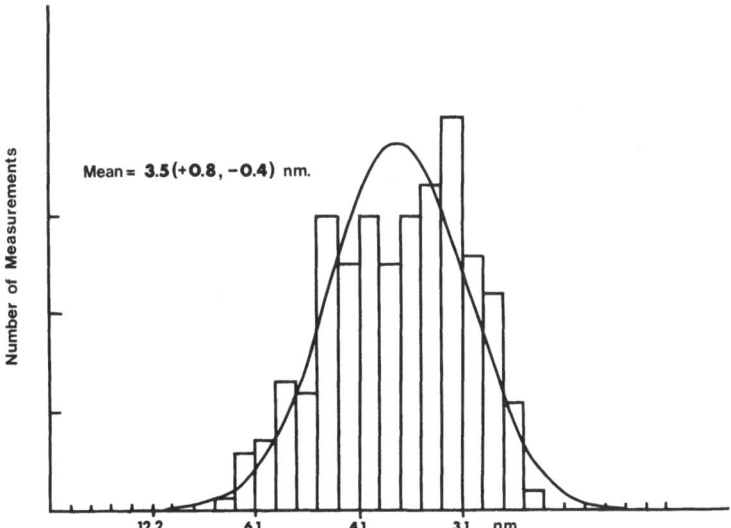

Figure 1. Distribution of lattice spacings for subfilaments in transverse sections of the myosin filament. The spacings were measured from optical-diffraction patterns obtained from images of transverse sections of individual myosin filaments. The lateral muscles of the freshwater killifish (*Fundulus diaphanus*) were used, and the muscle was fixed by the multistage fixation procedure shown by Ashton and Pepe (1981) to give the best-preserved subfilament organization. The mean spacing of 3.5 nm is the spacing between hexagonal lattice planes, and this corresponds to a center-to-center spacing of about 4 nm between subfilaments.

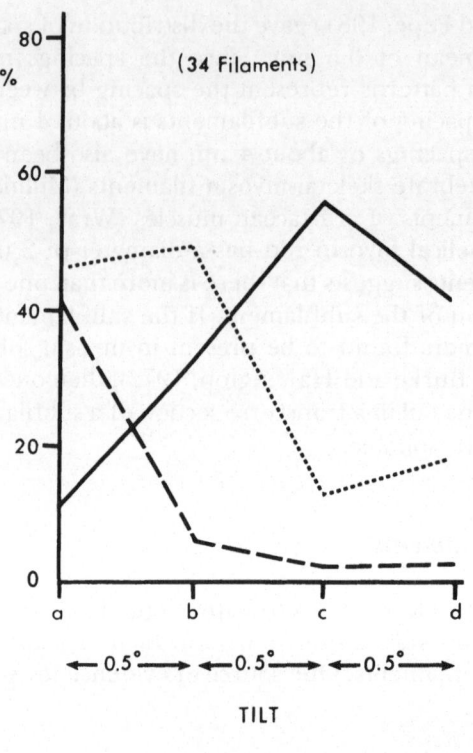

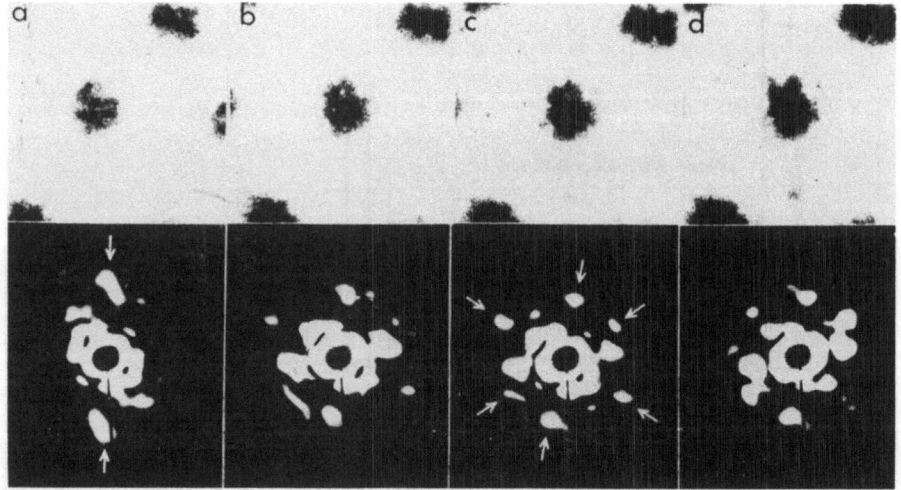

Figure 2. Evidence for parallel subfilaments in the shaft of myosin filaments from the change in optical-diffraction pattern (from single filaments) with tilt (sections 300 nm thick). *Bottom:* Optical-diffraction patterns obtained from the images of the single myosin filament directly above each pattern. Only the image of the filament in the center was used to produce the diffraction pattern. Each image from (a) through (d) differs in tilt by 0.5°, with the tilt axis oriented perpendicularly. The optical-diffraction pattern from (c) shows strong diffraction maxima in the three directions indicated by arrows. A tilt of 0.5° to (b) and another 0.5° to (a) results in strong diffraction maxima (indicated by arrows) in only one direction parallel to the axis of tilt.

is the change observed in the optical-diffraction pattern obtained from transverse sections at different tilts. Transverse sections of the filament, 140 or 300 nm thick, were tilted in intervals of 0.5–2°, and optical diffraction patterns were obtained from the images at each tilt interval. The changes in the optical-diffraction patterns that were obtained with tilt were as would be expected for hexagonally packed parallel rod subfilaments (Pepe *et al.*, 1981). In 140-nm-thick sections, a tilt of 1.6° with the tilt axis perpendicular to one of the lattice planes of the hexagonal lattice of subfilaments should change the optical-diffraction pattern from one that shows substructure spacings in three directions to one that shows substructure spacings in one direction, since the projection of the mass of the subfilaments in the tilted section will be concentrated in rows along the direction of tilt. With sections 300 nm thick, a tilt of only 0.75° should give the same result. Using 300-nm-thick sections, a tilt interval of 0.5° was used, whereas with 140-nm sections, a tilt interval of 2° was used. In transverse sections 300 nm thick, a tilt of 0.5° changed the optical-diffraction pattern from one with strong spacings in three directions to one with a strong spacing along the axis of tilt and two weaker spacings, and then to only one strong spacing along the axis of tilt after an additional 0.5° of tilt, just as expected for hexagonally packed parallel rods. In this case, an area of 34 filaments was studied, and as can be seen in Fig. 2, the number of filaments giving subfilament spacings in three directions declined sharply after 1° of tilt, while the number giving one or two directions increased sharply. In transverse sections 140 nm thick, a tilt of 2° gave a similar change in the optical-diffraction pattern (Fig. 3). These findings are what would be expected for hexagonally packed parallel rods spaced 4 nm apart.

Another source of evidence for subfilaments parallel to the long axis of the filament is the observation that vertebrate skeletal-myosin filaments can be made to splay into three groups of subfilaments along most of the length of the filament (Maw and Rowe, 1980). This effect was produced by putting a suspension of separated filaments onto a carbon film and washing with water prior to negative staining for electron microscopy. The splaying was not always obtained under these conditions. Using a 10 mM tris–citrate buffer, pH 8, which Reisler *et al.* (1980) have shown to produce minifilaments (discussed in section 7), we have found that we can get the splaying uniformly and consistently. An example of splayed filaments obtained in this way is shown in Fig. 4. As pointed out by Maw and Rowe (1980), there is no evidence for coiling of the three groups of subfilaments. They appear uniformly straight

---

Top: Graph indicating the percentages of filaments (total of 34 filaments) that give spacings in one direction (---), two directions (---), or three directions (———) in the optical-diffraction patterns obtained at each tilt interval of 0.5°. The number that gives three directions is maximum in (c) and decreases sharply in (a) after tilting by 1°. This is accompanied by a corresponding increase in the number of filaments that give one direction and two directions. These observations correspond to what is expected for hexagonally packed parallel rods 300 nm long and spaced 4 nm apart where a tilt of 0.75° along one of the lattice planes would change the optical-diffraction pattern from one with three directions to one with one direction (Pepe *et al.*, 1981).

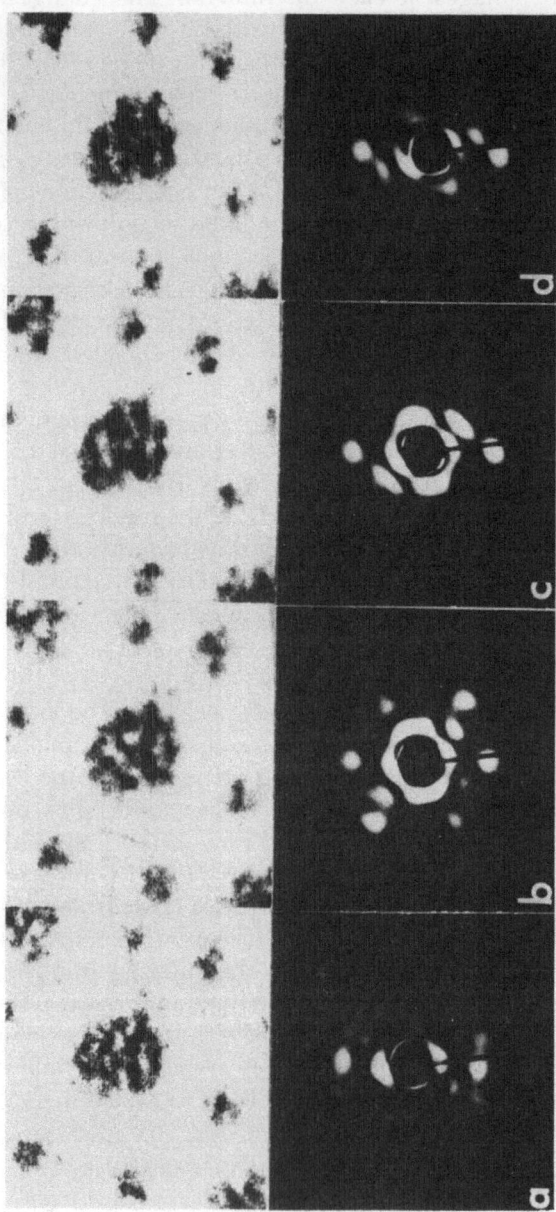

Figure 3. Evidence for parallel subfilaments in the shaft of myosin filaments from the change in optical-diffraction pattern (from single filaments) with tilt (sections 140 nm thick). Similarly to the 300-nm-thick sections in Fig. 2, tilting of 140-nm-thick sections changes the optical-diffraction pattern from one with strong diffraction maxima in three directions (b) to one with one direction parallel to the axis of tilt (a, c). The image of the filament is above the optical-diffraction pattern obtained from that image. In this case, the tilt interval was 2°. This change corresponds to what is expected for hexagonally packed parallel rods 140 nm long and spaced 4 nm apart, where a tilt of 1.6° along one of the lattice planes would be expected to change the optical-diffraction pattern from one with three directions to one with one direction (Pepe *et al.*, 1981).

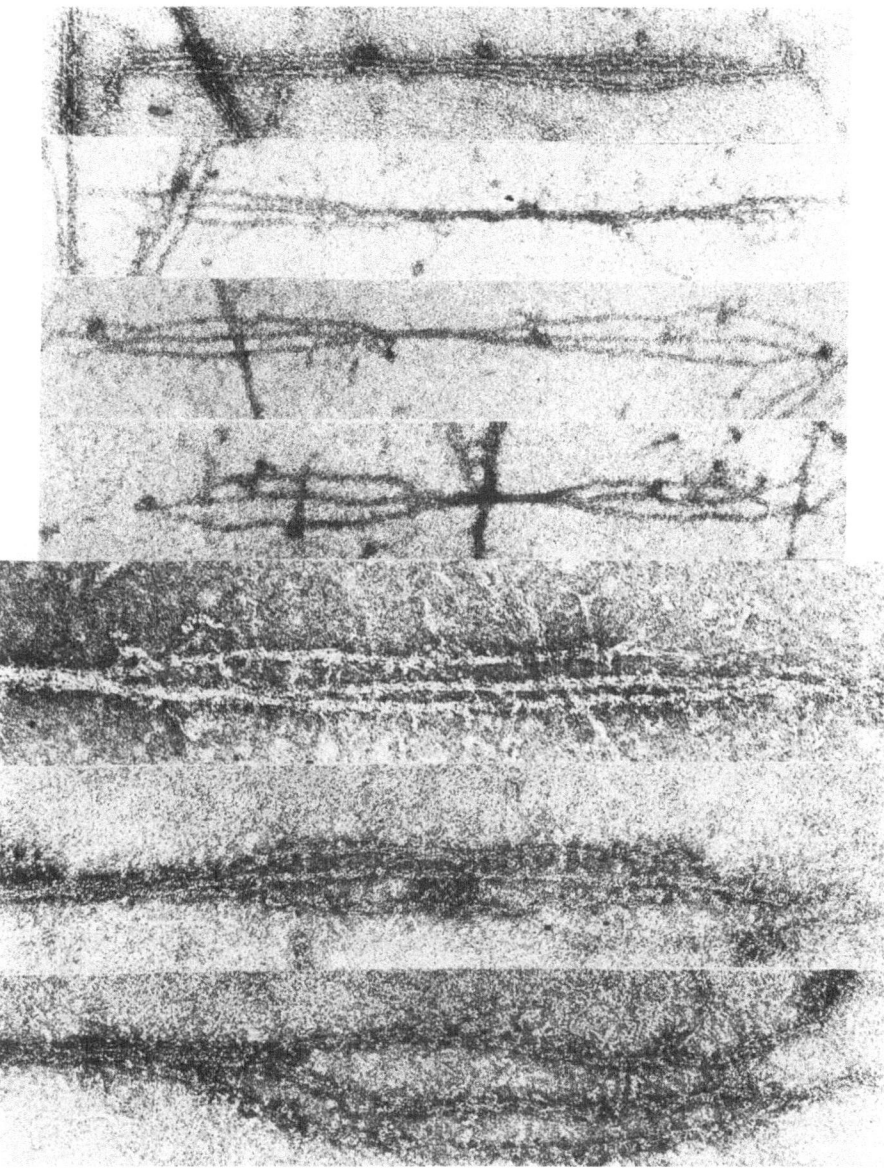

Figure 4. Splayed natural filaments. Natural filaments were washed with the tris–citrate buffer used by Reisler *et al.* (1980) to produce myosin minifilaments. This buffer caused the filaments to splay into three groups of subfilaments. This result was obtained consistently and uniformly with the tris–citrate buffer. The top four images show the entire splayed filament, and the bottom three show half the splayed filament at higher magnification. The cross-bridges are not preserved well enough to determine the interval at which they occur along each of the three groups of subfilaments into which the filament splays.

and therefore provide strong evidence for the presence of parallel subfilaments in the shaft of vertebrate skeletal-myosin filaments.

4. Threefold Symmetry

The early observations by Franzini-Armstrong and Porter (1964) of triangular transverse profiles indicated that the vertebrate skeletal myosin filament has threefold symmetry. Similarly, the triangular transverse profiles were observed in chicken muscle and formed one of the constraints in constructing a detailed model for the vertebrate skeletal-myosin filament (Pepe, 1966a, 1967a). If the sections are relatively thin (100 nm), the triangular transverse profiles are seen clearly only in sections through the bare zone region of the filament (Fig. 5). Therefore, it is possible that the filament symmetry may change outside the bare zone region. However, when the section thickness is increased to about 400 nm, where even if the bare zone region were included entirely it would comprise only 150 nm of the 400-nm length of the filament in the section, it becomes clear that the transverse profile along the entire length of the shaft of the filament is triangular (Fig. 5) and therefore that the shaft of the filament has threefold rotational symmetry in transverse projection. Another feature of the transverse profile becomes evident in these thicker sections: each apex of the triangle is blunted. Recently, Freundlich *et al.* (1980) have suggested that in these thick sections, this profile is imposed by the attachment of the myosin cross-bridges to the surrounding actin filaments. However, evidence that it is more likely due to the arrangement of the parallel subfilaments in the shaft of the filament will be presented later. The threefold rotational symmetry in transverse projection of the shaft along the entire length of the filament is further supported by the observation made by Maw and Rowe (1980) and discussed in Section 3 that the filament can be made to splay into three groups of subfilaments along almost the entire length of the filament (see Fig. 3). This and the triangular transverse profile of the filament have been cited as compelling evidence for the threefold symmetry of the filament as well as for the presence of parallel subfilaments. As will be described in detail in Section 10, although threefold rotational symmetry in transverse projection is consistent with *both* the two available detailed models of the myosin-filament structure (Pepe, 1966a, 1967a; Squire, 1973), the presence of parallel subfilaments is consistent only with the model proposed by Pepe (1966a, 1967a).

5. Subfilament Arrangement and Number

In earlier observations of substructure in transverse sections (Baccetti, 1965, 1966; Gilev, 1966a–c), the exact number and arrangement of subunits were elusive. Pepe and Drucker (1972), using transverse sections of vertebrate skeletal-myosin filaments, took advantage of the threefold rotational symmetry indicated by triangular transverse profiles in the bare zone region of the fila-

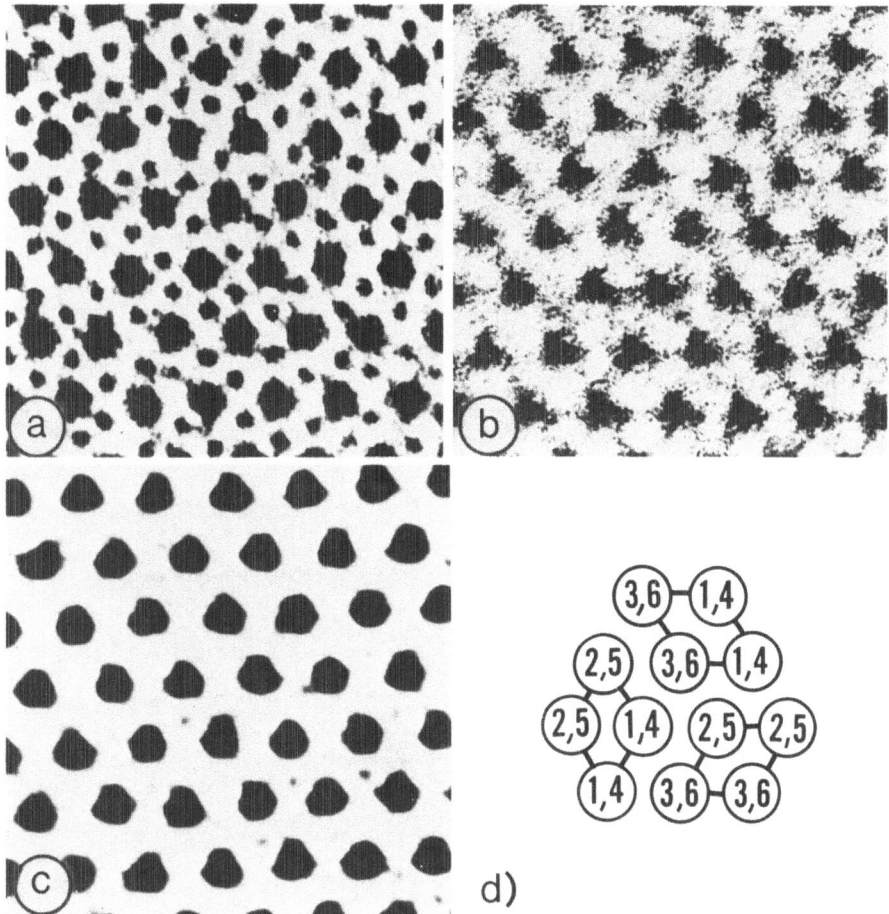

Figure 5. Transverse profiles of myosin filaments. (a) Transverse section less than 90 nm thick taken in the myosin- and actin-filament overlap region of the A-band. The filaments have a variety of profiles, some of which appear triangular. (b) Transverse section somewhat thinner than that in (a), taken in the bare zone region of the filaments immediately adjacent to the M-band region. The filaments all show clear triangular profiles that are all oriented similarly. This similar orientation is characteristic for fish muscle (Franzini-Armstrong and Porter, 1964; Pepe, 1975). (c) Transverse sections 400 nm thick. The density of the shaft of the myosin filament where the subfilaments are tightly packed is accentuated relative to the surround. The transverse profiles of the shaft of the filament can be described as triangular with each apex blunted. (d) Shaft of the myosin filament as predicted by the two-stranded model proposed by Pepe (1966a, 1967a). The transverse profile corresponds to that observed in the 400-nm-thick transverse sections (c).

ment and photographically enhanced the threefold symmetrical structures by Markham rotation. They concluded that there are 12 subfilaments hexagonally packed to form a triangle with 3 centrally located and 9 peripherally located (Fig. 6). This arrangement gives a transverse profile that is a triangle with each apex blunted consistent with what is seen in 400-nm-thick transverse sections (see Fig. 5). However, at the time that we obtained the photographi-

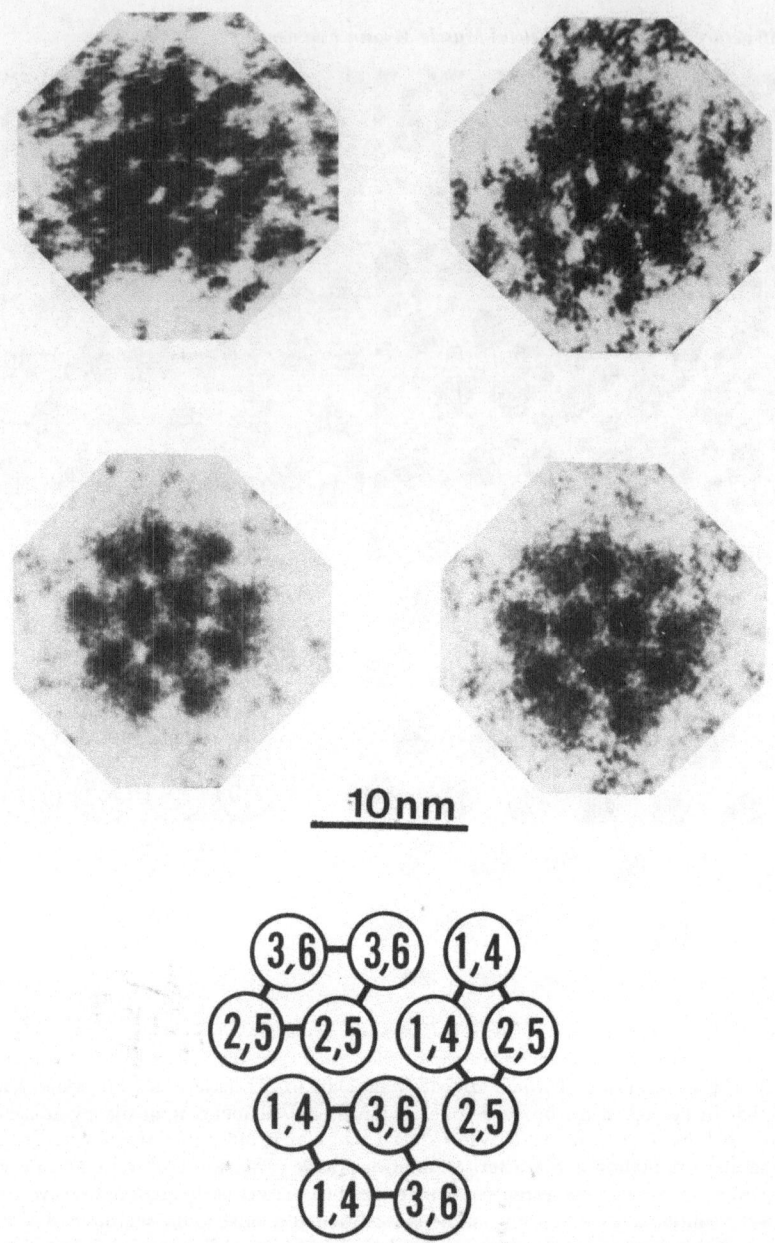

Figure 6. Photographic enhancement of the subfilaments in a transverse section of the myosin filament. Images of individual myosin filaments in transverse section were enhanced by Markham rotation taking advantage of the three-fold symmetry of the filament. One third of the exposure time was used for three exposures, differing by 120° rotation, which were superimposed. The top images are of the same filament without photographic enhancement. The bottom images are two rotational enhancements of the same filament. There are 12 subfilaments hexagonally packed with 3 in the center and 9 at the periphery forming a triangular profile with a subfilament missing at each apex of the triangle. The subfilaments are about 3 nm in diameter and spaced about 4 nm apart (Pepe and Drucker, 1972). The diagram at the bottom is of the two-stranded model proposed by Pepe (1966a, 1967a), in which the predicted arrangement and number of subfilaments are similar to those observed in the photographically enhanced images.

cally enhanced images (Fig. 6), we did not have evidence that made it unlikely for the images to be a result of artifact. More recently, using images characterized by optical-diffraction and tilting experiments, as already described (see Figs. 2 and 3), autocorrelation methods were also used to assure ourselves of the presence of real structure (Pepe *et al.*, 1981). The images characterized by these methods were then analyzed by a number of computer methods to circumvent the problems associated with photographic enhancement (Stewart *et al.*, 1981). This has enabled objective criteria to be employed and so enabled the number and arrangement of the subfilaments to be established clearly. The result of rotational averaging of four combined images is shown in Fig. 7. This shows the same arrangement of subfilaments as was observed previously by photographic enhancement (Fig. 6). This same arrangement of 12 subfilaments was also obtained by projection and linear-superposition methods, even though the assumptions made by these different methods of processing the images are different (Stewart *et al.*, 1981).

6. Sources of Artifacts

In considering the observations of subfilaments in transverse sections of the myosin filament, we have to keep in mind the possible sources of artifacts and how these might contribute to the observations. For instance, there might be artifacts related to the background granularity of the image, or artifacts related to focus or to multiple scattering, which can occur in the thicker sections (Stewart *et al.*, 1981).

It is unlikely that artifacts associated with background granularity or with focus make a substantial contribution to the image. First, the background-phase granularity is related to microscope defocus and would be expected to change with focus, which was not the case over a wide focal range; second, the position of phase granules would be expected to be essentially random, and therefore the frequency and consistency with which these optical-diffraction patterns were observed argue against this. Furthermore, a random process would not be expected to give the systematic changes observed in the optical-diffraction patterns on tilting. Therefore, it is unlikely that the filament structure could be a phase-granule artifact or an artifact of focus. At 200 kV, which was the accelerating voltage of the electron beam used in obtaining these images, multiple scattering is unlikely to be a problem for the 140-nm-thick sections, though it might be a problem for the 300-nm sections. However, since the substructure present in both the 140- and 300-nm-thick sections was so similar and changed so similarly and predictably on tilting (see Figs. 2 and 3), it is unlikely that the substructure could be a result of multiple scattering either.

On the basis of optical-diffraction studies of models of muscle transverse sections, Luther and Squire (1980) have suggested that filament fine structure could be an artifact associated with aperture truncation of high spatial fre-

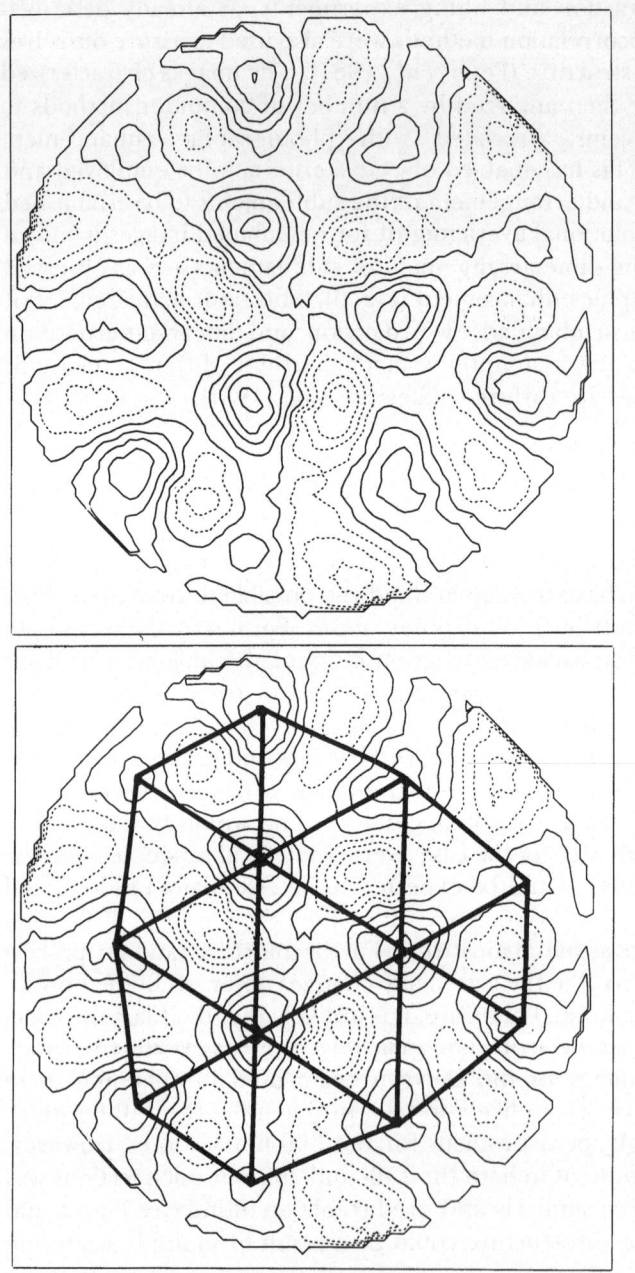

Figure 7. Computer processing of images of transverse sections of individual myosin filaments by rotational averaging. Contour density maps of the rotational averaging of four combined images are shown. At left, the bold straight lines join the centers of the subfilaments. At right is the same map without the lines. There are 12 subfilaments hexagonally packed with 3 centrally located and 9 peripherally located, forming a triangular profile with a subfilament missing at each apex. The diagram at the bottom is the two-stranded model proposed by Pepe (1966a, 1967a), in which the predicted arrangement and number of subfilaments are similar to those observed in the rotationally averaged images.

quencies. There are two reasons that this is unlikely. The first is that the effect observed by Luther and Squire (1980) was due primarily to their model's having very sharp edges on the myosin-filament profiles. This gave rise to an artifactually high content of high-frequency information in their optical-diffraction patterns, and this led to their observation of truncation "ripples" in their reconstructed images. Their "artifactual substructure" disappears when models with smoothed edges are employed (Ashton and Stewart, unpublished results). Second, the actual truncation frequency of the electron-microscope objective apertures employed in our study was substantially higher than that expected to introduce artifactual detail at the frequencies of about 4 nm^{-1} observed in the micrographs.

The high correspondence between the various methods used to process the images (rotational filtering, projection, and superposition methods) argues very strongly that the pattern of 12 subfilaments (Fig. 7), observed in each case, is present in the original micrographs and is not an artifact associated with processing (Stewart *et al.*, 1981). The rotationally filtered images (Fig. 7) showed clearly the rows of subunits, which are completely unrelated to the assumption of the threefold symmetry, and the translational methods showed clearly the threefold symmetry, which is not related to the assumption of linear repeats.

Therefore, we are now reasonably confident of the presence of 12 subfilaments parallel to the long axis of the filament, hexagonally packed with a center-to-center spacing of about 4 nm and arranged with 3 centrally located and 9 peripherally located as shown in Figs. 6 and 7.

7. Minifilaments

The experiment that resulted in the finding that natural filaments could be splayed easily and consistently by washing the filaments with 10 mM tris–citrate buffer, pH 8 (see Fig. 4), was suggested by some recent experiments of Reisler *et al.* (1980). They showed that a myosin solution dialyzed into 10 mM Tris–citrate, pH 8, will form myosin minifilaments that are highly uniform in size. These minifilaments have a length of 300 nm, a diameter of about 8 nm, and a bare zone in the middle of the filaments about 180–200 nm in length with projections along the rest of the length (Fig. 8). From hydrodynamic studies, the minifilaments were found to contain from 16 to 18 myosin molecules. It is tempting to speculate that these short filaments with small diameters may be related to an organizational unit for the myosin filament that is higher than the myosin dimer. This possibility led to trials of the use of the buffer in which minifilaments form to attempt to splay natural filaments. The fact that this buffer is so successful at splaying the natural filaments (Fig. 4) suggests that the organizational units of which the minifilaments are built may be related to the three units into which the natural filament splays. A possibility for this relationship is discussed in Section 10 in relation to available models for the vertebrate skeletal myosin filament.

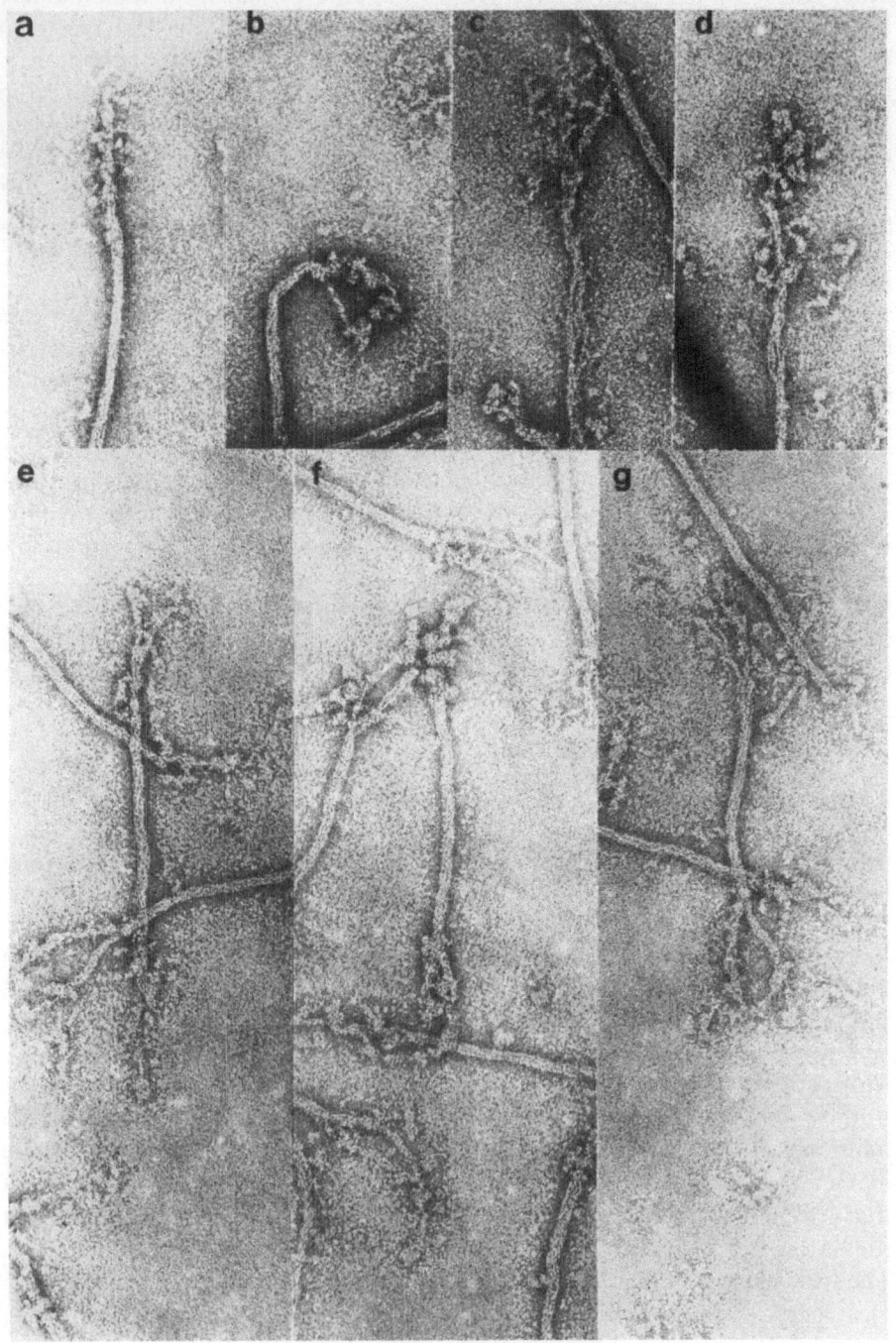

Although they are not the minifilaments described by Reisler *et al.* (1980), short filaments have also been observed at KCl concentrations of 0.3 M. These have been formed by reducing the KCl concentration of a myosin solution from 0.6 to 0.3M (synthetic filaments) as well as by exposing a natural filament suspension to 0.3 M KCl. At this KCl concentration, Kaminer and Bell (1966) measured synthetic-filament lengths of about 0.2 μm, which is about the same length as was obtained by Trinick and Cooper (1980) with natural filaments.

In addition to the short synthetic filaments observed at high concentrations of KCl (0.3 M), long synthetic filaments have also been observed in 0.3 M KCl, but only after standing in the cold for several days (Eaton and Pepe, 1974). These filaments have a periodicity of 43 nm and are smaller in diameter than natural filaments or synthetic filaments formed by reducing the KCl concentration to 0.15 M KCl. It is conceivable that these filaments are related to the units into which native filaments splay on exposure to water (Maw and Rowe, 1980) or tris–citrate buffer (see Fig. 4), although there is no strong evidence for this at present.

8. Tapered Ends of the Filaments

Longitudinal sections through the A-band of vertebrate skeletal-muscle myofibrils in many cases show a less dense gap in the A-band very close to the A–I junction (Fig. 9). This gap can be seen with (Fig. 9b) or without (Fig. 9a) antimyosin staining (Pepe, 1979). Craig and Offer (1976a) have attributed this gap to the absence of myosin cross-bridges at the position corresponding to the third 14-nm interval in from the tapered end of the filament. This missing cross-bridge interval is probably related in some way to how the myosin filament tapers and therefore is an important structural characteristic of the filament. Its relationship to available models for myosin-filament structure is discussed in Section 10.

9. Differences in Structure along the Length of the Filaments

Until recently, there has been no evidence for structural differences along the length of the myosin filament other than the evidence that in the middle of the filament, myosin molecules are oppositely oriented tail-to-tail, and that along the rest of the filament, they are similarly oriented head-to-tail in each half of the filament (Huxley, 1963). Evidence for structural differences along each half of the filament wherein myosin molecules are simi-

Figure 8. Minifilaments of Reisler *et al.* (1980). This micrograph shows the minifilaments that Reisler *et al.* (1980) produced on dialysis of myosin against 10 mM tris–citrate buffer, pH 8. These minifilaments have a very uniform length distribution. They are 300 nm long with a diameter of about 8 nm and a bare zone region in the middle of the filaments about 180–200 nm long with projections along the rest of the filament. Kindly provided by Dr. Emil Reisler.

Figure 9. Gap in myosin cross-bridges in the tapering region of the myosin filament close to the A–I junction in the A-band. Along each A–I junction, a short distance in from the edge of the A-band, is a gap that can be seen easily in unstained (a) as well as antimyosin-stained A-bands (b). This is observed in both fish (a) and chicken (b) muscles. This gap has been attributed to the absence of myosin cross-bridges at the position corresponding to the third 14-nm level in from the tapered end by Craig and Offer (1976a), who used anti-S-1 staining.

larly oriented has come from (1) optical-diffraction studies of transverse sections taken at different positions along the filament (Pepe *et al.*, 1981; Ashton and Pepe, 1981); (2) localization of C-protein binding to the myosin filament (Pepe and Drucker, 1975; Craig and Offer, 1976b) and restriction of binding to only a portion of the myosin filament (Pepe and Drucker, 1975); (3) the restriction of specific anti-L staining to only a portion of the myosin filament (Pepe, 1975); and (4) observations of filament length as a function of KCl concentration obtained with natural (Trinick and Cooper, 1980) and synthetic myosin filaments (Pepe and Drucker, unpublished results).

9.1. Optical-Diffraction Studies of Transverse Sections

Optical-diffraction patterns obtained from images of transverse sections of myosin filaments show subfilament spacings of about 4 nm (Pepe and Dowben, 1977; Pepe *et al.*, 1981), which is consistent with the subfilament spacings obtained from X-ray diffraction studies of both vertebrate (Millman, 1979) and invertebrate muscles (Wray, 1979a,b), as has already been discussed. Three types of optical-diffraction patterns were obtained—those with diffraction maxima in three directions, two directions, or only one direction (Pepe and Dowben, 1977), as shown in Fig. 10. As was discussed above, these

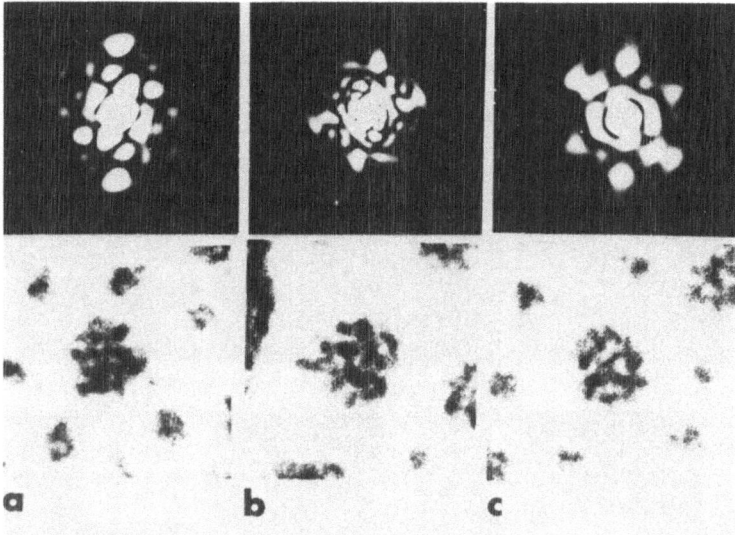

Figure 10. Types of optical-diffraction patterns observed from images of transverse sections of individual filaments. The image of the myosin filament is on the bottom, and the optical-diffraction pattern obtained from it (without including the surrounding actin filaments) is shown above. Optical-diffraction patterns with strong maxima in one (a), two (b), or three (c) directions were observed. The subfilament spacing represented by these patterns is about 4 nm.

different patterns can be obtained by tilting a filament that gives subfilament spacings in three directions. However, it is possible that tilt is not the sole source of these different patterns. For instance, Fig. 11 illustrates two possibilities other than tilt that could give rise to these patterns from a transverse section through 12 parallel subfilaments. A compression of the hexagonal lattice or a change in the orientation of elongated subfilament profiles could produce these patterns. The consequent possibility that these different patterns arise from structurally different myosin filaments or from structurally different portions of the same myosin filament was considered. To investigate these possibilities and to get an idea of the extent to which tilt contributes to the patterns, serial transverse sections of the filaments were studied (Pepe *et al.*, 1981). If the different patterns arise from structurally different filaments, the proportion of each pattern that is observed should not change from section to section along the length of the filament. If the patterns arise from structurally different portions of the same filament, the proportion of each pattern that is observed should change with position along the length of the filament. Serial transverse sections about 140 nm thick were taken through the A-band, and a group of about 25 filaments were followed from section to section. What we found was that near the tapered end of the filaments, most of the filaments gave optical-diffraction patterns that showed subfilament spacings in three directions, very few showed spacing in one direction, and some showed spacing in two directions (Fig. 12). In sections

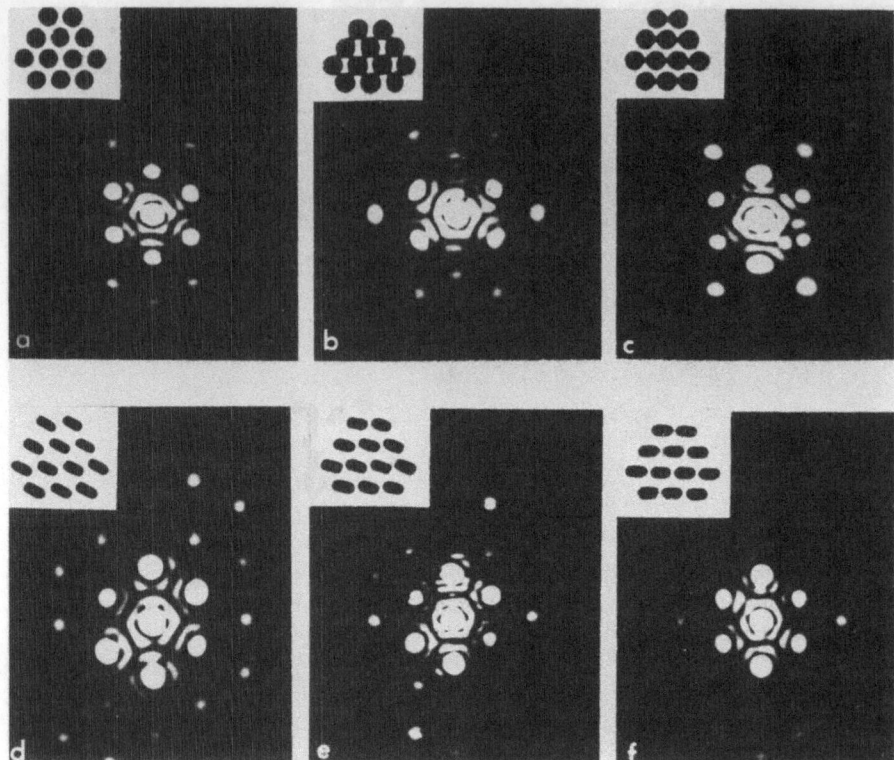

Figure 11. Two possibilities for differences in subfilament organization that could produce optical-diffraction patterns similar to those observed. (a–c) By compression of a hexagonal lattice of circular subfilament profiles (*insets*), the optical-diffraction pattern can be changed from one giving strong spacings in three directions (a) to one giving strong spacings in two directions (b) or in one direction (c).(d–f) If the subfilament profiles are elongated, a change in orientation of the subfilament profiles without compression of the lattice (*insets*) can give similar differences in optical-diffraction patterns. A rotation of 15° in the subfilament profiles from the orientations in (d) (which gives strong spacings in three directions) will change the optical-diffraction pattern to one that gives strong spacings in two directions (e), and a 30° rotation will produce a pattern that gives strong spacings in only one direction (f)

farther away from the tapered ends of the filaments, the number that gave subfilament spacings in three directions decreased drastically, while the numbers that gave one direction and two directions increased, and on approaching the M-band region of the filament, there was an increase in three directions with a decrease in one and two directions. The two halves of the filament were symmetrical. Because the relative proportions of these patterns change so strikingly along the filament, the different patterns cannot come entirely from tilting, nor can they come from structurally different filaments, and therefore the structural factors responsible for these differences must be changing along the length of the same filament. However, from these data, it was not possible to distinguish among the different structural characteristics that could give rise to these patterns.

Figure 12. Differences in optical-diffraction patterns observed from serial transverse sections along the myosin filament (Pepe *et al.*, 1981) and speculations on a possibility for explaining the pattern of C-protein binding to the myosin filament. (A) Serial transverse sections 140 nm thick were taken, and optical-diffraction patterns were obtained from the same 25 filaments along their length. Five sections were obtained in each half of the filament, and the two halves were symmetrical. Near the tapered ends, a large proportion of the filaments gave optical-diffraction patterns with strong maxima in three directions (●), and the number decreased sharply on moving away from the tapered end and increased again on approaching the M-band region. Concomitantly, the number of filaments that give strong maxima in one direction (▲) or two directions (○) was minimal near the tapered ends, increased on moving away from the tapered ends, and decreased again on approaching the M-band. Although tilting (see Figs. 2 and 3) contributed to the patterns observed, the strong change in proportions of the patterns in different parts of the filament cannot result entirely from tilting (Pepe *et al.*, 1981). (B) The positions of C-protein binding (arrows at 43-nm intervals) relative to the intervals at which myosin cross-bridges are found (14-nm intervals) are indicated (see Fig. 13 for localization of C-protein). (C) Assuming the transverse profile of each of the 12 subfilaments to be elongated (con-

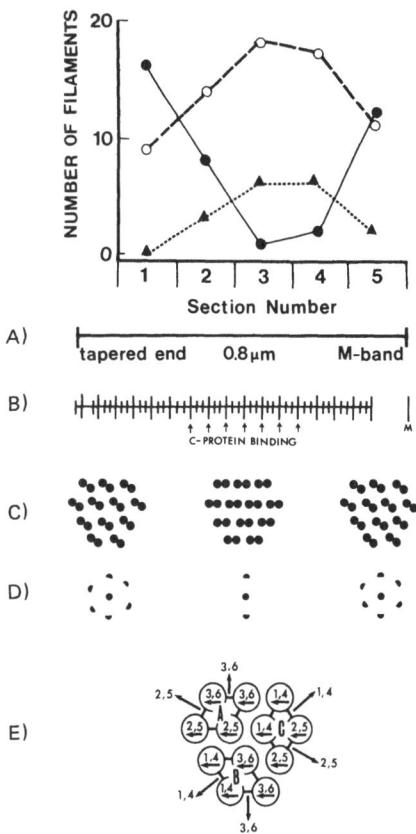

sistent with their being made up of myosin dimers and with a subfilament spacing of 4 nm), these relative orientations would give the optical-diffraction patterns shown diagrammatically in (D) consistent with the observations in (A). (E) Transverse section through the two-stranded model for the myosin filament proposed by Pepe (1966a, 1967a) made up of myosin dimers as the structural unit. The numbers indicate levels at which myosin molecules occur. The arrows projecting from the surface indicate the approximate position of the myosin cross-bridges occurring at the levels indicated by the numbers. This approximates a two-stranded helical arrangement with two myosin molecules at each position on the helix. The arrows in each subfilament indicate the orientation of the elongated profile in the C-protein binding region. Note that subfilaments with elongated profiles parallel to the surface of the filament occur only at levels 1, 3, 4, and 6, not at levels 2 and 5. This gives a 43-nm repeat, which could account for the binding of C-protein at 43-nm intervals only in the region of the filament where the elongated profiles are oriented closely to the orientation shown. See the text for details.

9.2. Restriction of C-Protein Binding to Specific Positions along Only a Portion of the Filament

Structural differences along the length of the filaments have also been suggested by observations of the localization of C-protein in the A-band of myofibrils (Pepe and Drucker, 1975; Craig and Offer, 1976b). In these studies, it was concluded from antibody localization in electron microscopy that the C-protein is bound to the surface of the filaments at seven positions

43 nm apart and located about halfway between the tapered ends and the M-band region of the filaments. The antibody used in these studies was not tagged with an electron-opaque marker; it was visualized as increased protein density appearing periodically in the A-band (Fig. 13b). If, however, the antibody was not bound periodically, it would be difficult to observe because it does not have an electron-opaque tag (Pepe, 1976). Therefore, the localization of the antibody midway between the tapered ends and the M-band region of the filaments as observed in Fig. 13b could represent only a part of the antibody staining, if, for instance, the C-protein was present in a nonperiodic fashion along the rest of the myosin filament. To show that the presence of C-protein was restricted to the periodic localization observed in electron microscopy, fluorescent-labeled antibody was used (Pepe and Drucker, 1975), and it was shown that the fluorescence is restricted to two bands, one in each half of the A-band (Fig. 13a), corresponding to the periodic localization observed in electron microscopy. The localization of C-protein in the A-band corresponds to the region of the myosin filament from which optical-diffraction patterns with diffraction maxima in only one direction were obtained maximally, as discussed above. This restricted localization of the C-protein to only a portion of the myosin filament and the 43-nm interval of C-protein binding are puzzling, since myosin cross-bridges occur at 14-nm intervals along the entire length of the myosin filament except for the central bare zone.

The strongest evidence that C-protein binds to the myosin molecules in the filament with the repeat of the underlying myosin molecules has come from studies of the binding of C-protein to light meromyosin (LMM) para-

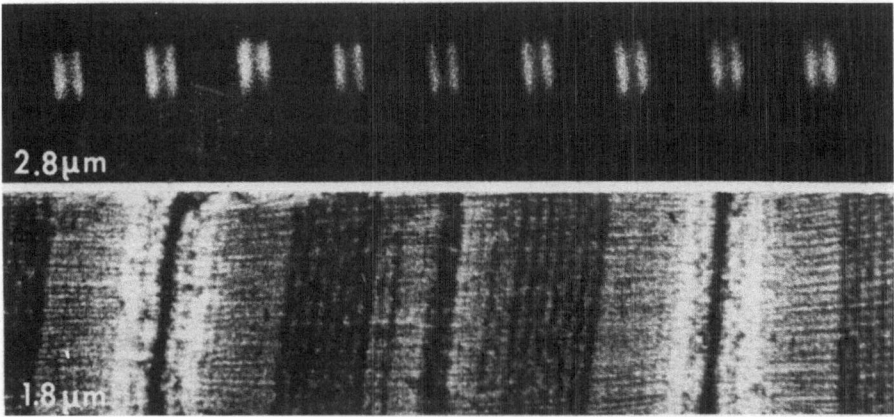

Figure 13. C-protein localization along the myosin filament. (a) Fluorescent anti-C-protein localization in the A-band (Pepe and Drucker, 1975). The anti-C-protein is restricted to two bands in the A-band that correspond in position to the groups of narrow dense bands observed in electron microscopy in (b) after anti-C-protein staining. (b) Localization of C-protein observed in electron microscopy with untagged antibody (Pepe and Drucker, 1975; Craig and Offer, 1976b). The seven bands farthest from the M-band region were shown to be specific for C-protein, and the other two bands were shown to represent the localization of two other proteins.

crystals. In the first studies done, the C-protein was shown to bind to LMM paracrystals with a 43-nm axial repeat, resulting in enhancement of the 43-nm repeat (Moos, 1972; Moos *et al.,* 1975). However, since the C-protein and the underlying LMM repeat were not both visualized in the same micrographs, it was not possible to conclude that the LMM repeat and the C-protein repeat were identical. A slight difference in the two repeats could be present but undetectable. Later studies of C-protein binding to an LMM paracrystal that had a 43-nm axial repeat of two narrow dark bands separated by 14 nm, showed that the C-protein was bound between the narrow dark bands 14 nm apart with visibility of both the C-protein and the underlying LMM repeat, demonstrating that the two repeats are the same (Chowrashi and Pepe, 1977). However, these studies are not strictly analogous to the binding of C-protein to myosin filaments, since the paracrystals all have a 43-nm repeat matching the repeat of C-protein binding, whereas in the myosin filament, there are cross-bridges at intervals of 14 nm. More recently, Safer and Pepe (1980) have studied the binding of C-protein to LMM paracrystals that have a 14-nm axial repeat and have found that the C-protein is still bound at 43-nm intervals, binding between every third pair of narrow dense bands along the paracrystals. This pattern of binding is as puzzling for the LMM paracrystal as it is for the myosin filament and must reflect some packing arrangement that makes C-protein binding sites available for C-protein binding either on the surface of the paracrystal or in the C-protein binding region of the myosin filament, only at 43-nm intervals. Outside the C-protein binding region of the myosin filaments, there must obtain a different structural arrangement such that C-protein binding sites are not available on the surface of the filament.

9.3. Restriction of Specific Anti-Light Meromyosin Staining to Only a Portion of the Filament

By staining myofibrils with an affinity-purified antibody that has been shown to be specific to the LMM fragment of myosin, we found that staining of the A-band is restricted to two bands, one along each A–I junction (Pepe, 1966b, 1967b). Therefore, LMM-specific antigenic sites are available for staining only at and near the tapered ends of the myosin filaments. This is outside the C-protein binding region. A possible explanation for this is that the C-protein and other proteins present between the C-protein region and the M-band might be binding to the LMM-specific antigenic sites and thereby preventing binding of the antibody in this region. Another, probably more likely is that the restricted anti-LMM staining of the A-band seen in Fig. 14 is a further reflection of differences in structural arrangement in different parts of the myosin filament.

9.4. Filament Length as a Function of KCl Concentration

Another observation that suggests some form of structural differences along the length of the filament has been reported recently by Trinick and

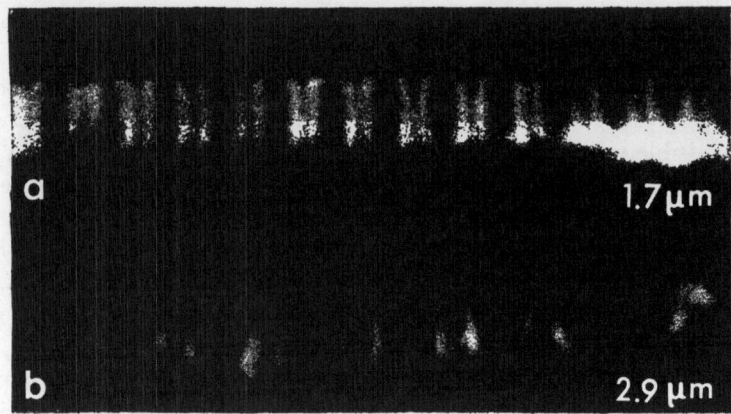

Figure 14. Anti-LMM localization along the myosin filament. Fluorescent anti-LMM staining of myofibrils with sarcomere lengths of 1.7 μm (a) and 2.9 μm (b). The availability of antigenic sites that are found only on the LMM fragment of myosin is restricted to the lateral regions of the A-band along each A-I junction.

Cooper (1980). They have reported that with increasing KCl concentration, the length of natural filaments in suspension drops gradually from about 1.5 to about 1.3 nm and then abruptly from 1.3 to 0.8 nm and again gradually to less than 0.8 nm. At the KCl concentration where the abrupt decrease in length occurred, there were two populations of filaments present simultaneously, i.e., those around 1.3 nm and those around 0.8 nm in length, whereas at all other KCl concentrations, only one population of lengths was observed. One possibility these workers suggested for the abrupt change in length from 1.3 to 0.8 nm was that it might be related to the presence of C-protein on the myosin filaments.

Recently, we have obtained similar results with synthetic myosin filaments (Pepe and Drucker, unpublished results) formed from column-purified myosin from which other proteins such as C-protein have been removed. Regardless of how the KCl concentration was arrived at, i.e., by dilution from 0.6 M KCl or by increasing the KCl concentration from 0.15 M KCl by dialysis, the lengths obtained were the same for a given KCl concentration. As can be seen in Fig. 15, there is clearly a difference in the length of the filaments as a function of KCl concentration for lengths greater and less then about 1.4 μm similar to the difference that has been observed with natural filaments. C-protein cannot be involved in these observations, since it is not present in the myosin used to prepare the synthetic filaments, suggesting that structural differences along the filament may be responsible for these differences above and below the length of 1.4 μm.

Comparing the observations made by Trinick and Cooper (1980) with natural filaments and those made by Pepe and Drucker (Fig. 15) with synthetic filaments, it is noteworthy that the change in slope in both cases occurred when the filaments reached about 1.3–1.4 μm in length. However, the bimodal distribution obtained in this length range with natural filaments was not observed with the synthetic filaments, with which a unimodal distribution

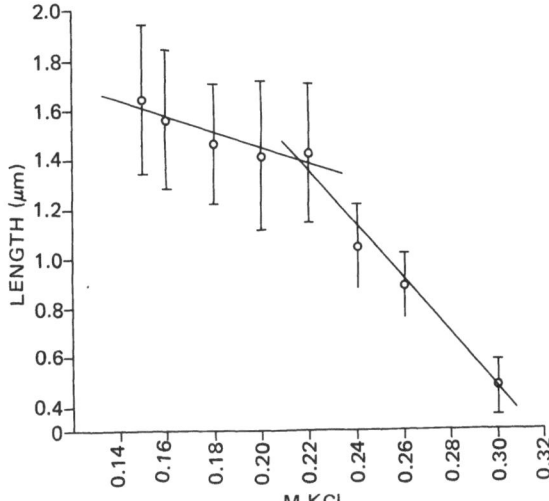

Figure 15. Length of synthetic myosin filaments as a function of KCl concentration. The change in length with KCl concentration was found to be different for filaments greater and less than 1.4 μm in length. A unimodal distribution of lengths was found at all KCl concentrations (Pepe and Drucker, unpublished results).

was observed at all KCl concentrations. The bimodal distribution at 1.3–1.4 μm with natural filaments might be the direct result of a stabilizing effect produced by the presence of C-protein. If a higher concentration of KCl is required to remove the C-protein than is required to shorten the filament, the result would be that on removal of the C-protein, there would be a sharp decrease in length to the length characteristic for the higher KCl concentration. This could produce a bimodal distribution at the critical KCl concentration where the C-protein begins to be dissociated from the filaments.

10. Modeling

There are two detailed models for myosin-filament packing: a three-stranded model with three myosin molecules per 14-nm interval (Squire, 1973) and a two-stranded model with four myosin molecules per 14-nm interval (Pepe, 1966a, 1967a). These are shown diagrammatically in Fig. 16.

In the *two-stranded model,* there are 12 parallel subfilaments hexagonally packed with 9 on the surface and 3 centrally located, giving a triangular profile with a subfilament missing at each apex. This arrangement of subfilaments has been observed (see Figs. 6 and 7), and the observed spacing of 4 nm between subfilaments are constructed of myosin dimers. The model is also consistent with the shape of the transverse profile observed for vertebrate skeletal filaments in thick sections (see Fig. 5). The 12 parallel subfilaments in this model form three subgroups, each of which has the same stagger relationship between the four subfilaments of the subgroup. Within each subgroup, there are staggers of 14 and 43 nm, while the stagger between the subgroups is 28.6 nm. The observed splaying of natural filaments shown in Fig. 3 and reported by Maw and Rowe (1980) is consistent with the presence of these three subgroups in the model. Therefore, threefold rotational symmetry in transverse projection is not inconsistent with this two-stranded model.

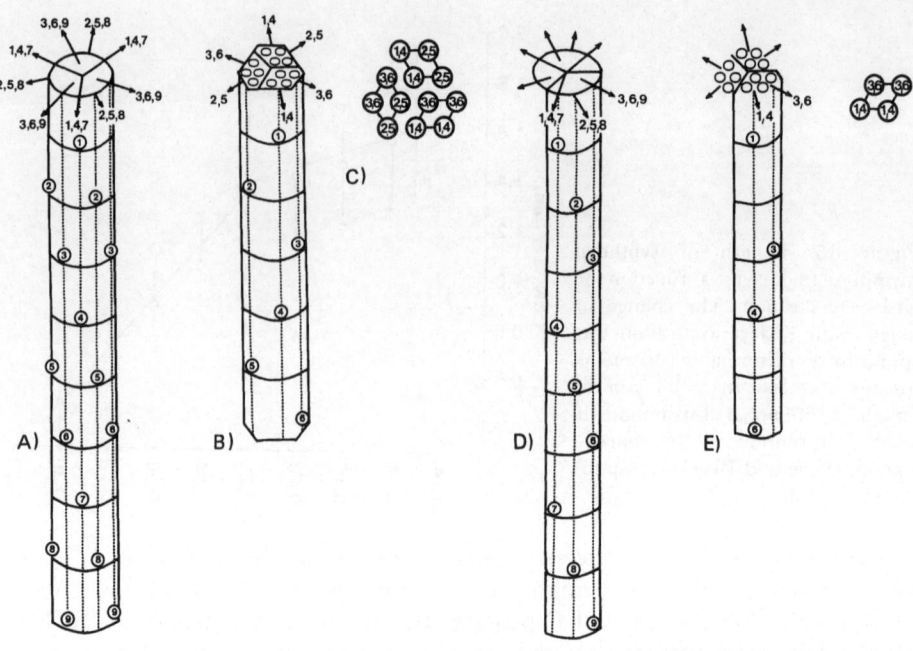

Figure 16. Models for the vertebrate skeletal-myosin filament. (A) Three-stranded model proposed by Squire (1973) for the myosin filament. The numbers indicate the levels at which myosin cross-bridges occur at intervals of 14 nm. There are three myosin molecules at each level. The myosin molecules are all equivalent and therefore are tilted with respect to the long axis of the filament. (B) Two-stranded model proposed by Pepe (1966a, 1967a) for the myosin filament, constructed from myosin dimers as the fundamental structural unit. The model is made up of 12 parallel subfilaments related as shown in transverse section to the right of the column. (C) The numbers indicate the levels at which myosin cross-bridges occur at 14-nm intervals. Each subfilament is a linear aggregate of myosin dimers. The 12 subfilaments consist of three groups of 4 subfilaments in which the relationships within a group are the same for all three. (D) If a three-stranded model can be constructed of parallel subfilaments to be consistent with the finding that the filament can be splayed into three groups of subfilaments (see Fig. 4), then along each group, myosin cross-bridges would occur at intervals of 14 nm as shown diagrammatically here for one of the three. (E) The two-stranded model is made up of parallel subfilaments consistent with the splaying of filaments into parallel bundles of filaments (Fig. 4). It is also made up of three groups of parallel subfilaments in which the relationships within a group are the same for all three consistent with the observation that the filaments splay into three groups of parallel subfilaments (Fig. 4). Along each of these groups, myosin cross-bridges would occur at intervals of 43 nm as shown diagrammatically here for one of the three. There would be bridges at two neighboring levels every 43 nm. The preservation of myosin cross-bridge order along the splayed filaments is not good enough yet to distinguish between a 14-nm and a 43-nm cross-bridge repeat along the separated three groups of subfilaments.

In the two-stranded model made up of myosin dimers, there would be four myosin molecules per 14-nm interval along the filament consistent with the data obtained from quantitative SDS–polyacrylamide gel electrophoresis when the actin band (used as a reference to quantitate the myosin) is resolved from the two closely neighboring protein bands (Pepe and Drucker, 1975). The arrangement of myosin heads (or cross-bridges) on the surface of the filament approximates a two-stranded 6/1 helix with two myosin molecules contributing to each position on the helix.

The *three-stranded model* is constructed of single myosin molecules that are in equivalent positions. This necessitates that the molecules be tilted with respect to the long axis of the filament. This model is clearly not consistent with the observations of a 4-nm spacing between subfilaments (see Fig. 1), with the presence of 12 subfilaments arranged as shown in Figs. 6 and 7, or with the subfilaments being parallel to the long axis of the filament as shown by both the tilt experiments (see Figs. 2 and 3) and the splaying of filaments (see Fig. 4) as discussed earlier. It may be possible to construct a three-stranded model with parallel subfilaments; however, it is difficult to conceive how 12 of these could be reconciled easily with a three-stranded model.

An important difference between the two models in Fig. 16 appears when the filament is splayed into three units as observed in Fig. 3. Note in Fig. 16E that along one of the three units for the two-stranded model, myosin cross-bridge positions occur at levels 3 and 4 separated by 14 nm and then at level 6 for a separation of 28.6 nm and again at level 1 for a separation of 14 nm. Therefore, the two-stranded model predicts a 43-nm axial repeat along each of the three units into which the filament splays. If a three-stranded model with parallel subfilaments can be constructed, then, as can be seen in Fig. 16D, it would predict myosin cross-bridges every 14 nm along each of the three units into which the filament splays. As yet, it has not been possible to observe the cross-bridge repeat along the three units in splayed filaments (see Fig. 4). However, it is noteworthy that Eaton and Pepe (1974) observed filaments with a 43-nm axial repeat formed at a KCl concentration of 0.3 M. It is interesting to speculate that these filaments may correspond to one of the three units into which the natural myosin filaments can be splayed (see Fig. 4) (Maw and Rowe, 1980).

10.1. Organizational Units of Structure

The possibility that the basic structural unit for assembly of the myosin filament is a myosin dimer (Harrington and Burke, 1972; Burke and Harrington, 1972) has already been discussed and is consistent with a subfilament spacing of 4 nm (see Fig. 1). The minifilaments described by Reisler *et al.* (1980) and shown in Fig. 8 (Section 7) suggest that there may be another larger organizational unit. As already discussed, these minifilaments have a length of about 300 nm, a diameter of about 8 nm, and a bare zone region of about 180–200 nm, and on the basis of hydrodynamic studies, they are composed of 16–18 myosin molecules. In Fig. 17, I have represented a possible structure for the minifilaments made up from myosin dimers and based on the relationships existing in one of the three subgroups of four subfilaments in the two-stranded model. The length of this unit of 16 molecules is 315 nm, the bare zone is 200 nm, and the diameter is over 7 nm. All these parameters are consistent with those measured for the minifilaments. Presumably, the tris–citrate buffer in which the minifilaments are formed selectively inhibits

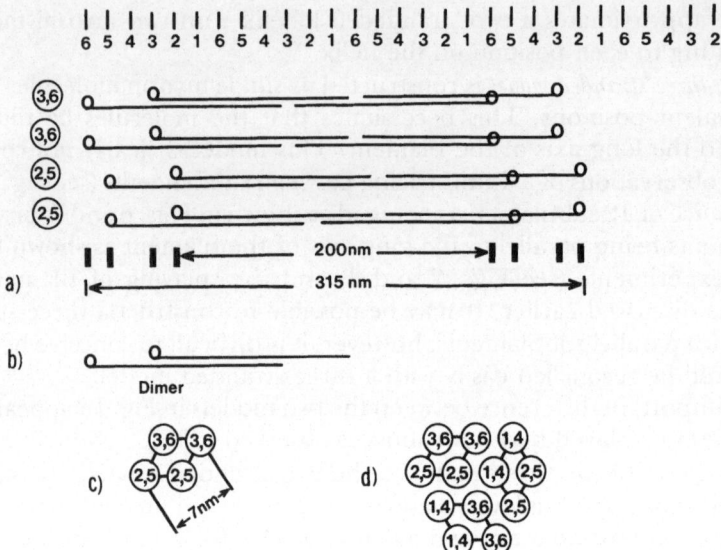

Figure 17. Possible structure for the minifilaments described by Reisler *et al.* (1980). The characteristics of the minifilaments described by Reisler *et al.* (1980) and shown in Fig. 8 are consistent
with a structure made up of myosin dimers (b) as the fundamental structural unit and with the
dimers related according to the relationship existing in each one of the three groups of four
subfilaments (c) in the two-stranded model proposed by Pepe (1966a, 1967a) with the myosin
dimer as the fundamental structural unit (d). (a) Two organizational units of four myosin dimers
related as in (c) are aggregated tail-to-tail. This possibility for minifilament structure would
contain 16 myosin molecules and have a diameter somewhat greater than 7 nm and a bare zone
region of 200 nm, all consistent with the parameters found for minifilaments (see Fig. 8). The
model would predict a gap in myosin cross-bridge intervals in each half of the minifilament as
shown in (a), but preservation of myosin cross-bridge order is not yet good enough to detect this.

side-to-side interactions between a group of four dimers but does not inhibit
the interactions within the group of four. This possible relationship between
the dimers in Fig. 17 would be consistent with the finding that the tris–citrate
buffer is very efficient at splaying filaments into three parallel strands (see
Fig. 4). Therefore, the units of four myosin dimers that are aggregated tail-
to-tail in the minifilaments could represent another organizational unit for
the myosin filament. The preservation of the myosin cross-bridge organization on the minifilaments is not good enough yet to determine whether or not
there is a gap in myosin cross-bridges as predicted by the model in Fig. 17, just
as it is not yet possible to determine the cross-bridge interval along splayed
filaments (see Fig. 16).

   Using the organizational unit of four myosin dimers as the unit by which
the filament tapers, it becomes possible to explain another observation, which
is the gap in myosin cross-bridges observed near the tapered end of the
filaments in the A-band near each A–I junction (see Fig. 9). This possibility is
shown in Fig. 18, in which the three groups of four subfilaments are shown
and the taper is in units of four myosin dimers. Note that there will be no

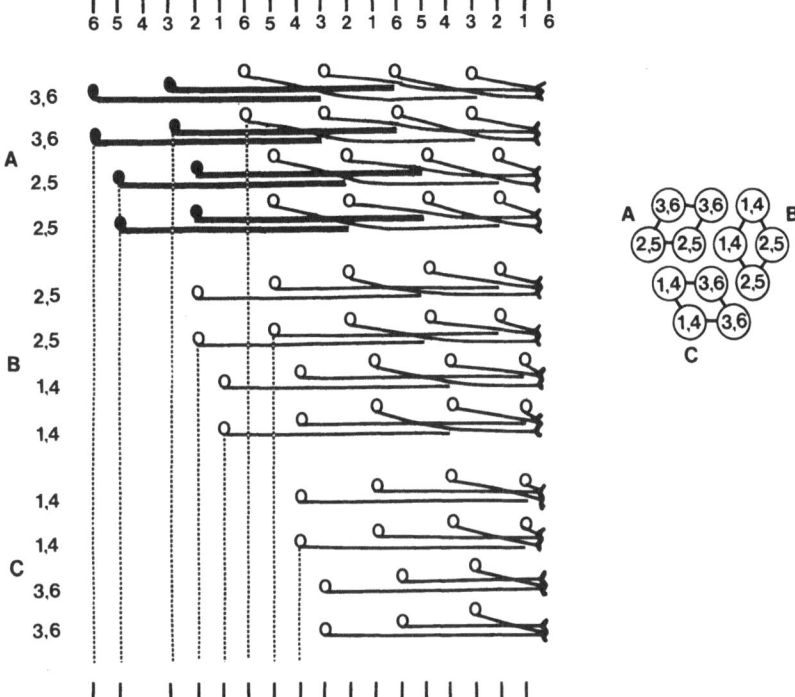

Figure 18. Possible structure for the tapered ends of myosin filaments. Craig and Offer (1976a) have concluded that there is a gap in myosin cross-bridges at the third level in from the tapered end of the myosin filament. If the two-stranded model proposed by Pepe (1966a, 1967a), using myosin dimers as the fundamental structural unit, is tapered by the next highest organizational level [suggested by Reisler's minifilaments (see Figs. 8 and 17)], which is a group of four myosin dimers related as in Fig. 17, then the missing level of myosin cross-bridges can be explained as shown diagrammatically here. An organizational unit of four myosin dimers is represented in bold outline in the group of four subfilaments labeled A. The levels at which myosin cross-bridges are present are indicated at the bottom.

myosin cross-bridges at the third 14-nm interval from the tapered end of the filament.

Therefore, to summarize, it appears very likely from the data already discussed that the filament is made up of 12 parallel subfilaments. They are hexagonally packed with a spacing of 4 nm and arranged so that there are 3 centrally located and 9 peripherally located. This structure can be constructed from myosin dimers (Harrington and Burke, 1972; Burke and Harrington, 1972), which are head-to-tail dimers with a stagger of 43 nm. The next organizational unit higher than the myosin dimer would be an aggregate of four dimers, and two of these aggregates tail-to-tail would give a structure corresponding to the minifilaments described by Reisler *et al.* (1980). The next higher organizational level of the filament would be a group of 4 subfilaments, and three of these groups make up the 12 subfilaments of the total filament. These three groups of 4 subfilaments would correspond to the three strands into which the myosin filament can be splayed. Each subfilament may be considered to be a linear aggregate of myosin dimers, and each group of 4 subfilaments may be considered as being constructed by linear aggregation of the organizational unit of four dimers. The gap in myosin cross-bridges in the A-band along each A–I junction can be accounted for by tapering of the filament by the organizational unit of four myosin dimers.

10.2. Speculations on Structural Differences along the Filament

One of the most puzzling observations related to possible differences in structure along the filament is the restricted binding of C-protein to only a portion of the filament and its binding to that portion at intervals of 43 nm instead of the 14-nm interval at which myosin cross-bridges occur (see Fig. 13b). As has already been discussed, this difference and the differences in optical-diffraction patterns observed from serial transverse sections (see Fig. 12) as well as the characteristics of LMM staining (see Fig. 14) and the curve of filament length as a function of KCl concentration (see Fig. 15) are probably all related in some way.

Though highly speculative, it is possible to show how structural differences along the length of the myosin filament could lead to the puzzling C-protein binding characteristics, using the two-stranded model in Fig. 16B. To do this, I will assume that the transverse profiles of the subfilaments are elongated as shown in Fig. 11D–F, although there is no evidence to indicate the shape of these profiles. This is what would be expected if each subfilament is a linear aggregate of myosin dimers in the two-stranded model as already discussed. The different orientations of the elongated profiles will be the source of the different optical-diffraction patterns observed in transverse sections from different portions of the filament. Beneath the graph in Fig. 12A, I have indicated the positions of C-protein binding along the filament in Fig. 12B and the positions where optical-diffraction patterns with spacings in one and three directions are seen maximally (Fig. 12D). The difference in orientation of elongated profiles from either the M-band region or the tapered end to the middle of the C-protein region represents a rotation of the elongated profiles of about 30° as indicated in Fig. 12C. Using this, there would be a 20° rotation from one edge of the C-protein region to the other. This is a rotation ±10° relative to that in the middle of the C-protein region. In this scheme, each subfilament twists with a pitch that is much longer than the length of the filament. The restriction of C-protein binding to this portion of the filament could therefore be explained if the C-protein binding sites are available for binding C-protein on the surface of the filament only for the range of orientations of elongated profiles present in the C-protein binding region, and this range of orientations would not occur anywhere else along the length of the filament. Below the C-protein region indicated in Fig. 12B, I have indicated the two-stranded model (Fig. 12E). The arrows in the circles indicate the orientations of the elongated profiles of the subfilaments. The numbers indicate the levels at which myosin molecules occur in the subfilaments, and the numbers associated with the arrows projecting from the surface indicate the levels and approximate positions of the cross bridges. Note that subfilaments with their elongated profile oriented parallel to the surface of the filament have myosin molecules at levels 3, 6, 1, and 4, but not at levels 2 and 5. Therefore, the C-protein binding sites available for binding in this region of the filament will have a 43-nm repeat. Thus, this simple scheme succeeds in accounting for the restricted binding of C-protein to only a portion of the filament and for the 43-nm interval of binding, rather than a 14-nm interval, which would correspond to the interval of myosin cross-bridges. This is highly speculative and is of interest only because it does present one possibility for explaining the puzzling binding characteristics of C-protein to the myosin filament.

ACKNOWLEDGMENT. This work was supported by U.S.P.H.S. Grant HL-15835 to the Pennsylvania Muscle Institute.

References

Ashton, F. T., and Pepe, F. A., 1981, The myosin filament. VIII. Preservation of subfilament organization, *J. Microsc.* **123**:93–108.

Baccetti, B., 1965, Perplexities and confirmations on the problem of the myosinic filament structure, *Boll. Soc. Ital. Biol. Sper.* **42**:1181–1184.

Baccetti, B., 1966, New observations on the ultrastructure of the myofilament, *J. Ultrastruct. Res.* **13**:245–256.

Burke, M., and Harrington, W. F., 1972, Geometry of the myosin dimer in high-salt media. II. Hydrodynamic studies on macromodels of myosin and its rod segments, *Biochemistry* **11**:1456–1462.

Chowrashi, P. K., and Pepe, F. A., 1977, Light meromyosin paracrystal formation, *J. Cell Biol.* **74**:136–152.

Craig, R., 1977, Structure of A-segments from frog and rabbit skeletal muscle, *J. Mol. Biol.* **109**:69–81.

Craig, R., and Offer, G., 1976a, Axial arrangement of cross bridges in thick filaments of vertebrate skeletal muscle, *J. Mol. Biol.* **102**:325–332.

Craig, R., and Offer, G., 1976b, The location of C-protein in rabbit skeletal muscle, *Proc. R. Soc. Lond. Ser. B* **192**:451–461.

Emes, C. H., and Rowe, A. J., 1978, Frictional properties and molecular weight of native and synthetic myosin filaments from vertebrate skeletal muscle, *Biochim. Biophys. Acta* **537**:125–144.

Eaton, B. L., and Pepe, F. A., 1974, Myosin filaments showing a 430 Å axial repeat periodicity, *J. Mol. Biol.* **82**:421–423.

Franzini-Armstrong, C., and Porter, K. R., 1964, Sarcolemmal invaginations constituting the T system in fish muscle fibers, *J. Cell Biol.* **22**:675–696.

Freundlich, A., Luther, P. K., and Squire, J. M., 1980, High voltage electron microscopy of cross bridge interactions in striated muscle, *J. Muscle Res. Cell Motil.* **1**:321–343.

Gilev, V. P., 1966a, Ultrastructure of thick filaments of muscle fibers, *Electron. Microsc.* **2**:689–690.

Gilev, V. P., 1966b, The ultrastructure of myofilaments. II. Further investigation of the thick filaments of crab muscles, *Biochim. Biophys. Acta* **112**:340–345.

Gilev, V. P., 1966c, On fine organization of "myosin" protofibrils of cross-striated muscle fibers, *Biofizika* **11**:274–278.

Harrington, W. F., and Burke, M., 1972, Geometry of the myosin dimer in high-salt media. I. Association behavior of rod segments from myosin, *Biochemistry* **11**:1448–1455.

Huxley, H. E., 1963, Electron microscopic studies on the structure of natural and synthetic protein filaments from striated muscle, *J. Mol. Biol.* **7**:281–308.

Huxley, H. E., and Brown, W., 1967, The low angle x-ray diagram of vertebrate striated muscle and its behaviour during contraction and rigor, *J. Mol. Biol.* **30**:383–434.

Kaminer, B., and Bell, A. L., 1966, Myosin filamentogenesis: Effects of pH and ionic concentration, *J. Mol. Biol.* **20**:391–401.

Katsura, I., and Noda, H., 1973, Assembly of myosin molecules into the structure of thick filaments of muscle, *Adv. Biophys.* **5**:177–202.

Lamvik, M. K., 1978, Muscle thick filament mass measured by electron scattering, *J. Mol. Biol.* **122**:55–68.

Luther, P. K., and Squire, J. M., 1980, Three-dimensional structure of the vertebrate muscle A-band. II. The myosin filament superlattice, *J. Mol. Biol.* **141**:409–439.

Marston, S. B., and Tregear, R. T., 1972, Evidence for a complex between myosin and ADP in relaxed muscle fibers, *Nature (London) New Biol.* **235**:23–24.

Maruyama, K., and Weber, A., 1972, Binding of adenosine triphosphate to myofibrils during contraction and relaxation, *Biochemistry* **11**:2990–2998.

Maw, M. C., and Rowe, A. J., 1980, Fraying of A-filaments into three subfilaments, *Nature (London)* **286**:412–414.

Millman, B. M., 1979, X-ray diffraction from chicken skeletal muscle, in: *Motility in Cell Function* (F. A. Pepe, J. W. Sanger, and V. T. Nachmias, eds.), pp. 351–354, Academic Press, New York.

Moos, C., 1972, Discussion: Interaction of C-protein with myosin and light meromyosin, *Cold Spring Harbor Symp. Quant. Biol.* **37**:93–95.

Moos, C., Offer, G., Starr, R., and Bennett, P., 1975, Interaction of C-protein with myosin, myosin rod and light meromyosin, *J. Mol Biol.* **97**:1–9.

Morimoto, K., and Harrington, W. F., 1974, Substructure of the thick filament of vertebrate striated muscle, *J. Mol. Biol.* **83**:83–97.

Pepe, F. A., 1966a, Organization of myosin molecules in the thick filament of striated muscle as revealed by antibody staining in electron microscopy, *Electron Microsc.* **2**:53–54.

Pepe, F. A., 1966b, Some aspects of the structural organization of the myofibril as revealed by antibody-staining methods, *J. Cell Biol.* **28**:505–525.

Pepe, F. A., 1967a, The myosin filament. I. Structural organization from antibody staining observed in electron microscopy, *J. Mol. Biol.* **27**:203–225.

Pepe, F. A., 1967b, The myosin filament. II. Interaction between myosin and actin filaments observed using antibody staining in fluorescent and electron microscopy, *J. Mol. Biol.* **27**:227–236.

Pepe, F. A., 1975, Structure of muscle filaments from immunohistochemical and ultrastructural studies, *J. Histochem. Cytochem.* **23**:543–562.

Pepe, F. A., 1976, Detectability of antibody in fluorescence and electron microscopy, in: *Cell Motility* (R. Goldman, T. Pollard, and J. Rosenbaum, eds.), pp. 337–346, Cold Spring Harbor Laboratory, New York.

Pepe, F. A., 1979, The myosin filament: Molecular structure, in: *Motility in Cell Function* (F. A. Pepe, J. W. Sanger, and V. T. Nachmias, eds.), pp. 103–116, Academic Press, New York.

Pepe, F. A., and Dowben, P., 1977, The myosin filament. V. Intermediate voltage electron microscopy and optical diffraction studies of the substructure, *J. Mol. Biol.* **113**:99–218.

Pepe, F. A., and Drucker, B., 1972, The myosin filament. IV. Observation of the internal structural arrangement, *J. Cell Biol.* **52**:255–260.

Pepe, F. A., and Drucker, B., 1975, The myosin filament. III. C-protein, *J. Mol. Biol.* **99**:609–617.

Pepe, F. A., and Drucker, B., 1979, The myosin filament. VI. Myosin content, *J. Mol. Biol.* **130**:379–393.

Pepe, F. A., Ashton, F. T., Dowben, P., and Stewart, M., 1981, The myosin filament. VII. Changes in internal structure along the length of the filament, *J. Mol. Biol.* **145**:421–440.

Potter, J., 1974, The content of troponin, tropomyosin, actin, and myosin in rabbit skeletal muscle myofibrils, *Arch. Biochem. Biophys.* **162**:436–441.

Reedy, M. K., 1976, Preservation of x-ray patterns from frog sartorius muscle prepared for electron microscopy, *Biophys. J.* **16**:126a.

Reisler, E., Smith, C., and Seegan, G., 1980, Myosin minifilaments, *J. Mol. Biol.* **143**:129–145.

Safer, D., and Pepe, F. A., 1980, Axial packing in light meromyosin paracrystals, *J. Mol. Biol.* **136**:343–358.

Squire, J. M., 1973, General model of myosin filament structure. III. Molecular packing arrangement in myosin filaments, *J. Mol. Biol.* **77**:291–323

Stewart, M., Ashton, F. T., Lieberson, R., and Pepe, F. A., 1981, The myosin filament. IX. Determination of subfilament positions by computer processing of electron micrographs, *J. Mol. Biol.* (in press).

Tregear, R. T., and Squire, J. M., 1973, Myosin content and filament structure in smooth and striated muscle, *J. Mol. Biol.* **77**:279–290.

Trinick, J., and Cooper, J., 1980, Sequential disassembly of vertebrate muscle thick filaments, *J. Mol. Biol.* **141**:315–321.

Weber, A., Heiz, R., and Reiss, I., 1969, The role of magnesium in the relaxation of myofibrils, *Biochemistry* **8**:2266–2271.

Wray, J., 1979a, X-ray diffraction studies of myosin filament structures in crustacean muscles, in: *Motility in Cell Function* (F. A. Pepe, J. W. Sanger, and V. T. Nachmias, eds.), pp. 347–350, Academic Press, New York.

Wray, J., 1979b, Structure of the backbone in myosin filaments of muscle, *Nature (London)* **270**:37–40.

12

Cross-Bridge Movement and the Conformational State of the Myosin Hinge in Skeletal Muscle

Hitoshi Ueno and William F. Harrington

Our recent experiments (Chiao and Harrington, 1979) on glycerinated skeletal-muscle fibers in rigor using the cross-linking method showed that the myosin heads could be cross-linked to the backbone of the thick filament while they are still attached to the thin filament. We further demonstrated that the heads could be made to move out from the thick filament by a small change in pH from 7.4 to 8.0. This behavior suggests that relatively small changes in the local ionic environment can release the cross-bridge from the thick-filament surface. Similar experiments (Sutoh *et al.,* 1978a) with synthetic myosin filaments showed that cross-linking of the subfragment-1 (S-1) subunits was markedly depressed at high pH (8.0–8.3). The myosin heads in synthetic filaments showed pH-sensitive association with the filament backbone similar to those in muscle fibers, which suggests that release of the heads from the thick-filament surface in muscle fibers at high pH is due to the intrinsic property of myosin rather than to an ionic effect between filaments. The possibility that such a release could lead to force generation through an α-helix–random coil transition in the subfragment-2 (S-2) region of myosin during a cross-bridge cycle has been considered in earlier papers (Harrington, 1971, 1979; Tsong *et al.,* 1979) and is supported by thermal melting experiments on the S-2 fragments isolated from rabbit skeletal myosin. Most of the melting in this structure appears to occur in the light meromyosin–heavy meromyosin (LMM–HMM) "hinge" region (Harrington *et al.,* unpublished). Hence, if this segment were to be stabilized in the α-helical state when the

Hitoshi Ueno and William F. Harrington • Department of Biology and McCollum–Pratt Institute, The Johns Hopkins University, Baltimore Maryland 21218.

head is close to the filament backbone, but became destabilized following attachment of the head to the thin filament during a cross-bridge cycle, the consequent partial melting of the S-2 segment would produce enough short-ening force to account for the tension developed in active muscle (Harrington, 1979).

In the study reported herein (Ueno and Harrington, 1981), we have reinvestigated the question of the effect of local charge on the cross-bridge release of glycerinated myofibrils in rigor and extended our earlier studies on cross-linking myosin segments in the thick filaments of muscle. Myofibrils were cross-linked at 5°C and $\mu = 0.05$ over the pH range 7.0–8.5, then di-gested with α-chymotrypsin at 20°C and $\mu = 0.6$ to form HMM and LMM subfragments. This subfragment mixture was subjected to further digestion with trypsin to yield short S-2 fragments (Sutoh *et al.*, 1978b). The extent of cross-linking LMM and short S-2 fragments was determined by densitometry of the bands corresponding to these species following electrophoresis on sodium dodecyl sulfate (SDS) gels. We could eliminate the contribution of intramolecular cross-linking in the analysis of the SDS gels by oxidizing LMM and short S-2 subfragments before electrophoresis to form the disulfide cross-linked species. We followed the time–course of cross-linking the S-2 and LMM as a function of pH to detect the disposition of cross-bridges, if any, away from the thick-filament surface. We compared the ratio of cross-linking rates of the two fragments (k_{S-2}/k_{LMM}) at each pH. This ratio eliminates the effect of pH on the cross-linking reaction. The normalized rate of cross-linking falls over a narrow pH range (7.4–8.4) with a sigmoidal profile (Fig. 1) closely similar to that observed earlier for the pH dependence of cross-linking of the S-1 subunits. This result suggests that the S-2 segment, like the S-1 subunit, is released and swings away from the thick-filament surface when the

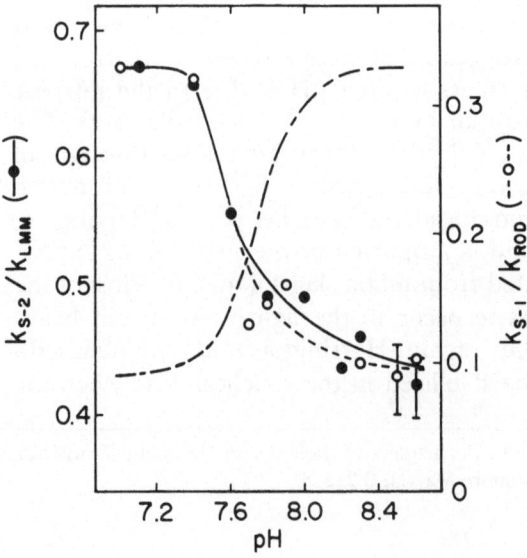

Figure 1. Titration curves of the rel-ative rate of cross-linking myosin short S-2 ($\bullet$), k_{S-2}^{0} ($= k_{S-2}/k_{LMM}$), and myosin heads ($\bigcirc$), k_{S-1}^{0} ($= k_{S-1}/k_{rod}$). Cross-linking was carried out in 40 mM imidazole-HCl (pH 7.0–7.4) or in 40 mM triethanolamine-HCl (pH 7.4–8.5) at 5°C. The cross-linking rate of myosin heads, k_{S-1}^{0}, is from Chiao and Harrington (1979) and includes additional data collected in this study. (---) Relative cleavage rate at the HMM–LMM hinge region of the myosin heavy chains in myofib-rils with α-chymotrypsin.

pH is raised. The earlier conclusion (Sutoh *et al.*, 1978a; Chiao and Harrington, 1979) that a major fraction of the S-2 link is immobilized on the surface when the pH is raised is corrected in this study. We also probed for conformational transitions in the HMM–LMM hinge region, based on the susceptibility of this region to proteolytic cleavage. Myofibrils were digested with α-chymotrypsin at different pH values over the range 7.0–8.6 in the presence of Mg^{2+} (0.1 mM) at 5°C. The time–course of the digestion was measured by determining the density of the myosin heavy-chain band on SDS gels at various digestion times and showed single exponential behavior when log (heavy-chain density) was plotted vs. digestion time. Observed rate constants thus obtained were corrected for the intrinsic pH dependence of α-chymotrypsin. This corrected rate constant showed a sharp sigmoidal increase when the pH was raised (with a transition midpoint at pH 7.8) just over the pH span where the cross-linking rates of S-1 and also S-2 showed transitions. We treated this cooperative transition process as a two-state equilibrium and estimate that about two protons are released from myosin when the cross-bridge is detached from the filament surface. Thus, the release of the cross-bridge appears to be a highly cooperative process and is accompanied by a conformational transition to a more open, proteolytically sensitive structure within the LMM–HMM hinge region, which is linked to a shift in local charge in the myosin molecule.

ACKNOWLEDGMENT. This research was carried out during the tenure of a Post-doctoral Fellowship to H.U. from the Muscular Dystrophy Association.

References

Chiao, Y. C., and Harrington, W. F., 1979, Cross-bridge movement in glycerinated rabbit psoas muscle fibers, *Biochemistry* **18**:959–963.

Harrington, W. F., 1971, A mechanochemical mechanism for muscle contraction, *Proc. Natl. Acad. Sci. U.S.A.* **68**:685–689.

Harrington, W. F., 1979, On the origin of the contractile force in skeletal muscle, *Proc. Natl. Acad. Sci. U.S.A.* **76**:5066–5070.

Sutoh, K., Chiao, Y. C., and Harrington, W. F., 1978a, Effect of pH on the cross-bridge arrangement in synthetic myosin filaments, *Biochemistry* **17**:1234–1239.

Sutoh, K., Sutoh, K., Karr, T., and Harrington, W. F., 1978b, Isolation and physico-chemical properties of a high molecular weight subfragment-2 of myosin, *J. Mol. Biol.* **126**:1–22.

Tsong, T. Y., Karr, T., and Harrington, W. F., 1979, Rapid helix–coil transitions in the S-2 region of myosin, *Proc. Natl. Acad. Sci. U.S.A.* **76**:1109–1113.

Ueno, H., and Harrington, W. F., 1981, Cross-bridge movement and the conformational state of the myosin hinge in skeletal muscle, *J. Mol. Biol.* **149**:619–640.

13

The Regulation of Myosin-Light-Chain Synthesis in Heterokaryons between Differentiated and Undifferentiated Myogenic Cells

Woodring E. Wright

Embryonic myoblasts from a variety of species are able to undergo a process of terminal differentiation *in vitro*. Freshly isolated myoblasts are morphologically and biochemically undifferentiated. After dividing several times, they become postmitotic, spontaneously fuse to form multinucleated myotubes, and initiate the synthesis of a variety of muscle-specific proteins. The synthesis of these proteins is coordinately controlled (Devlin and Emerson, 1978; Garrels, 1979), and although some translational regulation may be involved (Buckingham *et al.*, 1976; Bag and Sarkar, 1976), the major control appears to be primarily transcriptional (John *et al.*, 1977; Strohman *et al.*, 1977; Perriard, 1979). Little is known concerning the mechanisms by which the myoblasts shift from a program of cell division to one of terminal differentiation. In the study reported herein, polyethylene-glycol-induced fusion products (heterokaryons) between terminally differentiated skeletal myocytes and undifferentiated myoblasts were used to determine whether the formation of muscle-specific protein is under positive or negative control. Our results suggest that the synthesis of myosin is under positive inductive regulation, since undifferentiated rat myoblasts were observed to synthesize rat skeletal-myosin light chains when fused to differentiated chick myocytes.

Chick and rat myoblasts were chosen for these fusion experiments because the species-specific skeletal-myosin light chains can be resolved on two-

Woodring E. Wright • Department of Cell Biology and Internal Medicine, The University of Texas Health Science Center at Dallas, Dallas, Texas 75235.

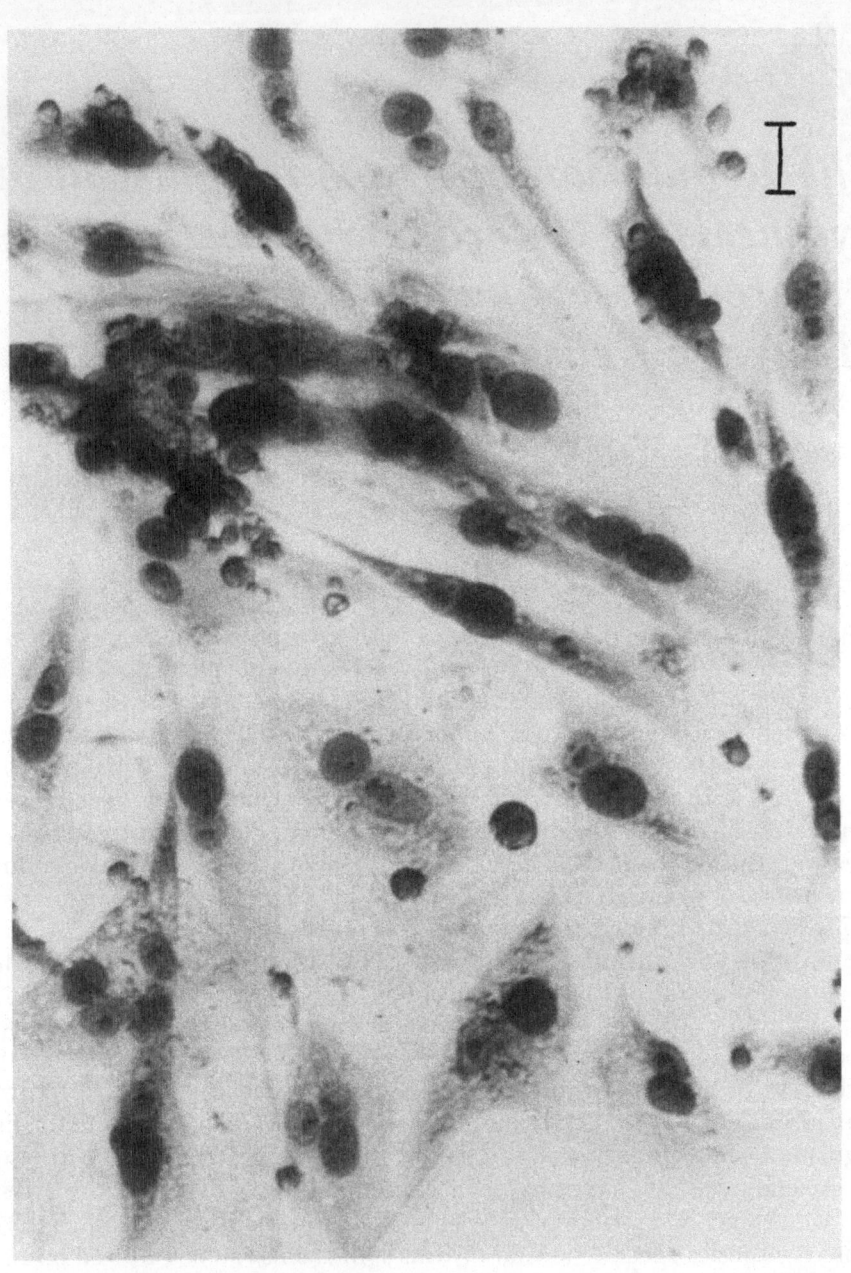

dimensional gels. Gene-dosage effects (Fougére *et al.*, 1972; Davidson, 1972; Peterson and Weiss, 1972; Brown and Weiss, 1975) would occur if multinucleated myotubes were fused to undifferentiated myoblasts. To avoid these effects, mononucleated differentiated cells (myocytes) were obtained by growing embryonic chick skeletal myoblasts in the presence of ethyleneglycol bis(aminoethyl ether)N,N'-tetraacetic acid (EGTA) and cytochalasin B, agents that prevent the spontaneous fusion of differentiating cells into multinucleated myotubes (Sanger, 1974; Vertel and Fischman, 1976). The near-diploid established line of L6 rat myoblasts also has the capacity to differentiate and form myotubes if they are allowed to become confluent (Yaffe, 1968). However, differentiated L6 cells synthesize only the embryonic form of rat skeletal-myosin light chain 1 (Whalen *et al.*, 1978; Garrels, 1979). L6 myoblasts were used in these studies as the undifferentiated parent because of the ease of maintaining dividing cultures in an undifferentiated state.

The traditional cell-hybridization technique is to isolate and analyze dividing hybrid clones. This approach is inappropriate for myocyte × myoblast fusions for several reasons. The myoblast differs from the myocyte by only a few cell divisions, and it is unlikely that the initial molecular interactions would be maintained during the many cell generations before the formation of a hybrid clone. Most important, the differentiated myocyte is postmitotic. Restricting the analysis to only those fusion products capable of giving rise to a dividing hybrid clone would clearly bias the results against the expression of differentiated functions. Therefore, in these experiments, we performed our analysis on heterokaryons, the initial fusion products before cell division has begun. Heterokaryons were obtained by selection with irreversible biochemical inhibitors (Wright and Hayflick, 1975; Wright, 1978). In this selective

---

Figure 1. Chick myocyte × rat myoblast heterokaryons. Twelve-day embryonic chick thigh muscle was mechanically dissociated, preplated to remove contaminating fibroblasts, and then distributed among gelatinized dishes in a medium consisting of 4 parts Dulbecco's Modified Eagle's Medium (DMEM) to 1 part Medium 199, 5% horse serum, 1% chick embryo extract, and antibiotics. EGTA, 1.75 mM, was added 24 hr later to prevent myotube formation, and 10^{-5} M cytosine arabinoside was added the following day to further reduce fibroblast contamination. Since many myocytes detach from the dish after several days' treatment with EGTA, the medium was changed on the 3rd day of culture to one containing 2 μg/ml cytochalasin B as the agent preventing spontaneous fusion to form myotubes. Seven-day-old chick cultures were trypsinized and subjected to an additional differential adhesion to further decrease fibroblast contamination, and the cells remaining in suspension were immediately used for heterokaryon experiments. L6 rat myoblasts, grown in medium (4 : 1 DMEM–Medium 199) supplemented with 15% fetal bovine serum, were maintained in continuous log-phase growth by subcultivation three times per week. Heterokaryons were isolated as previously described (Wright, 1978). Typical doses of the irreversible inhibitors used to kill these parental cells were 4.5 mM iodoacetamide or 0.004% diethylpyrocarbonate for chick myocytes and 5.5 mM iodoacetamide or 0.0055% diethylpyrocarbonate for L6 rat myoblasts. To increase the density of surviving heterokaryons, the day after fusion the population was enriched for viable cells by centrifuging the cells onto a cusion of Ficoll–sodium diatrizoate (Wright, 1981). The cells floating at the interface were harvested, washed, and plated in 6-mm microtiter wells. Aliquots were stained with Giemsa and counted the following day. The nuclei of chick myocytes stain more lightly than the L6 rat nuclei, so the heterokaryon nature of the binucleated cells is evident. Scale bar: 20 μm.

system, each parental population is treated with a lethal dose of a different irreversible agent, after which unreacted inhibitor is washed away. The parental cells are then mixed and finally fused with polyethylene glycol. Since a different lesion is produced in the parental cells, complementation and rescue can occur in heterokaryons, whereas parental cells and homokaryons die (Fig. 1). Populations in which greater than 95% of the cells were chick myocyte × rat myoblast heterokaryons were labeled with [^{35}S]methionine, and total cell extracts were analyzed on two-dimensional gels (O'Farrel, 1975).

Figure 2A shows the relative positions of chick and rat light chains on two-dimensional gels. Figure 2B demonstrates that chick myocyte × rat myoblast heterokaryons synthesized three types of skeletal myosin light chain 1: chick fast (c), rat adult fast (a), and rat embryonic (e). This pattern was not dependent on the specific biochemical treatments, since (iodoacetamide-treated chick) × (diethylpyrocarbonate-treated rat) heterokaryons gave the same result as (diethylpyrocarbonate-treated chick) × (iodoacetamide-treated rat) heterokaryons. L6 myoblast self × self homokaryons were constructed to control for the effects of the full biochemical selection procedure. Both these homokaryons and L6 cells cultured alone were negative for skeletal-myosin light chains. Chick myocyte × chick myocyte homokaryons did not show any new spots appearing near the location of rat myosin light chains, and they continued to synthesize chick skeletal light chains. As an additional control, L6 × L6 homokaryons were cocultivated with chick myocyte × myocyte homokaryons. Extracts of these cells showed the synthesis of only chick myosin light chains. Figure 3 confirms that the proteins made in the heterokaryons behave like myosin light chains in that they copurify with actomyosin on repeated cycles of high-ionic-strength extraction and low-ionic-strength precipitation.

These experiments demonstrate that chick myocytes can induce undifferentiated rat myoblasts to synthesize skeletal-myosin light chains when the two nuclei exist in a common cytoplasm. Furthermore, the experimental controls suggest that this result is not an artifact of the selective system or of simple contact between chick and rat cells. Finally, since this result is based on the behavior of heterokaryons before any cell division has begun, it is not a consequence of the rapid segregation of either parental set of chromosomes.

These observations imply that factors provided by the chick cell initiate the rat program of differentiation or at least the synthesis of rat myosin light chains. This factor(s) may be a specific positive inductive regulatory molecule. Alternatively, it could function through a variety of secondary mechanisms. For example, the presence of chick membrane receptors capable of responding to the culture environment, the alteration of critical metabolic pathways, or the inhibition of DNA synthesis in the L6 nucleus by the postmitotic chick nucleus could all potentially stimulate the expression of L6 rat myogenic functions.

The synthesis of adult rat light chain by chick myocyte × L6 myoblast heterokaryons suggests that the chick cell contains factors that either release the block of adult light-chain synthesis in L6 myotubes or dilute an L6 suppressive factor. Although differentiated L6 myotubes have been reported to

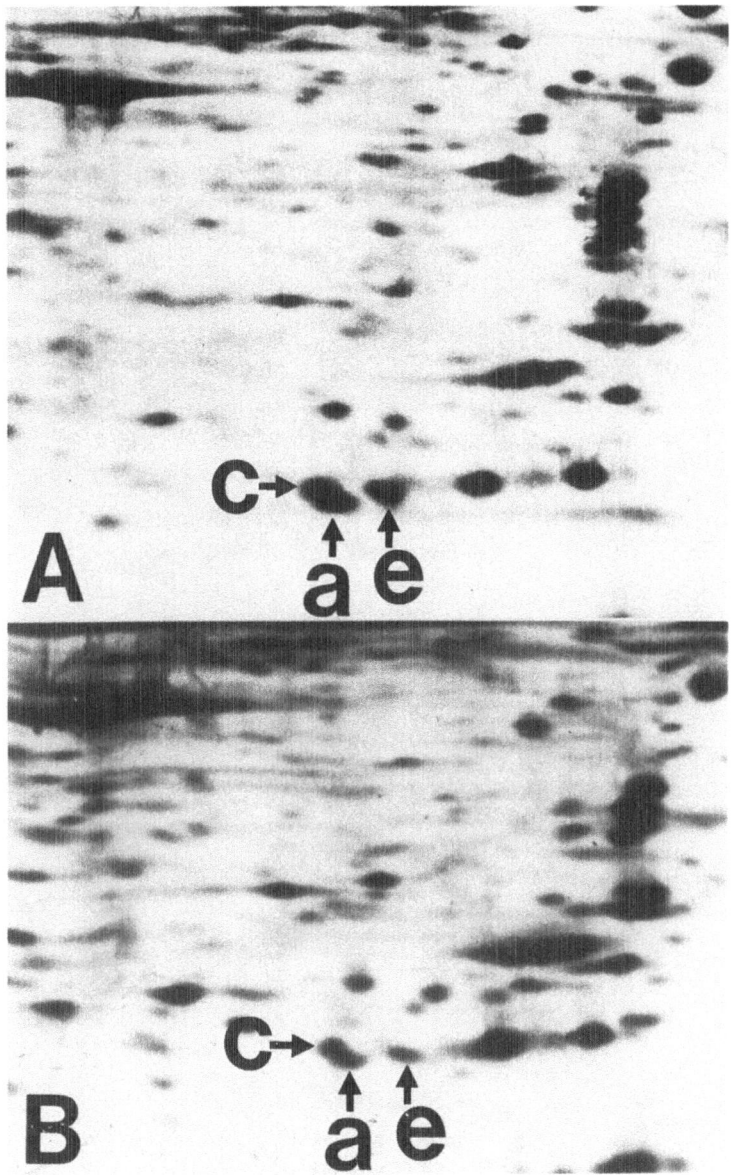

Figure 2. Two-dimensional polyacrylamide gels of heterokaryons and control. (A) Separate primary cultures of embryonic chick thigh and neonatal rat thigh myoblasts were allowed to differentiate and form multinucleate myotubes. Labeled cell extracts were coelectrophoresed to demonstrate the relative positions of light chain 1 from chick and rat. The identity of the light chains was established by comparison with published data (Devlin and Emerson, 1978; Whalen *et al.*, 1978) and their comigration with the myosin light chains of partially purified myofibrils from chick and rat tissue. (B) Chick myocyte × rat myoblast heterokaryons isolated as in Fig. 1. Cells were labeled overnight on the 4th day following fusion in methionine-free medium supplemented with 10% undialyzed fetal bovine serum and 300 μCi/ml [^{35}S]methionine. (c) Chick fast skeletal light chain 1; (a) adult rat fast skeletal-myosin light chain 1; (e) embryonic rat skeletal light chain 1.

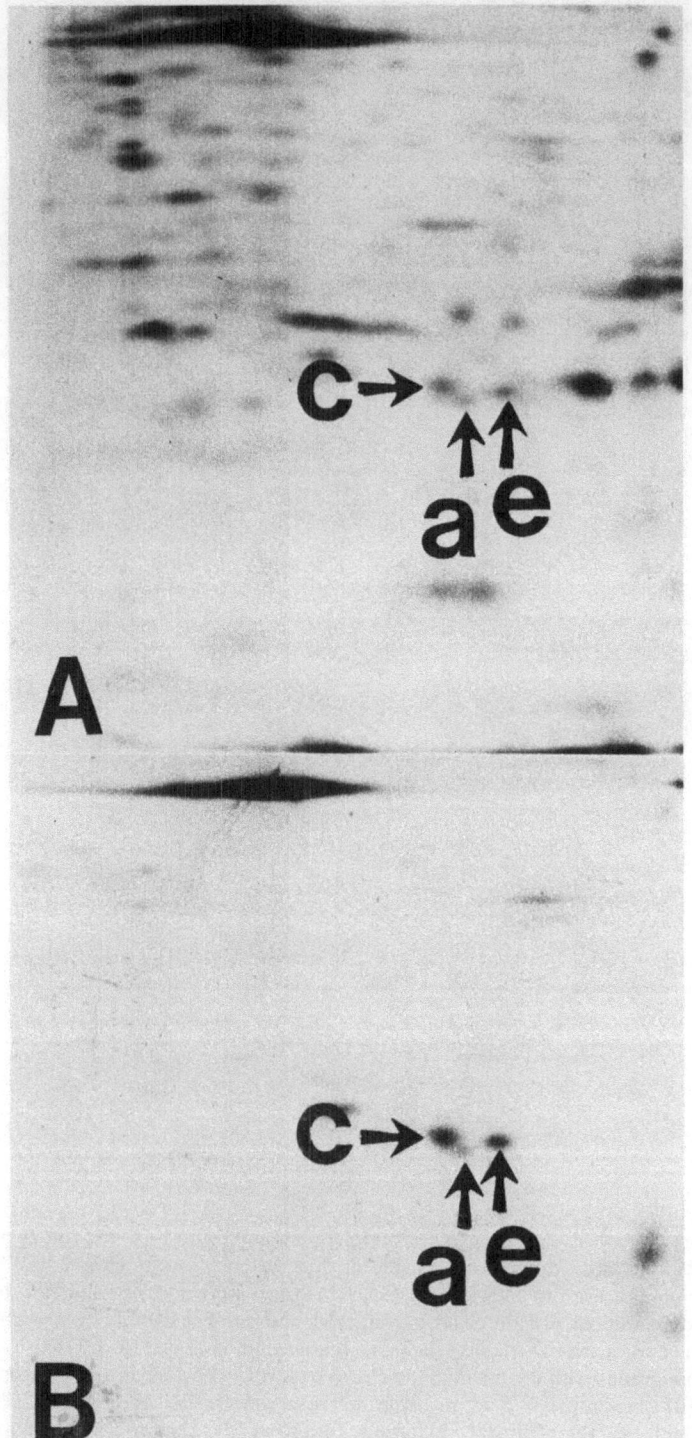

synthesize only the embryonic form of skeletal myosin light chain 1 (Whalen *et al.*, 1978; Garrels, 1979), we have been able to induce differentiated L6 myotubes to make very small amounts of adult light chain following trypsinization, purification of the myotubes, and subsequent cultivation in the presence of cytosine arabinoside. The consistent and rapid appearance of adult rat light chain in the chick myocyte × L6 myoblast heterokaryons probably represents a specific activation of the adult rat form. However, the ability of differentiated L6 cells to make small quantities of adult light chain under some conditions implies that its increased synthesis in the heterokaryons could be due to secondary effects.

Studies using hepatoma hybrids have clearly shown that liver functions can be activated in the nonhepatic parent if the hybrid contains twice as many hepatoma as nonhepatic chromosomes (Peterson and Weiss, 1972; Brown and Weiss, 1975). Although the majority of chick myocyte × rat myoblast heterokaryons in this study are binucleates, significant numbers of tri- and tetranucleated cells are present. It is possible that the light-chain synthesis exhibited by the total population of heterokaryons occurs only in those heterokaryons that contain more chick than rat nuclei. The fact that many binucleated heterokaryons are positive for chick myosin by indirect immunofluorescence (data not shown) argues against this possibility. However, chick myosin is present in the chick myocytes before heterokaryon formation. The proof that binucleated heterokaryons synthesize myosin awaits the availability of an antibody specific for rat myosin that does not cross-react with chick myosin.

Previous somatic-cell-hybridization experiments between cells of different histiotypic origins have led to the concept that "luxury" or differentiated functions are suppressed in hybrids between differentiated and "undifferentiated" cells (Ephrussi, 1972; Ringertz and Savage, 1976). The experiments reported herein were designed in part to test the hypothesis that alternate regulatory controls might be exhibited if the "undifferentiated" parent was a precursor cell within the same developmental lineage.

We have shown that differentiated chick myocytes are able to induce the synthesis of both fetal and adult rat skeletal-myosin light chains when fused to undifferentiated L6 rat myoblasts. This suggests that the control of terminal myogenesis is under positive inductive regulation. To our knowledge, this is the first report on the expression of differentiated functions in purified populations of heterokaryons. It is hoped that further experiments exploiting this system will lead to the isolation of the regulatory molecules responsible for these effects.

Figure 3. Partial purification of heterokaryon myosin light chains. Chick myocyte × rat myoblast heterokaryons were prepared as in Fig. 1. Actomyosin was partially purified by first lysing the cells with 0.5% Nonidet P-40 in 30 mM KCl in the presence of 30 μg cold carrier chick actomyosin, extracting the insoluble material in 0.6 M KCl, then precipitating the actomyosin by dilution to 30 mM KCl. (A) Total extract; (B) partially purified extract.

References

Bag, J., and Sarkar, S., 1976, Studies on a nonpolysomal ribonucleoprotein coding for myosin heavy chains from chick embryonic muscles, *J. Biol. Chem.* **251**:7600.

Brown, J. E., and Weiss, M. C., 1975, Activation of production of mouse liver enzymes in rat hepatoma–mouse lymphoid cell hybrids, *Cell* **6**:481.

Buckingham, M. E., Cohen, A., and Gros, F., 1976, Cytoplasmic distribution of pulse-labeled poly (A)-containing RNA, particularly 26S RNA, during myoblast growth and differentiation, *J. Mol. Biol.* **103**:611.

Davidson, R. L., 1972, Regulation of melanin synthesis in mammalian cells: Effects of gene dosage on the expression of differentiation, *Proc. Natl. Acad. Sci. U.S.A.* **69**:951.

Devlin, R. B., and Emerson, C. P., 1978, Coordinate regulation of contractile protein synthesis during myoblast differentiation, *Cell* **13**:599.

Ephrussi, B., 1972, *Hybridization of Somatic Cells*, Princeton University Press, Princeton, New Jersey.

Fougére, C., Ruiz, F., and Ephrussi, B., 1972, Gene dosage dependence of pigment synthesis in melanoma × fibroblast hybrids, *Proc. Natl. Acad. Sci. U.S.A.* **69**:330.

Garrels, J. I., 1979, Changes in protein synthesis during myogenesis in a clonal cell line, *Dev. Biol.* **73**:134.

John, H. A., Patrinou-Georgonlas, M., and Jones, K. W., 1977, Detection of myosin heavy chain mRNA during myogenesis in tissue culture by *in vitro* and *in situ* hybridization, *Cell* **12**:501.

O'Farrel, P., 1975, High resolution two-dimensional electrophoresis of proteins, *J. Biol. Chem.* **250**:4007.

Perriard, J.-C., 1979, Developmental regulation of creatine kinase isozymes in myogenic cell cultures from chicken: Levels of mRNA for creatine kinase subunits M and B, *J. Biol. Chem.* **254**:7036.

Peterson, J. A., and Weiss, M. C., 1972, Expression of differentiated functions in hepatoma cell hybrids: Induction of mouse albumin production in rat hepatoma–mouse fibroblast hybrids, *Proc. Natl. Acad. Sci. U.S.A.* **69**:571.

Sanger, J. W., 1974, The use of cytochalasin B to distinguish myoblasts from fibroblasts in cultures of developing chick striated muscle, *Proc. Natl. Acad. Sci. U.S.A.* **71**:3621.

Strohman, R. C., Moss, P. S., Micou-Eastwood, J., and Spector, D., 1977, Messenger RNA for myosin polypeptides: Isolation from single myogenic cell cultures, *Cell* **10**:265.

Vertel, B. M., and Fischman, D. A., 1976, Myosin accumulation in mononucleated cells of chick muscle cultures, *Dev. Biol.* **48**:438.

Whalen, R. G., Butler-Browne, G. S., and Gros, F., 1978, Identification of a novel form of myosin light chain present in embryonic muscle tissue and cultured muscle cells, *J. Mol. Biol.* **126**:415.

Wright, W. E., 1978, The isolation of heterokaryons and hybrids by a selective system using irreversible biochemical inhibitors, *Exp. Cell Res.* **112**:395.

Wright, W. E., 1981, The recovery of heterokaryons at high cell densities following isolation using irreversible biochemical inhibitors, *Som. Cell Genet.* (in press).

Wright, W. E., and Hayflick, L., 1975, Use of biochemical lesions for selection of human cells with hybrid cytoplasms, *Proc. Natl. Acad. Sci. U.S.A.* **72**:1812.

Yaffe, D., 1968, Retention of differentiation potentialities during prolonged cultivation of myogenic cells, *Proc. Natl. Acad. Sci. U.S.A.* **61**:477.

14

Phosphorylation of Myosin and the Regulation of Smooth-Muscle Actomyosin

David J. Hartshorne

1. Introduction

It is accepted that the contractile event in all types of muscle is initiated by an increase in the intracellular concentration of Ca^{2+}. One can consider the regulatory effect of Ca^{2+} under two broad categories. The first involves the regulation of the level of free Ca^{2+} within the cell and is usually linked to a membranous system, such as the sarcoplasmic reticulum; the second category implicates the contractile apparatus and more specifically is concerned with the effect that a given free Ca^{2+} level has on the function of the actomyosin system.

In smooth muscle, the structures that are primarily responsible for the regulation of the intracellular Ca^{2+} concentration are the sarcoplasmic reticulum and the plasma membrane. It has been suggested that the action potential might represent some influx of Ca^{2+} (Bolton, 1979), and if this is accepted, then the question that must be answered is how much of the Ca^{2+} required for the activation of the contractile apparatus enters the cell during the action potential and how much is derived from internal sources, such as the sarcoplasmic reticulum. This topic is still controversial, but there is a consensus that many smooth muscles contain a sarcoplasmic reticulum that functions in a manner similar to that in skeletal muscle. In support of this, it has been found that the volume of sarcoplasmic reticulum varies in different muscles and that this is related to the ability of a given muscle to contract in

David J. Hartshorne • Muscle Biology Group, Department of Nutrition and Food Science, University of Arizona, Tucson, Arizona 85721.

the absence of external Ca^{2+} (Johansson and Somlyo, 1980). Also, a membrane fraction from bovine vascular smooth muscle was isolated, and it was calculated that this had the capability or capacity to serve as both a sink and a source of the Ca^{2+} required for activation (Ford and Hess, 1975). Although many of the details concerning the sarcoplasmic reticulum in smooth muscle are missing, it is reasonable to assume that this organelle does control, at least partly, the level of Ca^{2+} within the cell [For a more detailed discussion, see the reviews by Somlyo (1980) and Johansson and Somlyo, 1980).]

Once the concentration of Ca^{2+} increases above a threshold level, then the contractile system is activated. Obviously, there are components present in the contractile apparatus that can detect the intracellular fluctuations in the Ca^{2+} level and interpret these to result in either a contraction at higher Ca^{2+} levels (on the order of 5×10^{-6} M) or a relaxation at lower Ca^{2+} levels ($<1 \times 10^{-6}$ M). These components are termed the regulatory proteins. In skeletal and cardiac muscle, the regulatory proteins are troponin and tropomyosin, which are located on the thin filaments, and a considerable amount of information has been accumulated on their mechanism of action (see other chapters). However, in smooth muscle, the regulatory system has not been unequivocally identified, and the various viewpoints are discussed in this review. Before the regulatory components are considered, it is worthwhile to outline the system on which the influence of regulation is exerted, namely, the contractile apparatus.

2. Contractile Apparatus

The protein components of the cell that respond to the Ca^{2+} fluctuations and utilize the chemical energy of ATP to result in either shortening or tension development are termed collectively the contractile apparatus. In smooth muscle, the major contractile proteins are myosin, actin, and tropomyosin, and minor components of the contractile apparatus include the proteins that are involved in the Ca^{2+}-dependent regulatory mechanisms. There are several aspects of the contractile apparatus in smooth muscle that are of interest, and some of these are considered below.

2.1. Content of Contractile Proteins

It is important to establish the content of the major components of the contractile apparatus, since this will form a basis for understanding much of the data derived from biochemical and physiological studies. It has been known for many years (Csapo, 1948; Needham and Cawkwell, 1956) that smooth muscles contain actomyosin, and it was the general impression from these early studies that the amount of actomyosin was less than in skeletal muscle. A qualitative distinction was also made, and it was claimed that smooth muscles contained a specialized form of actomyosin, termed tonoactomyosin, that was soluble at low ionic strength (Hamoir and Laszt, 1962;

Laszt and Hamoir, 1961). One suggestion was that this type of actomyosin was involved predominantly in tonic contractions (Laszt, 1964). An increased solubility of actomyosin at low ionic strengths seems to be a general feature associated with all smooth muscles, although the existence of a specialized form of actomyosin is no longer accepted (Rüegg *et al.*, 1965). One explanation for the increased solubility of the actomyosin might be the higher content of actin that is found in smooth as compared to skeletal muscle.

The values for the contractile protein contents that are accepted today are derived mainly from sodium dodecyl sulfate (SDS) electrophoresis. The most comprehensive study is that carried out by Cohen and Murphy (1978), who found that smooth muscles can be classified roughly into two groups, based on their content of actin. Arterial muscle contains more actin than the second category, which was listed as nonarterial and was represented by muscle from the esophagus, trachea, intestine, and uterus. Average values for the two categories of smooth muscle are given in Table 1. For comparative purposes, approximate values obtained with skeletal muscle are also listed in Table 1. Within the two categories of smooth muscle, the significant difference is the increased content of actin in arterial muscle. The higher content of tropomyosin in the latter is merely a consequence of the increased actin content, since tropomyosin binds to actin at a fixed stoichiometry to form the thin filament. What the increased actin concentration means in terms of muscle function is not clear, but it appears that arterial smooth muscle develops more force than most other smooth muscles, and the increased actin concentration might be related to this finding (Murphy, 1979). When smooth and skeletal muscles are compared, the most striking difference is the content of myosin. The approximate concentrations of myosin in the two muscle types are listed in Table 1, and it is apparent that the molarity of myosin in skeletal muscle is about 3-fold higher than in smooth muscle. This finding is particu-

Table 1. Content of the Major Contractile Proteins in Smooth and Skeletal Muscle

Contractile protein	Smooth[a]		Skeletal[b]
	Arterial	Nonarterial	
Myosin			
Content (mg/g cell wet wt.)	19.6 ± 0.7	19.5 ± 1.3	62
Molarity (mM)[c]	0.04	0.04	0.13
Actin			
Content (mg/g cell wet wt.)	50.0 ± 4.7	27.5 ± 1.8	22
Molarity (mM)[c]	1.2	0.66	0.5
Tropomyosin			
Content (mg/g cell wet wt.)	14.0 ± 1.2	7.7 ± 0.6	5
Molarity (mM)[c]	0.2	0.11	0.07

[a] Values taken from Cohen and Murphy (1978).
[b] Representative values.
[c] Molarities were calculated using a cell density of 1. The molecular weights used were: myosin, 470,000; actin, 42,000; tropomyosin, 68,000.

larly interesting, since it was shown (Murphy *et al.*, 1974) that smooth muscle can generate tension approximately equal to that in skeletal muscle, and it is known that the force developed is proportional to the number of cross-bridges acting in parallel (A. F. Huxley, 1957), which in turn is proportional to the myosin content. Thus, the relatively low myosin content in smooth muscle must be compensated for in some way, and several possibilities have been proposed. These include longer thick filaments in smooth muscle (Ashton *et al.*, 1975; Somlyo *et al.*, 1977), the higher actin content, and possibly different cross-bridge kinetics for the smooth-muscle system (Murphy, 1979).

The three proteins discussed above form the bulk of the contractile apparatus, but obviously some of the minor constituents are essential for its normal functioning. These include the regulatory components, which must be incorporated into the preceding tabulation when the data become available. At this time, only one additional component can be added, namely, calmodulin. It was estimated (Grand *et al.*, 1979) that calmodulin represents 0.42% of the total protein of rabbit uterus. If it is assumed that the total protein constitutes 15% of the wet weight of the cell, the content of calmodulin would be 0.63 mg/g wet weight, and using a molecular weight of 17,000 (Cheung, 1980), the concentration would be approximately 0.037 mM. It has been shown recently (reviews by Cheung, 1980; Means and Dedman, 1980) that calmodulin regulates many enzymatic mechanisms, and thus only a fraction of the total calmodulin would be associated with the contractile apparatus.

2.2. Properties of the Contractile Proteins

A detailed discussion of the contractile proteins in smooth muscle has been given previously (Hartshorne and Gorecka, 1980), and in the following sections, only a few points will be abstracted.

2.2.1. Myosin

In general, the isolation of myosin from smooth muscle is more difficult than from skeletal muscle. The major difficulty is associated with the higher actin content and the resistance of some actin–myosin interactions to dissociation by ATP. Proteolysis of the myosin (noticed in our laboratory using chicken gizzard) is also a potential complication. However, methods have been developed, and myosin has been isolated from many different smooth muscles (for a review, see Hartshorne and Gorecka, 1980). The physical parameters of smooth-muscle myosin are similar to those of myosins obtained from all other sources, in that the molecule consists of two large subunits, M_r about 200,000, and four small subunits, called light chains, two of M_r about 20,000 and two of M_r about 17,000. (The two 20,000-dalton light chains are the sites of phosphorylation and are discussed in Section 3.2.) In general, the smooth-muscle myosin molecule can be dissected by proteolysis into subfragments following a pattern similar to that established with skeletal-muscle myosin. Heavy meromyosin has been isolated (Onishi and Watanabe, 1979;

Seidel, 1978, 1980), and this reflects, to some extent, the properties of the parent myosin. Heavy meromyosin subfragment-1, isolated following papain digestion, lacks the 20,000-dalton light chain (Kendrick-Jones, 1973), but retains its activation of Mg^{2+}-ATPase activity by actin (Marston and Taylor, 1978; Mrwa and Rüegg, 1977; Seidel, 1978; Sobieszek and Small, 1976), although the activity is usually Ca^{2+}-insensitive. Digestion with α-chymotrypsin (Onishi and Watanabe, 1979; Seidel, 1978) or trypsin (Okamoto and Sekine, 1978) results in the disappearance of the 20,000-dalton light chain, although this is concomitant with the loss of actin activation. The difference in the ability of actin to activate the subfragments produced using different proteolytic enzymes is not understood. A potential complication in the interpretation of the proteolysis experiments is the possibility of heavy-chain cleavage in the region of the active site, which could affect the ATPase characteristics. Despite the difficulty in interpretation, this type of experimental approach is worthy of additional effort, since it could contribute to our knowledge of light-chain function. The question that remains to be answered is whether the dephosphorylated 20,000-dalton light chains inhibit or prevent the activation of ATPase activity by actin, and inhibition is relieved by phosphorylation, or whether the dephosphorylated light chains are dormant and activated by phosphorylation. In the first case, removal of the 20,000-dalton light chains would result in activation of ATPase activity; in the second case, in the loss of actin activation.

The aggregation of myosin to form filaments has been known for many years and was found initially with skeletal-muscle myosin (H. E. Huxley, 1963) and later demonstrated with smooth-muscle myosin (Hanson and Lowy, 1964; Kaminer, 1969; Shoenberg, 1965, 1969). The shape and size of the filaments, however, were found to be extremely variable and depended on several factors (e.g., ionic strength, pH, concentration of myosin), and it was therefore difficult to assign any one filament type as an *in vivo* model. In more recent experiments, longer filaments have been assembled that do not show a central bare zone. Sobieszek (1977a) suggested that the building block for this type of filament is an antiparallel myosin dimer with an overlap of about 600 Å and with the tails twisted around each other by a total of 180°. The dimer is aggregated in a six-stranded helix to form the filament. The same basic structure was suggested for several types of myosin (Hinssen *et al.*, 1978), and it was pointed out that this mode of assembly for smooth and nonmuscle myosins is distinct from that observed with skeletal-muscle myosin. Wachsberger and Pepe (1974), using rabbit uterine myosin, found that short, tapered bipolar filaments with a central bare zone were formed initially and that in the presence of Mg^{2+}, these aggregated by an end-to-end overlap of the tapered ends to form longer filaments with no bare-zone region. Both bipolar and nonpolar filaments were obtained by Suzuki *et al.* (1978), who felt that the type of filament produced depended on the speed of dilution of the protein solution. The short bipolar filaments were observed by Craig and Megerman (1977), and a longer side-polar filament was also found. It is proposed that in the latter, the cross-bridges on one side of the filament are of polarity opposite to those on the other side of the filament. This myosin aggregate is unusual in that it does not show a helical arrangement of myosin molecules, but presents a rectangular profile where the cross-bridges are found only on the two opposite faces. The advantage of this type of assembly is that it would allow the interaction with actin along the entire filament length, and thus might accommodate or facilitate the wide range of shortening of which smooth muscle is capable. The same advantage could be suggested for the other nonpolar structures, although with a helical arrangement of the myosin molecules, some constraints might be imposed by the geometry of the helix.

It is apparent from the foregoing discussion that the way in which the myosin molecules are assembled to form the thick filaments in smooth muscle is not established and must be clarified in the future. One of the complications is that unlike the situation in skeletal muscle, where the morphology of the *in vivo* thick filament is well documented, the natural form of the myosin filaments in smooth muscle has not been described adequately. Indeed, it was not until the late 1960s that the existence of thick filaments in smooth muscle was established (Kelly and Rice, 1968; Nonomura, 1968). Part of the problem appears to be that the thick filaments in smooth muscle are more labile than their skeletal-muscle counterparts and were lost during the processing of the tissue for electron microscopy. Ribbonlike aggregates were observed (Lowy and Small, 1970; Small and Squire, 1972), and these are also thought to be artifacts produced during the fixation procedures (Jones *et al.*, 1973; Shoenberg and Haselgrove, 1974; Somlyo *et al.*, 1973). However, thick filaments are now accepted as a component of the ultrastructure of the smooth-muscle cell, and they have been seen in a variety of smooth muscles (for a review, see Somlyo, 1980). Despite this, there are several features about the thick filaments that are not established, one of which is the filament length. Ashton *et al.* (1975) determined a length of 2.2 μm for the filaments in rabbit portal anterior mesenteric vein, a length greater than that of the filaments in rabbit skeletal muscle (H. E. Huxley, 1963). Filaments up to 8 μm long were reported in homogenates of taenia coli (Small, 1977), although it is not clear whether these were aggregates of shorter filaments. Another feature that has not been clarified is whether the smooth-muscle thick filaments possess a central bare zone. This is significant, since it reflects the mode of packing of the myosin molecules and thus the arrangement of the cross-bridges on the filament. Cross-bridges have been observed on smooth-muscle thick filaments (Ashton *et al.*, 1975; Devine and Somlyo, 1971; Somlyo *et al.*, 1973), and their distribution in a helical arrangement is consistent with the observations (Somlyo, 1980). There is one other factor that might affect the thick-filament morphology in smooth muscle, and that is the level of myosin phosphorylation. Suzuki *et al.* (1978) showed that phosphorylation stabilized the filamentous structure to the dissociating effect of ATP. Thus, one might predict that in the relaxed state, the thick filaments (i.e., with most of the myosin in the dephosphorylated form) would be more labile and possibly subject to the generation of artifactual structures.

The next property of myosin that will be considered is its ATPase activity. It may be stated that in general, the ATPase activities of smooth-muscle myosin are lower than those of skeletal-muscle myosin. The characteristics of the ATPase activities are also different, and both types of muscles exhibit a pattern of behavior that is characteristic of the muscle type. Usually, three different ATPase activities are measured, these being when Ca^{2+}-ATP is the substrate, when Mg^{2+}-ATP is the substrate, and when K^+-ATP is the substrate (often termed the K^+-EDTA ATPase activity). Obviously, only the Mg^{2+}-ATPase activity is of physiological relevance, but the other activities are useful to characterize a given myosin. The K^+-EDTA activity is usually about 30% of the fast-skeletal-muscle myosin values, but otherwise its properties (i.e., the inhibition by actin and by sulfhydryl modification) are similar. However, the Ca^{2+}-ATPase activities are quite different. It was noted several years ago for uterine actomyosin (Needham and Cawkwell, 1956) that the Ca^{2+}-ATPase activity showed a marked activation when the ionic strength was increased. Subsequently, it was found, using gizzard myosin, that a marked activation occurred on tryptic digestion or the addition of various denaturing agents (Bárány *et al.,* 1966). More recently, it was shown that the optimum value of the Ca^{2+}-ATPase activity of stomach myosin is at 0.3 M KCl (Katoh and Kubo, 1977). This type of behavior is not characteristic of skeletal-muscle myosin, in which the effect of ionic strength on the Ca^{2+}-ATPase activity is not marked. However, nonmuscle myosin appears to be similar to smooth-muscle myosin, and it was found that the Ca^{2+}-ATPase activity of myosin from fibroblasts was activated by KCl (Yerna *et al.,* 1978). Another difference observed with the Ca^{2+}-ATPase activities is that smooth-muscle myosin does not exhibit the activation of ATPase activity associated with limited sulfhydryl modification (Wachsberger and Kaldor, 1971; Yamaguchi *et al.,* 1970). Skeletal-muscle myosin is activated on reaction of the sulfhydryl 1 residue (Sekine *et al.,* 1962).

The Mg^{2+}-ATPase activities of pure myosins from both skeletal and smooth muscles are similar and are on the order of 2 nmoles phosphate liberated/min per mg myosin at 25°C. Both myosins show a similar pH–activity profile, and maxima are observed at both acidic (pH≈5) and alkaline (pH≈10) pH values. The pH optimum of actin activation for both myosin types is in the neutral pH range (Driska and Hartshorne, 1975), and this is consistent with the optimum pH for tension development in glycerinated arterial fibers (Mrwa *et al.,* 1974). The most significant difference in the Mg^{2+}-ATPase activity of smooth- and skeletal-muscle myosins is seen in the effect of actin. The Mg^{2+}-ATPase activity of skeletal-muscle myosin is activated over 100-fold by actin, and additional protein components are not necessary for this effect. In contrast, the Mg^{2+}-ATPase activity of smooth-muscle myosin is activated much less by actin (Bárány *et al.,* 1966; Driska and Hartshorne, 1975; Yamaguchi *et al.,* 1970), and in fact, if both proteins are pure, the activation is negligible. It is clear, therefore, that the activation requires additional factors; these contribute the primary requirements for the regulatory system in smooth muscle and are discussed in Section 3. However, at this point, it is convenient to discuss the Mg^{2+}-ATPase activity of smooth-muscle actomyosin with the understanding that proteins in addition to myosin and actin are involved. In general, the ATPase activity of skeletal-muscle actomyosin falls within the range of 0.5–1.0 μmole P_i/min per mg actomyosin, and the values reported for smooth-muscle actomyosin are all lower than this. However, the values obtained for the latter are extremely variable and fall between 0.01 and 0.3 μmole P_i/min per mg actomyosin (Driska and Hartshorne, 1975; Mrwa and Rüegg, 1975; Russell, 1973; Sobieszek, 1977b; Sobieszek and Bremel, 1975; Sobieszek and Small, 1976, 1977; Sparrow *et al.*, 1970). The majority of reports favor the lower end of this range. At present, there is no adequate explanation to account for such a wide divergence of activities. Some contributing factors could be variations in the assay conditions (e.g., Mg^{2+} and protein concentration, ionic strength) and also the content of regulatory proteins, i.e., the activating components that a given actomyosin contains. The extent of myosin phosphorylation may also be an important variable (see Section 3). In addition, the kinetics of actomyosin ATPase activity is usually nonlinear and exhibits a rapid initial phase followed by a slower rate of hydrolysis (Hartshorne and Persechini, 1980). Thus, if the ATPase activity is calculated over a short time period, it is usually higher than one estimated over a longer period (Murphy *et al.*, 1969). It is interesting that the higher Mg^{2+}-ATPase activities (Sobieszek, 1977b; Sobieszek and Bremel, 1975; Sobieszek and Small, 1976, 1977) were obtained after only a brief reaction time and therefore were reflective of the faster initial rate. Whether the faster or the slower phase of ATP hydrolysis more accurately reflects the *in vivo* situation has not been established, although it is possible that both phases could be incorporated into the contractile mechanism. In terms of steady-state cross-bridge cycling, the slower phase is certainly more consistent with the physiological date. It was calculated (Mrwa *et al.*, 1975) that for arterial muscle at 37°C, the cross-bridge cycling rate is on the order of 1 sec^{-1}. This would correspond to an actomyosin ATPase activity at 25°C of about 0.01 μmole P_i/min per mg myosin.

It is customary for muscle biochemists to use actomyosin as the model for the contractile apparatus, and one example of this is in the investigations of the Ca^{2+}-dependent regulatory system. To reflect the *in vivo* situation, it is required that the Mg^{2+}-ATPase activity of smooth-muscle actomyosin be activated by Ca^{2+} (on the order of 5×10^{-6} M) and inhibited when the Ca^{2+} level is reduced. Historically, this proved to be an elusive requirement, and it was not until 1970 (Sparrow *et al.*, 1970) that a Ca^{2+}-dependent actomyosin was isolated from arterial smooth muscle. Since this time, however, several groups have isolated Ca^{2+}-sensitive actomyosins from a variety of smooth muscles (for a review, see Hartshorne and Gorecka, 1980), and these have been instrumental in furthering our understanding of the regulatory mechanism. The essential constituents of a Ca^{2+}-sensitive actomyosin have not been defined, although it can be stated that subunits similar to those of skeletal-muscle troponin are not essential for Ca^{2+}-regulation (Bremel *et al.*, 1977; Driska and Hartshorne, 1975; Sobieszek and Bremel, 1975).

2.2.2. Actin

Of the three major contractile proteins, actin isolated from a variety of sources shows the least variation. Most of the well-known properties of skeletal-muscle actin are similar to those of smooth-muscle actin (for a review, see Hartshorne and Gorecka, 1980). However, it has recently been demonstrated that slight differences do exist, and three isoelectric variants of actin have been identified. The most acidic form, α, is found in skeletal and cardiac muscle, nonmuscle actins contain the β and γ forms, and smooth-muscle actin comigrates with the γ form, which is the most basic of the three (Rubenstein and Spudich, 1977; Vanderkerckhove and Weber, 1978a; Whalen *et al.,* 1976). The difference in isoelectric points between the α- and γ-actins is restricted to a few amino acids at the N terminus (Vandekerckhove and Weber, 1978b; Zechel, 1979), and the rest of the sequence is similar. Recently it has been suggested that γ-actin binds Mg^{2+}-ADP more strongly than β-actin (Anderson, 1979). However, these differences are relatively minor, and most investigators feel that the biological properties of skeletal- and smooth-muscle actins are very similar, and in general, the two actin can be used interchangeably in hybrid systems with myosin. An exception to this generalization is found in the work of Ebashi and coworkers (1975, 1976), who report a requirement for smooth-muscle actin in their studies on the regulatory mechanism of smooth muscle.

In smooth muscle, as in skeletal muscle, filamentous actin is organized into the thin filaments. These are numerous, and in general, the higher content of actin relative to myosin in smooth muscle is reflected by the ratio of thin to thick filaments; for example, ratios were obtained of 10–14:1 (Heumann, 1969) and 12:1 (Bois, 1973) in intestinal muscle and 12–15:1 (Devine and Somlyo, 1971; Somlyo *et al.,* 1973) in vascular smooth muscle. In rabbit skeletal muscle, there are about 2 thin filaments per thick filament. The length of the smooth-muscle thin filaments has not been established, although Nonomura *et al.* (1980) reported that they are shorter than in skeletal muscle. Troponin is not found in smooth-muscle thin filaments (Driska, 1976; Nonomura *et al.,* 1980; Sobieszek and Small, 1976), and the only other major constituent is tropomyosin, which is found at a stoichiometry of about 1 tropomyosin molecule per 7 actins. It is interesting that despite the absence of troponin, X-ray diffraction data indicate that the position of tropomyosin on the thin filament is altered on activation of the muscle (Parry and Squire, 1973; Vibert *et al.,* 1972). In striated muscle, the movement of tropomyosin from a blocking to a nonblocking position, corresponding to the relaxed and contracted states, respectively, has been proposed as an essential component of the regulatory mechanism (see other chapters). However, in striated muscle, this requires the participation of troponin, which is thought to direct and stabilize the tropomyosin positions. Thus, the significance of the tropomyosin shift in smooth muscle is not understood, although it may reflect some aspect of the cross-bridge–actin interaction.

One end of the thin filament is associated with an amorphous structure called a dense body. Dense bodies are located both within the cytoplasm and attached to the plasma membrane (for a review, see Somlyo, 1980). It is thought that the dense bodies serve a function analogous to that of the Z-lines in striated muscle and anchor the thin filaments and thereby aid in the transmission of tension evenly throughout the cell. Also associated with the dense bodies are the 10-nm, or intermediate, filaments (see Goldman volume this series). Their function is not established, although it is clear that they do not participate directly in the contractile process, and they may form a cytoskeletal network (Ashton *et al.*, 1975; Cooke, 1976; Cooke and Fay, 1972; Small and Sobieszek, 1977a), which again could aid in the distribution of tension within the cell.

2.2.3. Tropomyosin

It has been known for many years that tropomyosin is a component of smooth muscle (for a review, see Hartshorne and Gorecka, 1980). In terms of their gross physical parameters, tropomyosins from all muscle sources are similar. The molecule is asymmetric, almost entirely α-helical, and composed of two subunits arranged in register in a coiled-coil configuration. At low ionic strength, tropomyosin forms an end-to-end polymer, and this is assumed to be its structure on the thin filament, where one strand of tropomyosin is associated with each of two actin strands (Murray and Weber, 1974). The stoichiometry is 1 tropomyosin molecule per 7 actin molecules, and this is maintained in hybrid systems where the actin and tropomyosin are isolated from different muscle sources.

It was shown a few years ago that contrary to earlier beliefs, the tropomyosin subunits in skeletal muscle are not identical. Two subunits, α and β, can be identified in rabbit psoas muscle (Cummins and Perry, 1973), and the α-subunit is present in greater amounts (Cummins and Perry, 1973; Johnson, 1974). Subsequently, it was shown that in slow (red) skeletal muscle (Cummins and Perry, 1974) and in fetal muscle (Amphlett *et al.*, 1976), the proportion of β is increased, whereas in rabbit and avian cardiac muscle, only the α-subunit is found (Cummins, 1979; Cummins and Perry, 1974; Mak *et al.*, 1979). In the larger mammalian hearts, the β-subunit is also found (Leger *et al.*, 1976; Ookubo *et al.*, 1975). The original observations that allowed a discrimination between the two subunits were based on a difference in electrophoretic mobilities. Despite this, the molecular weights for the two subunits are similar, and each subunit contains the same number of amino acids. The complete sequence of both α- (Stone and Smillie, 1978; Stone *et al.*, 1974) and β-subunits (Mak *et al.*, 1979) has been determined and found to differ in 39 amino acid substitutions. So far, there is no indication of a functional difference between the two subunits (Leger *et al.*, 1976).

The tropomyosin of smooth muscles has not been as extensively characterized as that of striated muscle. Different subunits can be detected; for example, tropomyosin from chicken gizzard shows two components on SDS–

polyacrylamide electrophoresis (Bremel *et al.*, 1977; Cummins and Perry, 1974; Driska and Hartshorne, 1975), although neither is equivalent to the skeletal-muscle subunits (Cummins and Perry, 1974). The apparent molecular weights of the two gizzard subunits are 36,000 and 39,000 based on electrophoretic mobilities. Since the α- and β-subunits were shown to have similar molecular weights, but different electrophoretic mobilities, these values must be confirmed by other techniques. It was suggested (Strasburg and Greaser, 1976) that with gizzard tropomyosin, only the homodimer (i.e., 2 × 36,000 or 2 × 39,000) is formed and the heterodimer is not found. Tropomyosin has been identified in other smooth muscles, rabbit and pig uterus (Cummins and Perry, 1974), hog carotid artery (Murphy *et al.*, 1974), and calf aorta (Fine and Blitz, 1975) and is thought to consist of a single subunit.

The function of tropomyosin in smooth muscle is not understood. The consensus, as already outlined, is that a troponinlike mechanism is not functional in smooth muscle, though recently Marston *et al.* (1980) have identified components from aorta thin filaments that were similar to the skeletal troponin subunits. It is interesting that although troponin is not usually found in smooth-muscle preparations, smooth-muscle tropomyosin has retained a troponin binding site (Ebashi *et al.*, 1976; Sparrow and van Bockxmeer, 1972) and is able to function in the skeletal-muscle system. Tropomyosin does not bind to leiotonin (Nonomura and Ebashi, 1980), but it is essential for the functioning of the leiotonin system (Ebashi *et al.*, 1978). Thus, it is difficult to imagine that tropomyosin in smooth muscle is functional via a stoichiometric interaction with another regulatory protein. It is a consistent observation, however, that tropomyosin is required for full activation of smooth-muscle actomyosin (Chacko *et al.*, 1977; Hartshorne *et al.*, 1977b; Nonomura and Ebashi, 1980; Sobieszek and Small, 1977), and yet the mechanism of activation is obscure. Possibly tropomyosin is involved in a potentiation along the thin filament of some effects generated by a separate mechanism, and this hypothesis could be applied to either the phosphorylation or the leiotonin theory of regulation. The function of tropomyosin in smooth muscle (and also in nonmuscle systems) must be established by future research, although one may assume as a working hypothesis that its effect is secondary to another mechanism and that the tropomyosin may potentiate an imposed influence via interactions with its thin-filament partner, actin.

2.2.4. Summary Statement

The preceding sections should now be condensed to state what is thought to be the contractile mechanism in smooth muscle. It has been shown that both thick and thin filaments are present in smooth muscle and that in some instances cross-bridges were identified on the thick filaments. Thus, it is assumed that the sliding-filament model, which was developed for skeletal muscle, also operates in smooth muscle. This is made more attractive by the finding that Z-line analogues, called the dense bodies, are also present in smooth muscle. In the sliding-filament model, length changes are achieved by a relative sliding of the two sets of filaments, and neither the thin nor the thick filaments alter in length. The focus of tension development is the cross-bridge–actin interaction, and for each cross-bridge cycle, one molecule of ATP is hydrolyzed. Very little is known about the kinetics of the cross-bridge cycle in smooth muscle, although usually it is acknowledged to be slower than in skeletal muscle. To some extent, the reduced rate of cross-bridge cycling is reflected by the specific ATPase activity of the smooth-muscle actomyosin. In its simplest concept, the regulatory mechanism, which is part of the contractile apparatus, functions by controlling the cross-bridge–actin interactions. If these interactions are prevented, the muscle relaxes. This feature is common to the regulation of all muscle types, but the mechanism by which this is achieved in smooth muscle is quite distinct from its striated-muscle counterpart.

3. Regulation

The activity of the contractile apparatus in its native state is governed by the level of free Ca^{2+}. It is known that components associated with the contractile proteins are responsible for the regulatory effect, and this section will outline some of the ideas that have been proposed to account for regulation in smooth muscle.

Historically, directions in smooth-muscle biochemistry followed patterns previously established with skeletal muscle, and the initial efforts to identify the regulatory mechanism in smooth muscle were no exception. Following the discovery of troponin in skeletal muscle by Ebashi and his colleagues (Ebashi, 1963; Ebashi *et al.*, 1968), smooth muscle was analyzed for troponinlike components. In several instances, success was reported (Carsten, 1971; Ebashi *et al.*, 1966; Ito and Hotta, 1976; Ito *et al.*, 1976; Shibata *et al.*, 1973; Sparrow and van Bockxmeer, 1972). However, the current opinion is that since troponinlike components are not detected in Ca^{2+}-sensitive actomyosin, they could not constitute the primary regulatory system. This does not exclude a troponinlike system in smooth muscle, but merely suggests that if troponin does exist, its function must be secondary to, or complementary to, another regulatory mechanism. In this regard, it is interesting to note that Marston *et al.* (1980) have reported dual regulation in smooth muscle.

The next discovery that was critical to the smooth-muscle field came from Szent-Györgyi and his colleagues, who were working with molluscan muscle. They discovered that the regulation of molluscan actomyosin was due to the binding of Ca^{2+} to two of the myosin light chains (Kendrick-Jones *et al.*, 1970, 1976; Szent-Györgyi *et al.*, 1973) and was therefore linked to the myosin molecule. An elegant test was devised to distinguish between myosin-linked and actin-linked systems (Lehman *et al.*, 1972; Lehman and Szent-Györgyi, 1975), and using this procedure, Bremel (1974) found that the control system in chicken-gizzard actomyosin was myosin-linked. However, the system in smooth muscle proved not to be as simple as that in molluscan muscle, because as the smooth muscle myosin is purified, its actin-activated ATPase activity is reduced. It became apparent that additional factors are required and that these are removed during the purification procedures. These factors are now recognized by most investigators to be the regulatory components, since activation occurs only in the presence of Ca^{2+}. Despite this complication, the system still retains its classification as myosin-linked, because it is thought that the activating process is focused on the myosin molecule. The existence of a myosin-linked system in chicken gizzard was confirmed (Bremel *et al.*, 1977; Hartshorne *et al.*, 1977a; Ikebe *et al.*, 1977; Sobieszek and Small, 1976) and extended to actomyosins from other smooth muscles (Borejdo and Oplatka, 1976; Frederiksen, 1976; Mrwa and Rüegg, 1975; Takeuchi and Tonomura, 1977). Although most investigators accept that the regulatory mechanism in smooth muscle is myosin-linked, this opinion is not unanimous (Ebashi *et al.*, 1975).

3.1. Requirement for Regulation

Although there is some disparity as to the focus of the regulatory system, there is universal agreement on the basic requirements for regulation in smooth muscle. In the skeletal-muscle system, a mixture of pure myosin and actin will possess an Mg^{2+}-ATPase activity that is close to maximal, and the function of the troponin–tropomyosin complex is to inhibit activity in the absence of Ca^{2+}. In the presence of Ca^{2+}, there is no inhibition, and sometimes even a slight activation. In the smooth-muscle system, the mode of regulation is quite distinct. A mixture of pure myosin and actin from smooth muscle has negligible Mg^{2+}-ATPase activity, and the function of the regulatory proteins is to activate ATPase activity, but only in the presence of Ca^{2+}. Obviously, a dormant complex cannot be regulated, and therefore the primary requirement for regulation in smooth muscle is an activation of the actomyosin complex. This much is accepted, and the controversy that exists is centered on the identity of the activating factor. There are basically two theories. The most popular theory is that activation is achieved as a result of the phosphorylation of the 20,000-dalton light chains of myosin. The minority opinion, held by Ebashi and his colleagues, is that regulation is due to an actin-linked system called leiotonin (see Section 3), which does not involve the phosphorylation of myosin.

Conceptually, the simplest mode of regulation is that found in many invertebrates (Lehman and Szent-Györgyi, 1975) where the Ca^{2+}-free light chain is inhibitory. In skeletal muscle, the Ca^{2+}-binding site is shifted to the thin filament, and here it is the Ca^{2+}-free troponin complex that is inhibitory. If one considers only the phosphorylation theory of regulation in smooth muscle, then it may be proposed that the dominant role of regulation by Ca^{2+} is lost and phosphorylation acts as its substitute. The dephosphorylated light chain inhibits the Mg^{2+}-ATPase activity of actomyosin, and the inhibition is relieved by phosphorylation. Thus, in this sense, phosphorylation acts to derepress the actomyosin complex rather than as a true activator. Superficially, the leiotonin system would appear to be similar to the classic skeletal-muscle mode of regulation wherein Ca^{2+}-binding to some component of the thin filament relieves the inhibition. However, the distinction between the two mechanisms is that actin plus myosin forms an active complex in skeletal muscle and an inactive complex in smooth muscle. Thus, leiotonin is called upon both to activate or derepress the myosin molecule and to regulate the thin filament.

3.2. Phosphorylation Theory of Regulation

The hypothesis that the phosphorylation of the myosin molecule forms the key event in the activation of the contractile apparatus is the more popular of the two theories. Sobieszek (1977a) and Bremel *et al.* (1977) first reported that chicken-gizzard myosin and actomyosin were phosphorylated on the 20,000-dalton light chains of the myosin molecule and that this event regulated the actin–myosin interaction. Also, the Ca^{2+} requirements for phosphorylation and activation of ATPase activity were shown to be similar. Since these observations were made, there have been many reports of the phosphorylation of smooth-muscle myosin (Aksoy *et al.*, 1976; Barron *et al.*, 1979; Chacko *et al.*, 1977; DiSalvo *et al.*, 1978; Frearson *et al.*, 1976; Gorecka *et al.*, 1976; Huszar and Bailey, 1979; Ikebe *et al.*, 1977, 1978; Sobieszek, 1977b; Sobieszek and Small, 1977). A few key facts may be abstracted from these studies and used to outline the phosphorylation theory: (1) the two 20,000-dalton light chains of the myosin molecule are phosphorylated by a specific enzyme, the myosin-light-chain kinase (MLCK); (2) 1 mole of phosphate is incorporated per light chain; (3) the MLCK is active only in the presence of Ca^{2+} at concentrations similar to those required to initiate contraction; (4) phosphorylation is thought to be a prerequisite for the activation by actin of the Mg^{2+}-ATPase activity of myosin; and (5) an additional enzyme is required to "deactivate" the myosin, and this is a myosin-light-chain phosphatase (MLCP).

In this theory, an increase in the intracellular Ca^{2+} concentration activates the MLCK, which phosphorylates the myosin light chains and initiates the contractile process. Cross-bridge cycling will continue as long as Ca^{2+} is present and tension development, or shortening, occurs. When the Ca^{2+} level is descreased, the MLCK becomes inactive, the MLCP dephosphorylates the myosin, and the muscle relaxes.

There are several components of this simple cyclic mechanism that should be considered in more detail, and these are presented below.

3.2.1. Myosin-Light-Chain Kinase

During the earlier phase of the work on the regulatory mechanism of smooth muscle, it was necessary to identify and isolate the proteins associated with the MLCK. It was concluded from the studies of Dabrowska *et al.* (1977)

that the MLCK of chicken gizzard is composed of two subunits of molecular weights approximately 105,000 and 17,000. Neither subunit alone possesses any kinase activity. Subsequently, it was discovered (Dabrowska *et al.*, 1978) that the smaller subunit is calmodulin, which is known to bind Ca^{2+} and has recently been shown to regulate several processes (Cheung, 1980; Means and Dedman, 1980). The calmodulin dependence of the MLCK was confirmed by Adelstein *et al.* (1978) using turkey-gizzard smooth muscle, and this was also reported for the skeletal-muscle system by Yazawa *et al.* (1978). Many reports have now appeared that document the calmodulin dependence of kinases from different sources (Dabrowska and Hartshorne, 1978; Hathaway and Adelstein, 1979; Nairn and Perry, 1979; Waisman *et al.*, 1978; Walsh *et al.*, 1979; Yerna *et al.*, 1979). A striking difference among the various MLCKs is the size of the larger subunit. Several examples are listed in Table 2. One factor suggested as being at least partly responsible for the range of observed molecular weights is limited proteolysis (Adelstein *et al.*, 1978; Walsh *et al.*, 1979).

Some of the kinetic parameters associated with the MLCKs are summarized in Table 2. In general, the V_{max} values are similar, except for that of the MLCK isolated from cardiac muscle, which is considerably less active than the other kinases. However, this value could be reduced as a result of proteolysis and may not be a valid reflection of the native enzyme (Walsh *et al.*, 1979). Several of the listed V_{max} values were estimated using isolated myosin light chains as the phosphate acceptor, and at least in the case of chicken-gizzard MLCK, the rate of phosphorylation is slower when whole myosin is used (Mrwa and Hartshorne, 1980). The choice of substrate may also be a factor in determining the K_m for the light chains. When whole myosin was used, with a partially purified MLCK from trachea, the K_m was found to be 8 μM; with skeletal-muscle myosin and MLCK, the K_m was 2 μM (Stull *et al.*, 1978). (The concentration of myosin in the smooth-muscle cell is about 40 μM (see Table 1) and the concentration of light chains about 80 μM.)

One of the reasons for examining the kinetic properties of the MLCK from chicken gizzard was to search for features in common with the actomyosin ATPase characteristics. To date, no marked similarities have emerged. The temperature coefficient (Q_{10}) for actomyosin ATPase activity is close to 3 (Driska and Hartshorne, 1975), whereas that for the MLCK is 2 (Mrwa and Hartshorne, 1980); actomyosin ATPase activity requires a relatively high free Mg^{2+} concentration, and the MLCK does not (Hartshorne *et al.*, 1980); the pH profiles for the two activities are distinct (Driska and Hartshorne, 1975; Mrwa and Hartshorne, 1980); and the effect of sulfhydryl modification is also different for the two systems (Hartshorne *et al.*, 1980). Thus, a link between the two enzymatic activities was not apparent from these studies. However, one might not expect the ATPase behavior to reflect the phosphorylation characteristics unless the rate of phosphorylation is slower than, or equal to, the subsequent rate of ATP hydrolysis of actomyosin. Assuming a value of 20 μmoles P_i transferred/min per mg kinase as representative of V_{max} for the MLCK, and assuming an M_r of 105,000, a turnover

Table II. Properties of Myosin-Light-Chain Kinases Isolated from Different Tissues

Source	Molecular weight	K_m (μM ATP)	V_{max} (μmoles·min⁻¹· mg⁻¹ kinase)[a]	K_m (μM light chain)	K_d (nM calmodulin)	References
Chicken gizzard	105,000	60-70	5-15	—	—	Dabrowska et al. (1977), Mrwa and Hartshorne (1980), Hartshorne et al. (1980)
Turkey gizzard	125,000-130,000	—	33	—	—	Adelstein et al. (1978), Conti and Adelstein (1980)
Skeletal muscle	77,000-80,000	200-400	15-30	40-50	—	Nairn and Perry (1979), Pires and Perry (1977)
	80,000	280	4.3	24	6.6	Yazawa and Yagi (1978)
	155,000	—	—	—		Walsh et al. (1981)
Cardiac muscle	85,000	175	0.03	21	1.3	Walsh et al. (1979)
	155,000	—	—	—	2.2	Walsh et al. (1981)
Blood platelets[b]	105,000	121	≮3.1	18	—	Hathaway and Adelstein (1979)

[a] Values given for 25°C and adjusted to this temperature if necessary, assuming a Q_{10} of 2.
[b] Data obtained using a partially purified enzyme.

number of about 35 sec^{-1} can be calculated. For an actomyosin-specific ATPase activity of 0.15 μmole P$_i$/min per mg myosin, the turnover number is about 1.2 sec^{-1}. Thus, even though the rate of phosphorylation with intact myosin is slower than that chosen here, it is unlikely to be rate-limiting to the hydrolysis of ATP by actomyosin. The amount of MLCK in the muscle cell is pertinent to this conclusion, since obviously the phosphorylation rate could be reduced if it were first-order with respect to the kinase concentration. This value is not established, but in our laboratory, the yield of the purified MLCK from chicken gizzards is approximately 2 mg from 100 g tissue. A minimum concentration of the MLCK (assuming no preparative losses) would be about 0.2 μM, which is at least 100-fold higher than the concentrations of MLCK used to estimate the value of V_{max}, and thus the *in vivo* level of the MLCK is probably not rate-limiting.

Another factor that could affect the level of light-chain phosphorylation is the phosphorylation of the larger subunit of the MLCK. Adelstein *et al.* (1978) and Conti and Adelstein (1980) found that this is achieved by the cyclic AMP (cAMP)-dependent protein kinase and that phosphorylation results in a decreased affinity for calmodulin and hence in a reduction in the MLCK rate. More recently, two sites of phosphorylation have been identified, termed A and B (Adelstein *et al.*, 1981). Both sites are phosphorylated in the absence of calmodulin, but only the B site is labeled when calmodulin is bound to the kinase. Only the phosphorylation of the A site results in the reduction of the affinity for calmodulin. Originally, it was suggested (Adelstein *et al.*, 1978) that an increase in the cAMP level within the cell might exert a dual physiological function, part of which would be to lower the concentration of available Ca^{2+} and the other part to modify the contractile apparatus via the cAMP-dependent kinase. With respect to the latter suggestion, some support is derived from the following: the addition of the catalytic subunit of the cAMP-dependent kinase resulted in a decrease in the Ca^{2+}-activated tension that was reversed on the addition of calmodulin (Hoar and Kerrick, 1980), and an inhibition of the Mg^{2+}-ATPase activity (Mrwa *et al.*, 1979; Silver and DiSalvo, 1979) and of myosin phosphorylation (Silver and DiSalvo, 1979) was obtained following the addition of cAMP to smooth-muscle actomyosins that contained the cAMP-dependent kinase.* Despite the apparent acceptance of this idea, there are some difficulties associated with it. The most obvious is that phosphorylation of the A site must occur in order to reduce the affinity for calmodulin, and this site is blocked by the Ca^{2+}-calmodulin complex. Obviously, at saturating levels of calmodulin, phosphorylation of the A sites would be prevented, and for a reduction in the activity of the MLCK to occur, the calmodulin concentration must be submaximal. The level of available calmodulin at different stages of the contraction cycle is not established, but even allowing that it could be less than stoichiometric with the larger subunit of the MLCK [the stoichiometry of the MLCK complex is 1 : 1 (Hartshorne *et al.*,

---

*In the absence of the cAMP-dependent protein kinase, neither cAMP nor cGMP (up to 100 μM) affected the MLCK activity (Hartshorne *et al.*, 1980).

1980)], there remains a problem associated with the rate constants for the formation of the complex. In simple terms, Ca^{2+}–calmodulin must dissociate from its partner before phosphorylation of the A site can occur. The reassociation of Ca^{2+}–calmodulin must therefore be slow enough to allow phosphorylation by the cAMP-dependent kinase. The "on" and "off" rates have not been calculated for the MLCK complex, but judging from the constants given in Table 2, a slow "on" rate would not be expected. Finally, the relationship between myosin phosphorylation and the contractile cycle in smooth muscle, especially in the relaxation phase (see Section 3.2.7), has not been established, and it is therefore difficult to assess the "cAMP theory" from a physiological viewpoint.

3.2.2. Site of Phosphorylation by the MLCK

Each of the two 20,000-dalton light chains on the smooth-muscle myosin molecule is phosphorylated, and 1 mole of phosphate can be incorporated per light chain. The sequence around the phosphorylated serine is given in Fig. 1, and this is compared to corresponding sequences from rabbit and chicken skeletal-muscle light chains [data are given for the L2, or 5,5'-dithiobis-(2-nitrobenzoic acid) (DTNB)–light chain]. The sequence to the C-terminal side of the serine is very similar, although the sequence toward the N terminus shows some variation. One common feature, however, is the presence on the N-terminal side of an adjacent hydroxylamino acid, either serine or threonine. The substrate requirements for the MLCK, though not established, are clearly distinct from those of the cAMP-dependent protein kinase (Fig. 1), which prefers probably two basic residues one or two amino acids on the N-terminal side of the serine (Glass and Krebs, 1980; Yeaman *et al.*, 1977).

Also shown in Fig. 1 are the sites of hydrolysis by papain and α-chymotrypsin for the gizzard light chain. Both enzymes release an N-terminal peptide and leave a core of about 17,000 daltons. In the case of α-chymotrypsin, the phosphorylation site is removed with the peptide, but this is retained by the larger fragment when papain is used. It is interesting, however, that the "17,000 papain fragment" cannot be phosphorylated by the MLCK (Jakes *et al.*, 1976), which indicates that the N-terminal region that is removed may contain substrate determinants.

The specificity of the MLCK is quite high, and it will not phosphorylate histone, casein, phosphorylase b, or phosphorylase kinase (Adelstein and Eisenberg, 1980). An apparent exception to this is found with the kinase described by Waisman *et al.* (1978), which phosphorylates light chains, histone, and phosphorylase kinase. It remains to be determined, however, whether or not this kinase is indeed the MLCK or another calmodulin-dependent protein kinase that has yet to be characterized. In general, the light chains from different muscles can be phosphorylated by an MLCK, although a preference is noted for the homologous system (Adelstein and Eisenberg, 1980). This tendency is more marked when myosin, instead of isolated light chains, is used as the substrate.

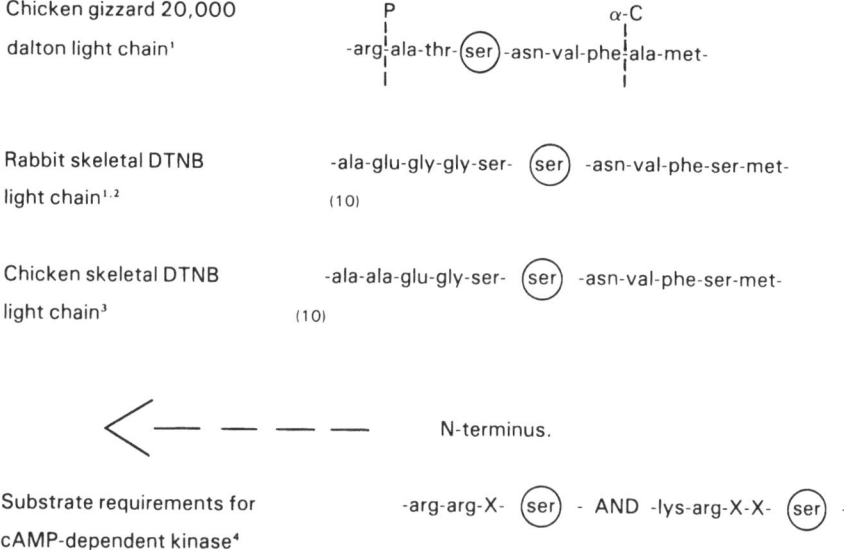

Figure 1. Comparison of the sequence around the phosphorylated serine residues for the chicken gizzard 20,000-dalton light chain, the rabbit skeletal DTNB–light chain, and the chicken skeletal DTNB–light chain. These are compared with the preferred sequence for the cAMP-dependent kinase. The phosphorylation sites are circled. The vertical dashed lines in the sequence for the gizzard light chain indicate the peptide bonds hydrolyzed by papain (P) and α-chymotrypsin (α-C). For the skeletal-muscle light chains, the 10th amino acid from the N terminus is indicated. References: (1) Jakes *et al.* (1976); (2) Perrie *et al.* (1973); (3) Matsuda *et al.* (1977); (4) Glass and Krebs (1980).

3.2.3. Effect of Phosphorylation

It is widely accepted that the phosphorylation of myosin allows the activation by actin of the Mg^{2+}-ATPase activity. An evaluation of this statement will be made in Section 3.2.7, but it is convenient at this stage to document the various observations. The initial report came from Sobieszek (1977a), who correlated the activation of the Mg^{2+}-ATPase activity of gizzard actomyosin with the phosphorylation of myosin. Subsequently, several investigators have shown a similar dependence using actomyosin (Aksoy *et al.*, 1976; Small and Sobieszek, 1977b; Sobieszek, 1977b), myofibrils (Sobieszek, 1977b), and partially purified components (Chacko *et al.*, 1977; Dabrowska *et al.*, 1977; Gorecka *et al.*, 1976; Huszar and Bailey, 1979; Ikebe *et al.*, 1978). A linear relationship between the extent of myosin phosphorylation and ATPase activity was found with gizzard actomyosin and myofibrils (Sobieszek, 1977b) and also from swine arterial actomyosin (Driska *et al.*, 1981). With a more physiologically oriented approach, a correlation between the development of tension and myosin phosphorylation has also been noted for various types of smooth-muscle-fiber preparations (Barron *et al.,* 1979, 1980; Driska *et al.,* 1981; Hoar *et al.,* 1979; Janis and Gualteri, 1978).

The effect of phosphorylation on the Ca^{2+}- and K^+-EDTA ATPase activities of myosin is not marked, although some effect has been noted for the Mg^{2+}-ATPase activity in the absence of actin (Suzuki *et al.,* 1978). As noted in Section 2.2.1, filaments of phosphorylated myosin are more resistant to dissociation by Mg^{2+}-ATP than filaments of dephosphorylated myosin, and it was suggested (Suzuki *et al.,* 1978) that the state of aggregation of myosin is related to ATPase activity, with the filamentous form showing higher activity. There is no direct evidence to indicate that the conformation of smooth-muscle myosin is altered on phosphorylation, although this has been reported with the light chains from skeletal-muscle myosin, and it was noted further that these changes were reflected in an alteration of the Ca^{2+}-binding properties of the light chains (Alexis and Gratzer, 1978). Other investigators, using whole myosin from cardiac muscle (Holroyde *et al.,* 1978; Kuwayama and Yagi, 1979) and skeletal muscle (Holroyde *et al.,* 1978), found no effect of phosphorylation on Ca^{2+}-binding properties.

Both skeletal and cardiac muscle contain an MLCK, although the function of myosin phosphorylation in these systems is not understood. Morgan *et al.* (1976) found that the phosphorylation of the light chains in striated muscle is not accompanied by a significant activation of the actin-activated ATPase activity. Recently, however, Pemrick (1980) found that the effect of phosphorylation of rabbit skeletal myosin is primarily to increase the affinity for actin-binding (i.e., a decrease in K_{app}) with no marked effect on the V_{max} of the actin-activated ATPase activity. Preliminary results with gizzard myosin (Siemankowski and Hartshorne, unpublished data) indicate that a similar situation exists with the smooth-muscle system. If these observations are accepted, this would mean that the activation of Mg^{2+}-ATPase activity of myosin as a result of phosphorylation would be partly dependent on the concentration of actin used in the assay (Pemrick, 1980). An effect that could be related to the change in K_{app} for actin is the finding of Manning and Stull (1979) that light-chain phosphorylation in rat extensor digitorum longus muscle is not obligatory for contraction, but may play a role in posttetanic potentiation.

Clearly, there is much to be learned, even at a very basic level, about the effects of phosphorylation on the properties of myosin. For example, with smooth-muscle myosin, it is not known whether the phosphorylation of one head or both heads is required for activation of ATPase activity. There is also the question of how the phosphorylation of the light chains can influence the ATPase sites, or actin-binding sites, which are thought to be located on the heavy chains. It is convenient to postulate that phosphorylation induces in the light chain a conformational change that is then transmitted to the heavy chains (in a sense, this effect is similar to that shown in allosteric systems), but there is no evidence for this and the molecular consequences of phosphorylation remain to be established.

3.2.4. Myosin-Light-Chain Phosphatase

In the phosphorylation theory, the role of the MLCP is to effect the deactivation of the contractile apparatus when the intracellular Ca^{2+} level is reduced. It is assumed that the MLCP is active both in the absence and in the presence of Ca^{2+}, although a net dephosphorylation of myosin will occur only when the MLCK is inhibited by the removal of Ca^{2+}. If correct, this assumption requires that the *in vivo* level of MLCP activity be considerably less than the level of MLCK activity. Although the *in vivo* activities have not been established, MLCP activity has been detected in several smooth-muscle systems (Aksoy *et al.*, 1976; Chacko *et al.*, 1977; Ikebe *et al.*, 1978; Onishi *et al.*, 1979; Sherry *et al.*, 1978), and in general, the phosphatase rates are lower than the kinase rates.

A MLCP was isolated first from skeletal muscle (Morgan *et al.*, 1976) and found to be a single subunit of M_r about 70,000. More recently, two phosphatases, termed I and II, have been isolated from turkey gizzard (Pato and Adelstein, 1980). Phosphatase I is composed of three subunits in an equimolar ratio of molecular weights, 60,000, 55,000, and 38,000, and phosphatase II is a single polypeptide with a molecular weight of 43,000. Both catalyzed the dephosphorylation of light chains at similar rates, 0.53 and 0.32 μmole/min per mg at 30°C for I and II, respectively. The substrate specificity of the two enzymes, however, is quite distinct. Phosphatase II dephosphorylates only myosin light chains (MLCK, histone IIA, and casein were also tested), whereas phosphatase I is less specific and dephosphorylates the light chains and the phosphorylated MLCK larger subunit (Pato and Adelstein, 1980).

At present, there is no evidence from *in vitro* data to indicate that the MLCP is regulated in its native state. Regulatory effects have been proposed with other protein phosphatases (for a review, see Lee *et al.*, 1980), and similar possibilities should not be excluded for the MLCP.

3.2.5. Role of Ca²⁺

Following excitation of the muscle, the level of intracellular Ca^{2+} increases, Ca^{2+} binds to calmodulin, and the Ca^{2+}–calmodulin complex activates the larger subunit of the MLCK. It has been proposed recently for the MLCK (Blumenthal and Stull, 1980) and for phosphodiesterase (Crouch and Klee, 1980) that three to four molecules of Ca^{2+} need to be bound by calmodulin before the latter will activate enzymatic activity. In the simplest concept, therefore, the role of Ca^{2+} is directed only at one site, namely, calmodulin. A slightly more complex situation was proposed (Chacko *et al.*, 1977) in which Ca^{2+} binding by the myosin molecule was also a requirement for the normal functioning of smooth muscle. This idea derives some support from the observations that several invertebrate myosins are regulated by the binding of Ca^{2+} to the light chains (Szent-Györgyi *et al.*, 1973), and it is also known that smooth-muscle myosin can bind Ca^{2+} (Bremel *et al.*, 1977; Sobieszek and

Small, 1976; Hirata *et al.*, 1980). Thus, the controversy that evolved was whether the phosphorylation–dephosphorylation of myosin is adequate to account for regulation or whether an additional component, i.e., the interaction of Ca^{2+} with myosin, is also required. To choose between the alternatives, the actin-activated Mg^{2+}-ATPase activity of phosphorylated myosin should be assayed in the absence and in the presence of Ca^{2+}. If phosphorylation alone is adequate, then no difference in the two assays would be observed; if Ca^{2+} binding by myosin is required, then the ATPase activity in the absence of Ca^{2+} would be inhibited. Although the idea is simple, its practical application proved more difficult. The major complication was the presence of contaminating kinase or phosphatase or both in the protein preparations. Thus, the approach that was adopted was to design experiments in which the effects of these contaminants would be reduced, and this led to the use of adenosine 5'-O(3-thiotriphosphate) (ATPγS). This ATP analogue can be utilized by the MLCK, and a thiophosphate group is attached to the 20,000-dalton light chain. The latter, however, serves as a poor substrate for phosphatases (Gergely *et al.*, 1976; Gratecos and Fischer, 1974; Morgan *et al.*, 1976). The net result is that myosin is "frozen" in the thiophosphorylated state and may be assayed without complications arising from contamination by phosphatases. It was found that as the extent of thiophosphorylation increased, the degree of Ca^{2+} sensitivity decreased and the ATPase activity of the actomyosin in the absence of Ca^{2+} approached that in the presence of Ca^{2+}; i.e., the myosin was locked in an active but unregulated state (Sherry *et al.*, 1978). The effects of ATPγS have also been tested with mechanically disrupted chicken-gizzard fibers (Hoar *et al.*, 1979) and skinned rabbit ileum strips (Cassidy *et al.*, 1979), and it was shown that an increase in the extent of thiophosphorylation results in the loss of Ca^{2+} regulation and the generation of an active state. The conclusion that was derived from these experiments is that phosphorylated myosin is active in both the absence and the presence of Ca^{2+} and that an additional regulatory component associated with the interaction of Ca^{2+} with myosin is not required (see Section 3.2.7 for an evaluation of this statement). Other investigators share this opinion, although their experimental procedures were different (Ikebe *et al.*, 1978; Small and Sobieszek, 1977a,b). It should be pointed out, however, that the work of Chacko *et al.* (1977), which suggested the importance of the Ca^{2+}–myosin interaction, was done using vas deferens, and this tissue has not been analyzed by other investigators.

Another factor that should be considered is the amount of Ca^{2+} that can be bound to myosin in the presence of Mg^{2+}, i.e., under conditions approaching those encountered *in vivo*. Numerous investigators have shown with skeletal and cardiac myosins that Mg^{2+} influences Ca^{2+} binding (Bremel and Weber, 1975; Holroyde *et al.*, 1979; Kuwayama and Yagi, 1979; Watterson *et al.*, 1979). In general, the effect of Mg^{2+} is to reduce Ca^{2+} binding. For example, Kuwayama and Yagi (1979) found that at 1×10^{-5} M free Ca^{2+}, in the absence and in the presence of 4.5 mM Mg^{2+}, the amount of Ca^{2+} bound to pig cardiac myosin was 1.3 and 0.2 moles per mole myosin, respectively. Using rabbit skeletal myosin, Watterson *et al.* (1979) calculated that at 1×10^{-5} M free Ca^{2+} and 1 mM Mg^{2+}, only about 50% of the Ca^{2+} sites would be occupied. Mg^{2+} competition with the Ca^{2+} binding of smooth-muscle myosin is not as well documented, although Hirata *et al.* (1980) found that with bovine aorta myosin, the amount of Ca^{2+} bound at 1×10^{-5} M Ca^{2+}, in the presence of 8 mM Mg^{2+}, was insignificant. These results contrast to those reported for molluscan myosin, where even in the presence of Mg^{2+}, significant amounts of Ca^{2+} are bound (Kendrick-Jones *et al.*, 1970). In summary, because the amount of Ca^{2+} bound to smooth-muscle myosin under physiological conditions is relatively low, it is unlikely that the interaction of Ca^{2+} with the myosin light chains is a dominant component of the regulatory mechanism in vertebrate smooth muscle as it is in many invertebrate muscles.

3.2.6. Summary of the Phosphorylation Theory

Much of the data discussed previously is incorporated in the cyclic mechanism shown in Fig. 2. If phosphorylation provides the *only* means of regulating the actin–myosin interaction, and hence cross-bridge cycling, then there are several features that are predictable. The first is that contraction of the muscle must be preceded by the phosphorylation of myosin; also, the cross-bridge cycling rate should be proportional to the degree of phosphorylation; and the rate of relaxation should correspond to the rate of myosin dephosphorylation. As will be indicated in the following section, these predictions are satisfied only in part by the experimental observations.

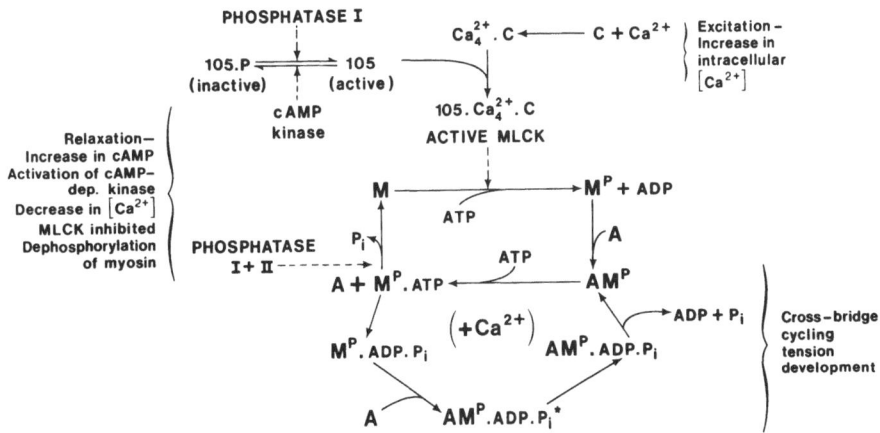

Figure 2. Summary scheme to illustrate the phosphorylation theory of regulation in smooth muscle. Abbreviations: (M, M^P) Dephosphorylated and phosphorylated myosin, respectively; (A) actin; (C) calmodulin; (105) apoenzyme of the myosin-light-chain kinase (MLCK). Some of the intermediates are shown for the hydrolysis of ATP by actomyosin. It is thought that the transition between the two intermediates for the actin–myosin-products complex represents a rate-limiting step (Marston and Taylor, 1978). Cross-bridge cycling, i.e., hydrolysis of ATP, will continue as long as Ca^{2+} is present. The possible relationship of the cAMP-dependent kinase and phosphatase I to the regulatory mechanism is also indicated (see the text).

3.2.7. Evaluation of the Phosphorylation Theory

The initial evidence suggesting a regulatory function for myosin phosphorylation was that as the extent of phosphorylation increased, the level of actin-activated ATPase activity also increased. This general relationship has been confirmed with actomyosins from several different smooth muscles and has been extended to the correlation of tension with phosphorylation in various muscle-fiber preparations (see Section 3.2.3.). Thus, the cumulative evidence indicates that the activation of ATPase activity, or the initiation of the contractile process, is associated with the phosphorylation of myosin. However, these studies are not sufficient to allow a conclusion that phosphorylation of myosin is the only factor involved. Primarily, this is because none of the preparations that were used was adequately defined with respect to its protein components, and an effect due to an "unknown" factor cannot be eliminated. Ebashi and collaborators have maintained for several years that the level of myosin phosphorylation does not necessarily parallel the extent of ATPase activity (Mikawa *et al.*, 1977), and to some extent data obtained in our laboratory agree with this statement. We observed that it is possible to prepare phosphorylated myosin that exhibits a very low activation by actin, and full activation is achieved only when additional components are added (Mrwa, Persechini, and Hartshorne, unpublished observations). Evidence from more physiologically oriented studies also suggests that additional factors might be involved. Using carotid-artery strips, Murphy and his colleagues (Aksoy *et al.*, 1980; Driska *et al.*, 1981; Murphy *et al.*, 1980) found that the tension developed on stimulation of the muscle was preceded by the phosphorylation of myosin, but the extent of phosphorylation declined significantly before maximum force was attained. Further, they found that maximum force could be maintained over long periods of time during which the extent of myosin phosphorylation was reduced to levels only slightly above those in the unstimulated tissue. The interpretation from these results was that phosphorylation regulated the maximum shortening velocity and thus the cross-bridge cycling rate, but that other factors could be involved in the maintenance of tonic isometric force. One factor that was considered is the formation of noncycling-cross-bridge–actin interactions that effectively could "freeze" the muscle at any given tension and maintain that tension with little energy cost (i.e., hydrolysis of ATP). This hypothesis is not new, and the Ca^{2+}-dependent formation of noncycling cross-bridges has been noted before (Siegman *et al.*, 1976a,b). What initiates the formation of the noncycling cross-bridge and how it is eventually dissociated is not known. In a study using tracheal smooth muscle, deLanerolle and Stull (1980) showed that myosin phosphorylation coincided temporally with the increase in isometric tension, although during relaxation, induced by atropine, the decrease in tension occurred faster than the decrease in the level of phosphorylation. These authors pointed out that a process other than the dephosphorylation of myosin may be important in mediating the relaxation response.

At this point, the general impression that is conveyed is that activation either of the Mg^{2+}-ATPase activity of actomyosin or of tension development seems to be intimately associated with the phosphorylation of myosin. However, there is also evidence to suggest that phosphorylation alone may not be adequate to fully account for the active state of actomyosin, and an additional system may be implicated.

It is interesting to test the validity of the preceding statement by considering it in light of other experimental observations that have been proposed as supportive of the phosphorylation scheme. One of these is the application of various phenothiazine derivatives, and the other is the use of ATPγS.

It was shown several years ago that phenothiazine derivatives were bound specifically by calmodulin (Levin and Weiss, 1977) and thus inhibited several calmodulin-dependent processes (Levin and Weiss, 1979), including MLCK (Hidaka *et al.*, 1979). When tested with smooth-muscle preparations, the phenothiazines inhibited tension development with skinned intestinal and arterial muscle (Kerrick *et al.*, 1980) and with intact arterial strips (Barron *et al.*, 1980) and also elicited the relaxation of Ca^{2+}-induced contractions in skinned taenia coli (Crosby and Diamond, 1980). The phosphorylation (Barron *et al.*, 1980) and thiophosphorylation (Kerrick *et al.*, 1980) of the myosin light chains were also prevented. With chicken-gizzard myosin and actomyosin, it was shown that these compounds inhibit both the phosphorylation of myosin and the Mg^{2+}-ATPase activity of actomyosin (Hidaka *et al.*, 1979, 1980; Sheterline, 1980). From these results, it would appear that the relationship between myosin phosphorylation and the activation of the contractile process is firmly established. The most reasonable explanation for the observed results is that the phenothiazines inhibit the MLCK and thus allow dephosphorylation by the MLCP. Once again, however, it should be pointed out that none of these experiments was performed using clearly defined systems, and one criticism that can be raised is that the site of action of the phenothiazine cannot be identified unequivocally. One is faced with the possibility that the phenothiazines may inhibit an unidentified process that is involved in the regulatory mechanism. In particular, it is important to establish what effect, if any, phenothiazines have on the leiotonin system (see Section 3.3).

The original objective motivating the use of ATPγS was to investigate the role of the Ca^{2+}-myosin interaction in the regulatory process (see Section 3.2.5). Once this was achieved, it was realized that these results were of additional significance, since they offered strong support for the phosphorylation theory. The use of ATPγS and the finding that the thiophosphorylation of myosin was correlated to the loss of Ca^{2+} sensitivity and the activation of ATPase activity in the absence of Ca^{2+} removed much of the uncertainty concerning the identity of the Ca^{2+}-dependent reaction that was involved in the regulatory process. It could be argued that the similarity of the Ca^{2+} requirements for the phosphorylation of myosin and for the activation of actomyosin ATPase activity (Aksoy *et al.*, 1976; DiSalvo *et al.*, 1978; Small and Sobieszek, 1977b) might be fortuitously similar and that the two processes are regulated by two independent systems that happen to possess similar Ca^{2+}

requirements (the multiplicity of reactions that are regulated by calmodulin lends credence to this possibility). The use of ATPγS removed most of this ambiguity and thus strengthened the hypothesis that the phosphorylation of myosin was a prerequisite for activation. One can speculate that if a second mechanism does exist, it might also be a kinase-linked reaction.

In summary, the evidence presented above is extremely strong in support of the involvement of a phosphorylation mechanism in the regulation of the contractile activity of smooth muscle, and it is most likely that light-chain phosphorylation is implicated in this mechanism. The evidence, however, cannot restrict the regulatory mechanism to only the phosphorylation and dephosphorylation of the myosin light chains, and there are suggestions that an additional mechanism is involved. The identity of the second mechanism is not established, but it should be inhibited by the phenothiazines, and if the ATPγS results are accepted, then it might also involve a kinase.

3.3. Leiotonin Theory of Regulation

The second, although less popular, system that has been proposed as the regulatory mechanism in smooth muscle is leiotonin. For comparative purposes, some properties of the phosphorylation and leiotonin theories are listed in Table 3. The most important difference between the two theories is that leiotonin is not functional via the phosphorylation of the myosin light chains (Mikawa *et al.*, 1977), and Ebashi and his colleagues (Ebashi *et al.*, 1977; Nonomura and Ebashi, 1980) are of the opinion that the phosphorylation of myosin is not involved in the regulatory mechanism of smooth muscle.

It would be convenient if the role of leiotonin could be compromised with that of the phosphorylation theory, possibly as the elusive second mechanism, although this cannot be achieved at present because of the lack of agreement concerning the role of myosin phosphorylation. Obviously, a priority for future research is to evaluate the roles of each system and to determine whether each is independent of the other or whether they interact to form a cooperative mechanism of regulation.

3.4. Summary of the Regulatory Mechanism in Smooth Muscle

It is accepted that contractile activity in smooth muscle is regulated by fluctuations in the intracellular concentration of Ca^{2+}. The increase or decrease of free Ca^{2+} is sensed by some component associated with the contractile apparatus, with the result that the cross-bridge–actin interactions are either promoted or dissociated. The identity of the system that fulfills this function is the subject of controversy, and two possibilities have been suggested. The most popular theory is that regulation of smooth-muscle actomyosin is achieved by the phosphorylation and dephosphorylation of the myosin molecule; the alternative viewpoint is that regulation is due to a system termed leiotonin.

The experimental evidence in support of the phosphorylation theory is

Table 3. Comparison of the Phosphorylation and Leiotonin Theories of Regulation in Smooth Muscle

Characteristics	Leiotonin[a]	Phosphorylation	Remarks
Regulatory components	Leiotonin A, M_r 80,000 Leiotonin C, M_r 17,000 Ca^{2+}-binding protein	MLCK (apoenzyme + calmodulin) MLCP (phosphatases I and II)	Leiotonic C similar to, but not identical with, calmodulin and troponin C.
Location	Thin filament	Unknown (probably soluble)	—
Stoichiometry (*in vitro* requirements)	1 Leiotonin 10 Tropomyosins	Unknown	If this reflects *in vivo* concentration, there would be about 5 leiotonin molecules per thin filament.
Mechanism	Actin-linked. Requires tropomyosin. Mechanism unknown; does not involve phosphorylation of myosin.	Myosin-linked. Involves phosphorylation and dephosphorylation of myosin.	Calmodulin can substitute for leiotonin C.

[a] Data taken from Nonomura and Ebashi (1980).

considerable, and one must conclude that the phosphorylation of the myosin light chains forms an integral part of the regulatory mechanism in smooth muscle. Some aspects of the contraction–relaxation cycle that are compatible with the phosphorylation theory are that contraction does not occur in the absence of phosphorylation, that phosphorylation precedes tension development, and also, it appears, that phosphorylation regulates the cross-bridge cycling rate (Aksoy *et al.*, 1980; Murphy *et al.*, 1980). However, there are flaws in the theory. There is no simple relationship between the extent of myosin phosphorylation and the level of isometric tension (Barron *et al.*, 1980; Driska *et al.*, 1981), and there are indications that during the relaxation phase, the decrease in tension is not correlated to the reduction of myosin phosphorylation (deLanerolle and Stull, 1980). Thus, it would appear that phosphorylation alone cannot account for all the experimental observations, and it is evident that other factors are implicated. Whether or not the complementary regulatory mechanism is related to the leiotonin system remains to be established.

ACKNOWLEDGMENT. The author is supported by Grant HL 23615 from the National Institutes of Health.

References

Adelstein, R. S., and Eisenberg, E., 1980, Regulation and kinetics of the actin–myosin–ATP interaction, *Annu. Rev. Biochem.* **49**:921.

Adelstein, R. S., Conti, M. A., Hathaway, D. R., and Klee, C. B., 1978, Phosphorylation of smooth muscle myosin light chain kinase by the catalytic subunit of adenosine 3':5'-monophosphate-dependent protein kinase, *J. Biol. Chem.* **253**:8347.

Adelstein, R. S., Pato, M. D., and Conti, M. A., 1981, The role of phosphorylation in regulating contractile proteins, *Adv. Cyclic Nucleotide Res.* **14** (in press).

Aksoy, M. O., Williams, D., Sharkey, E. M., and Hartshorne, D. J., 1976, A relationship between Ca^{2+} sensitivity and phosphorylation of gizzard actomyosin, *Biochem. Biophys. Res. Commun.* **69**:35.

Aksoy, M. O., Dillon, P. F., and Murphy, R. A., 1980, Phosphorylation of the 20,000 dalton myosin light chain (LC 20) regulates shortening velocity in vascular smooth muscle, *Fed. Proc. Fed. Am. Soc. Exp. Biol.* **39**:2042.

Alexis, M. N., and Gratzer, W. B., 1978, Interaction of skeletal myosin light chains with calcium ions, *Biochemistry* **17**:2319.

Amphlett, G. W., Syska, H., and Perry, S. V., 1976, The polymorphic forms of tropomyosin and troponin I in developing rabbit skeletal muscle, *FEBS Lett.* **63**:22.

Anderson, N. L., 1979, The β and γ cytoplasmic actins are differentially thermostabilized by Mg ADP; γ actin binds Mg ADP more strongly, *Biochem. Biophys. Res. Commun.* **89**:486.

Ashton, F. T., Somlyo, A. V., and Somlyo, A. P., 1975, The contractile apparatus of vascular smooth muscle: Intermediate high voltage stereo electron microscopy, *J. Mol. Biol.* **98**:17.

Bárány, M., Bárány, K., Gaetjens, E., and Bailin, G., 1966, Chicken gizzard myosin, *Arch. Biochem. Biophys.* **113**:205.

Barron, J. T., Bárány, M., and Bárány, K., 1979, Phosphorylation of the 20,000-dalton light chain of myosin of intact arterial smooth muscle in rest and in contraction, *J. Biol. Chem.* **254**:4954.

Barron, J. T., Bárány, M., Bárány, K., and Storti, R. V., 1980, Reversible phosphorylation and dephosphorylation of the 20,000-dalton light chain of myosin during the contraction-relaxation–contraction cycle of arterial smooth muscle, *J. Biol. Chem.* **255**:6238.

Blumenthal, D. K., and Stull, J. T., 1980, Activation of skeletal muscle myosin light chain kinase by Ca^{2+} and calmodulin, *Biochemistry* **19**:5608.

Bois, R. M., 1973, The organization of the contractile apparatus of vertebrate smooth muscle, *Anat. Rec.* **117**:61.

Bolton, T. B., 1979, Mechanisms of action of transmitters and other substances on smooth muscle, *Physiol. Rev.* **59**:606.

Borejdo, J., and Oplatka, A., 1976, Evidence for myosin-linked regulation in guinea pig taenia coli muscle, *Pfluegers Arch.* **366**:177.

Bremel, R. D., 1974, Myosin linked calcium regulation in vertebrate smooth muscle, *Nature (London)* **252**:405.

Bremel, R. D., and Weber, A., 1975, Calcium binding to rabbit skeletal myosin under physiological conditions, *Biochim. Biophys. Acta* **376**:366.

Bremel, R. D., Sobieszek, A., and Small, J. V., 1977, Regulation of actin–myosin interaction in vertebrate smooth muscle, in: *The Biochemistry of Smooth Muscle* (N. L. Stephens, ed.), pp. 533–549, University Park Press, Baltimore.

Carsten, M. E., 1971, Uterine smooth muscle: Troponin, *Arch. Biochem. Biophys.* **147**:353.

Cassidy, P. S., Hoar, P. E., and Kerrick, W. G. L., 1979, Irreversible thiophosphorylation and activation of tension in functionally skinned rabbit ileum strips by [^{35}S] ATPγS, *J. Biol. Chem.* **254**:11148.

Chacko, S., Conti, M. A., and Adelstein, R. S., 1977, Effect of phosphorylation of smooth muscle myosin on actin activation of Ca^{2+} regulation, *Proc. Natl. Acad. Sci. U.S.A.* **74**:129.

Cheung, W. Y., 1980, Calmodulin plays a pivotal role in cellular regulation, *Science* **207**:19.

Cohen, D. M., and Murphy, R. A., 1978, Differences in cellular contractile protein contents among porcine smooth muscles: Evidence for variation in the contractile system, *J. Gen. Physiol.* **72**:369.

Conti, M. A., and Adelstein, R. S., 1980, Phosphorylation by cyclic adenosine 3':5'-monophosphate-dependent protein kinase regulates myosin light chain kinase, *Fed. Proc. Fed. Am. Soc. Exp. Biol.* **39**:1569.

Cooke, P., 1976, A filamentous cytoskeleton in vertebrate smooth muscle fibers, *J. Cell Biol.* **68**:539.

Cooke, P. H., and Fay, F. S., 1972, Correlation between fiber length, ultrastructure, and the length–tension relationship of mammalian smooth muscle, *J. Cell Biol.* **52**:105.

Craig, R., and Megerman, J., 1977, Assembly of smooth muscle myosin into side-polar filaments, *J. Cell Biol.* **75**:990.

Crosby, N. D., and Diamond, J., 1980, Effects of phenothiazines on calcium induced contractions of chemically skinned smooth muscle, *Proc. West. Pharmacol. Soc.* **23**:335.

Crouch, T. H., and Klee, C. B., 1980, Positive cooperative binding of calcium to bovine brain calmodulin, *Biochemistry* **19**:3692.

Csapo, W., 1948, Actomyosin content of the uterus, *Nature (London)* **162**:218.

Cummins, P., 1979, The homology of the α-chains of cardiac and skeletal rabbit tropomyosin, *J. Mol. Cell. Cardiol.* **11**:109.

Cummins, P., and Perry, S. V., 1973, The subunits and biological activity of polymorphic forms of tropomyosin, *Biochem. J.* **133**:765.

Cummins, P., and Perry, S. V., 1974, Chemical and immunochemical characteristics of tropomyosins from striated and smooth muscle, *Biochem. J.* **141**:49.

Dabrowska, R., and Hartshorne, D. J., 1978, A Ca^{2+}- and modulator-dependent myosin light chain kinase from non-muscle cells, *Biochem. Biophys. Res. Commun.* **85**:1352.

Dabrowska, R., Aromatorio, D., Sherry, J. M. F., and Hartshorne, D. J., 1977, Composition of the myosin light chain kinase from chicken gizzard, *Biochem. Biophys. Res. Commun.* **78**:1263.

Dabrowska, R., Sherry, J. M. F., Aromatorio, D. K., and Hartshorne, D. J., 1978, Modulator protein as a component of the myosin light chain kinase from chicken gizzard, *Biochemistry* **17**:253.

DeLanerolle, P., and Stull, J. T., 1980, Myosin phosphorylation during contraction and relaxation of tracheal smooth muscle, *J. Biol. Chem.* **255**:9993.

Devine, C. E., and Somlyo, A. P., 1971, Thick filaments in vascular smooth muscle, *J. Cell Biol.* **49**:636.

DiSalvo, J., Gruenstein, E., and Silver, P., 1978, Ca^{2+} dependent phosphorylation of bovine aortic actomyosin, *Proc. Soc. Exp. Biol. Med.* **158**:410.

Driska, S. P., 1976, Calcium control of smooth muscle contractile proteins, Ph.D. thesis, Carnegie-Mellon University, Pittsburgh.

Driska, S., and Hartshorne, D. J., 1975, The contractile proteins of smooth muscle: Properties and components of a Ca^{2+}-sensitive actomyosin from chicken gizzard, *Arch. Biochem. Biophys.* **167**:203.

Driska, S., Aksoy, M. O., and Murphy, R. A., 1981, Myosin light chain phosphorylation associated with contraction in arterial smooth muscle, *Am. J. Physiol.* **240**:C222.

Ebashi, S., 1963, Third component participating in the superprecipitation of "natural actomyosin," *Nature (London)* **200**:1010.

Ebashi, S., Iwakura, H., Nakajima, H., Nakamura, R., and Ooi, Y., 1966, New structural proteins from dog heart and chicken gizzard, *Biochem. Z.* **345**: 201.

Ebashi, S., Kodama, A., and Ebashi, F., 1968, Troponin. I. Preparation and physiological function, *J. Biochem. (Tokyo)* **64**:465.

Ebashi, S., Toyo-oka, R., and Nonumura, Y., 1975, Gizzard troponin, *J. Biochem. (Tokyo)* **78**:859.

Ebashi, S., Nonomura, Y., Toyo-oka, T., and Katayama, E., 1976, Regulation of muscle contraction by the calcium–troponin–tropomyosin system, in: *Calcium in Biological Systems* (C. J. Duncan, ed.), pp. 349–360, Cambridge University Press, London.

Ebashi, S., Mikawa, T., Hirata, M., Toyo-oka, T., and Nonomura, Y., 1977, Regulatory proteins of smooth muscle, in: *Excitation–Contraction Coupling in Smooth Muscle* (R. Casteels, T. Godfraind, and J. C. Rüegg, eds.), pp. 325–334, Elsevier/North-Holland, Amsterdam.

Ebashi, S., Mikawa, T., Hirata, M., and Nonomura, Y., 1978, The regulatory role of calcium in muscle, *Ann. N. Y. Acad. Sci.* **307**:451.

Fine, R. E., and Blitz, A. L., 1975, A chemical comparison of tropomyosins from muscle and non-muscle tissues, *J. Mol. Biol.* **95**:447.

Ford, G. D., and Hess, M. L., 1975, Calcium-accumulating properties of subcellular fractions of bovine vascular smooth muscle, *Circ. Res.* **37**:580.

Frearson, N., Focant, B. W. W., and Perry, S. V., 1976, Phosphorylation of a light chain component of myosin from smooth muscle, *FEBS Lett.* **63**:27.

Frederiksen, D. W., 1976, Myosin-mediated Ca^{++}-regulation of actomyosin-adenosinetriphosphatase from porcine aorta, *Proc. Natl. Acad. Sci. U.S.A.* **73**:2706.

Gergely, P., Vereb, G., and Bot, G., 1976, Thiophosphate-activated phosphorylase kinase as a probe in the regulation of phosphorylase phosphatase, *Biochim. Biophys. Acta* **429**:809.

Glass, D. B., and Krebs, E. G., 1980, Protein phosphorylation catalyzed by cyclic AMP-dependent and cyclic GMP-dependent protein kinase, *Annu. Rev. Pharmacol. Toxicol.* **20**:363.

Gorecka, A., Aksoy, M. O., and Hartshorne, D. J., 1976, The effect of phosphorylation of gizzard myosin on actin activation, *Biochem. Biophys. Res. Commun.* **71**:325.

Grand, R. J. A., Perry, S. V., and Weeks, R. A., 1979, Troponin C-like proteins (calmodulins) from mammalian smooth muscle and other tissues, *Biochem. J.* **177**:521.

Gratecos, D., and Fischer, E. H., 1974, Adenosine 5′-O(3-thiotriphosphate) in the control of phosphorylase activity, *Biochem. Biophys. Res. Commun.* **58**:960.

Hamoir, G., and Laszt, L., 1962, Tonomyosin of arterial muscle, *Nature (London)* **193**:682.

Hanson, J., and Lowy, J., 1964, The problem of the location of myosin in vertebrate smooth muscle (discussion), *Proc. Roy. Soc. London Ser B* **160**:523.

Hartshorne, D. J., and Gorecka, A., 1980, The biochemistry of the contractile proteins of smooth muscle, in: *Handbook of Physiology*, Section 2, *The Cardiovascular System*, Vol. II, *Vascular Smooth Muscle* (D. F. Bohr, A. P. Somlyo, and H. V. Sparks, eds.), pp. 93–120, American Physiology Society, Bethesda, Maryland.

Hartshorne, D. J., and Persechini, A. J., 1980, Phosphorylation of myosin as a regulatory component in smooth muscle, *Ann. N. Y. Acad. Sci.* **356**:130.

Hartshorne, D. J., Abrams, L., Aksoy, M. O., Dabrowska, R., Driska, S., and Sharkey, E. M., 1977a, Molecular basis for the regulation of smooth muscle actomyosin, in: *The Biochemistry of Smooth Muscle* (N. L. Stephens, ed.), pp. 513–532, University Park Press, Baltimore.

Hartshorne, D. J., Gorecka, A., and Aksoy, M. O., 1977b, Aspects of the regulatory mechanism in smooth muscle, in: *Excitation–Contraction Coupling in Smooth Muscle* (R. Casteels, T. Godfraind, and J. C. Rüegg, eds.), pp. 377–384, Elsevier/North-Holland, Amsterdam.

Hartshorne, D. J., Siemankowski, R. F., and Aksoy, M. O., 1980, Ca regulation in smooth muscle and phosphorylation: Some properties of the myosin light chain kinase, in: *Regulatory Mechanism of Muscle Contraction* (S. Ebashi, K. Maruyama, and M. Endo, eds.), pp. 287–301, Japan Scientific Societies Press, Tokyo.

Hathaway, D. R., and Adelstein, R. S., 1979, Human platelet myosin light chain kinase requires the calcium-binding protein calmodulin for activity, *Proc. Natl. Acad. Sci. U.S.A.* **76**:1653.

Heumann, H.-G., 1969, Gibt es in glatten Vertebraten muskeln dicke Filamente? Elektronenmikroskopische Untersuchungen an der Darm-muskulatur der Hausmaus, *Zool. Anz (Suppl. BD)* **33**(Verh. Zool. Ges.):416.

Hidaka, H., Naka, M., and Yamaki, T., 1979, Effect of novel specific myosin light chain kinase inhibitors on Ca^{2+}-activated Mg^{2+}-ATPase of chicken gizzard actomyosin, *Biochem. Biophys. Res. Commun.* **90**:694.

Hidaka, H., Yamaki, T., Naka, M., Tanaka, T., Hayashi, H., and Kobayashi, R., 1980, Calcium-regulated modulator protein interacting agents inhibit smooth muscle calcium-stimulated protein kinase and ATPase, *Mol. Pharmacol.* **17**:66.

Hinssen, H., D'Haese, J., Small, J. V., and Sobieszek, A., 1978, Mode of filament assembly of myosins from muscle and nonmuscle cells, *J. Ultrastruct. Res.* **64**:282.

Hirata, M., Mikawa, T., Nonomura, Y., and Ebashi, S., 1980, Ca^{2+} regulation in vascular smooth muscle. II. Ca^{2+} binding of aorta leiotonin, *J. Biochem. (Tokyo)* **87**:369.

Hoar, P. E., and Kerrick, W. G. L., 1980, Catalytic subunit of c-AMP dependent protein kinase: Effect on contraction of functionally skinned muscle fibers, *Fed. Proc. Fed. Am. Soc. Exp. Biol.* **39**:1817.

Hoar, P. E., Kerrick, W. G. L., and Cassidy, P. S., 1979, Chicken gizzard: Relation between calcium-activated phosphorylation and contraction, *Science* **204**:503.

Holroyde, M. J., Potter, J. D., and Solaro, R. J., 1979, The calcium binding properties of phosphorylated and unphosphorylated cardiac and skeletal myosins, *J. Biol. Chem.* **254**:6478.

Huszar, G., and Bailey, P., 1979, Relationship between actin–myosin interaction and myosin light chain phosphorylation in human placental smooth muscle, *Am. J. Obstet. Gynecol.* **135**:718.

Huxley, A. F., 1957, Muscle structure and theories of contraction, *Prog. Biophys. Mol. Biol.* **7**:257.

Huxley, H. E., 1963, Electron microscope studies on the structure of natural and synthetic protein filaments from striated muscle, *J. Mol. Biol.* **7**:281.

Ikebe, M., Onishi, H., and Watanabe, S., 1977, Phosphorylation and dephosphorylation of a light chain of the chicken gizzard myosin molecule, *J. Biochem. (Tokyo)* **82**:299.

Ikebe, M., Aiba, T., Onishi, H., and Watanabe, S., 1978, Calcium sensitivity of contractile proteins from chicken gizzard muscle, *J. Biochem. (Tokyo)* **83**:1643.

Ito, N., and Hotta, K., 1976, Regulatory protein of bovine tracheal smooth muscle, *J. Biochem. (Tokyo)* **80**:401.

Ito, N., Takagi, T., and Hotta, K., 1976, Regulatory protein of vascular smooth muscle, *J. Biochem. (Tokyo)* **80**:899.

Jakes, R., Northrop, F., and Kendrick-Jones, J., 1976, Calcium binding regions of myosin "regulatory" light chains, *FEBS Lett.* **70**:229.

Janis, R. A., and Gualteri, R. T., 1978, Contraction of intact smooth muscle is associated with the phosphorylation of a 20,000 dalton protein, *Physiologist* **21**:59.

Johansson, B., and Somlyo, A. P., 1980, Electrophysiology and excitation–contraction coupling, in: *Handbook of Physiology*, Section 2, *The Cardiovascular System*, Vol. II, *Vascular Smooth Muscle* (D. F. Bohr, A. P. Somlyo, and H. V. Sparks, eds.), pp. 301–323, American Physiology Society, Bethesda, Maryland.

Johnson, L. S., 1974, Non-identical tropomyosin subunits in rat skeletal muscle, *Biochim. Biophys. Acta* **371**:219.

Jones, A. W., Somlyo, A. P., and Somlyo, A. V., 1973, Potassium accumulation in smooth muscle and associated ultrastructural changes, *J. Physiol. (London)* **232**:247.

Kaminer, B., 1969, Synthetic myosin filaments from vertebrate smooth muscle, *J. Mol. Biol.* **39**:257.

Katoh, N., and Kubo, S., 1977, Purification and some properties of rabbit stomach myosin, *J. Biochem. (Tokyo)* **81**:1497.

Kelley, R. E., and Rice, R. V., 1968, Localization of myosin filaments in smooth muscle, *J. Cell Biol.* **37**:105.

Kendrick-Jones, J., 1973, The subunit structure of gizzard myosin, *Philos. Trans. R. Soc. London Ser. B* **265**:183.

Kendrick-Jones, J., Lehman, W., and Szent-Györgyi, A. G., 1970, Regulation in molluscan muscles, *J. Mol. Biol.* **54**:313.

Kendrick-Jones, J., Szentkiralyi, E. M., and Szent-Györgyi, A. G., 1976, Regulatory light chains in myosins, *J. Mol. Biol.* **104**:747.

Kerrick, W. G. L., Hoar, P. E., and Cassidy, P. S., 1980, Ca²⁺-activated tension: The role of myosin light chain phosphorylation, *Fed Proc. Fed. Am. Soc. Exp. Biol.* **39**:1558.

Kuwayama, H., and Yagi, K., 1979, Ca²⁺ binding of pig cardiac myosin subfragment-1 and g₂ light chain, *J. Biochem. (Tokyo)* **85**:1245.

Laszt, L., 1964, Was ist Gefässtonus? Untersuchungen über die Beziehungen zwischen Gefässmuskel vontraktion und-volumen äunderung, *Angiologica* **1**:346.

Laszt, L., and Hamoir, G., 1961, Etude par electrophorèse et ultracentrifugation de la composition protéinique de la conche musculaire des carotides de bovidé, *Biochim. Biophys. Acta* **50**:430.

Lee, E. Y. C., Silberman, S. R., Ganapathi, M. K., Petrovi, S., and Paris, H., 1980, The phosphoprotein phosphatases: Properties of the enzymes involved in the regulation of glycogen metabolism, *Adv. Cyclic Nucleotide Res.* **13**:95.

Leger, J., Bouveret, P., Schwartz, K., and Swynghedauw, B., 1976, A comparative study of skeletal and cardiac tropomyosin: Subunits, thiol group content and biological activities, *Pfluegers Arch.* **362**:271.

Lehman, W., and Szent-Györgyi, A. G., 1975, Regulation of muscular contraction: Distribution of actin control and myosin control in the animal kingdom, *J. Gen. Physiol.* **66**:1.

Lehman, W., Kendrick-Jones, J., and Szent-Györgyi, A. G., 1972, Myosin-linked regulatory systems: Comparative studies, *Cold Spring Harbor Symp. Quant. Biol.* **37**:319.

Levin, R. M., and Weiss, B., 1977, Binding of trifluoperazine to the calcium-dependent activator of cyclic nucleotide phosphodiesterase, *Mol. Pharmacol.* **13**:690.

Levin, R. M., and Weiss, B., 1979, Selective binding of antipsychotics and other psychoactive agents to the calcium-dependent activator of cyclic nucleotide phosphodiesterase, *J. Pharmacol. Exp. Ther.* **208**:454.

Lowy, J., and Small, J. V., 1970, The organization of myosin and actin in vertebrate smooth muscle, *Nature (London)* **227**:46.

Mak, A. S., Lewis, W. G., and Smillie, L. B., 1979, Amino acid sequences of rabbit skeletal β- and cardiac tropomyosins, *FEBS Lett.* **105**:232.

Manning, D. R., and Stull, J. T., 1979, Myosin light chain phosphorylation and phosphorylase a activity in rat extensor digitorum longus muscle, *Biochem. Biophys. Res. Commun.* **90**:164.

Marston, S. B., and Taylor, E. W., 1978, Mechanism of myosin and actomyosin ATPase in chicken gizzard smooth muscle, *FEBS Lett.* **86**:167.

Marston, S. B., Trevett, R. M., and Walters, M., 1980, Calcium ion-regulated thin filaments from vascular smooth muscle. *Biochem. J.* **185**:355.

Matsuda, G., Suzuyama, Y., Maita, T., and Umegane, T., 1977, The L-2 light chain of chicken skeletal muscle myosin, *FEBS Lett.* **85**:53.

Means, A. R., and Dedman, J. R., 1980, Calmodulin—an intracellular calcium receptor, *Nature (London)* **285**:73.

Mikawa, T., Nonomura, Y., and Ebashi, S., 1977, Does phosphorylation of myosin light chain have direct relation to regulation in smooth muscle?, *J. Biochem. (Tokyo)* **82**:1789.

Morgan, M., Perry, S. V., and Ottaway, J., 1976, Myosin light-chain phosphatase, *Biochem. J.* **157**:687.

Mrwa, U., and Hartshorne, D. J., 1980, Phosphorylation of smooth muscle myosin and myosin light chains, *Fed. Proc. Fed. Am. Soc. Exp. Biol.* **39**:1564.

Mrwa, U., and Rüegg, J. C., 1975, Myosin-linked calcium regulation in vascular smooth muscle, *FEBS Lett.* **60**:81.

Mrwa, U., and Rüegg, J. C., 1977, The role of the regulatory light chain in pig carotid smooth muscle ATPase, in: *Excitation–Contraction Coupling in Smooth Muscle* (R. Casteels, T. Godfraind, and J. C. Rüegg, eds.), pp. 353–357, Elsevier/North-Holland, Amsterdam.

Mrwa, U., Achtig, I., and Rüegg, J. C., 1974, Influences of calcium concentration and pH on the tension development and ATPase activity of the arterial actomyosin contractile system, *Blood Vessels* **11**:277.

Mrwa, U., Paul, R. J., Kreye, V. A. W., and Rüegg, J. C., 1975, The contractile mechanism of vascular smooth muscle, *INSERM* **50**:319.

Mrwa, U., Troschka, M., and Rüegg, J. C., 1979, Cyclic AMP-dependent inhibition of smooth muscle actomyosin, *FEBS Lett.* **107**:371.

Murphy, R. A., 1979, Filament organization and contractile function in vertebrate smooth muscle, *Annu. Rev. Physiol.* **41**:737.

Murphy, R. A., Bohr, D. F., and Newman, D. L., 1969, Arterial actomyosin: Mg, Ca, and ATP ion dependencies for ATPase activity, *Am. J. Physiol.* **217**:666.

Murphy, R. A., Herlihy, J. T., and Megerman, J., 1974, Force-generating capacity and contractile protein content of arterial smooth muscle, *J. Gen. Physiol.* **64**:691.

Murphy, R. A., Aksoy, M. O., and Dillon, P. R., 1980, Regulation in vascular smooth muscle: Ca^{++}-dependent myosin light chain (LC) phosphorylation mediates cross-bridge cycling, *Fed. Proc. Fed. Am. Soc. Exp. Biol.* **39**:1817.

Murray, J. M., and Weber, A., 1974, The cooperative action of muscle proteins, *Sci. Am.* **230**:58.

Nairn, A. C., and Perry, S. V., 1979, Calmodulin and myosin light-chain kinase of rabbit fast skeletal muscle, *Biochem. J.* **179**:89.

Needham, D. M., and Cawkwell, J. M., 1956, Some properties of the actomyosin-like protein of the uterus, *Biochem. J.* **63**:337.

Nonomura, Y., 1968, Myofilaments in smooth muscle of guinea pig's taenia coli, *J. Cell Biol.* **39**:741.

Nonomura, Y., and Ebashi, S., 1980, Calcium regulatory mechanism in vertebrate smooth muscle, *Biomed. Res.* **1:1.**

Nonomura, Y., Mikawa, T., and Ebashi, S., 1980, Ca^{2+} sensitive thin filament from chicken gizzard smooth muscle, *Proc. Jpn. Acad.* **56**(B):178.

Okamoto, Y., and Sekine, T., 1978, Effects of tryptic digestion on the enzymatic activities of chicken gizzard myosin, *J. Biochem. (Tokyo)* **83**:1375.

Onishi, H., and Watanabe, S., 1979, Chicken gizzard heavy meromyosin that retains the two light-chain components, including a phosphorylatable one, *J. Biochem. (Tokyo)* **85**:457.

Onishi, H., Iijima, S., Anzai, H., and Watanabe, S., 1979, The possible role of myosin light-chain phosphatase in relaxation of chicken gizzard muscle, *J. Biochem. (Tokyo)* **86**:1283.

Ookubo, N., Ueno, H., and Ooi, T., 1975, Similarities and differences of the α and β components of tropomyosin, *J. Biochem. (Tokyo)* **78**:739.

Parry, D. A. D., and Squire, J. M., 1973, Structural role of tropomyosin in muscle regulation: Analysis of the x-ray diffraction patterns from relaxed and contracting muscles, *J. Mol. Biol.* **75**:33.

Pato, M. D., and Adelstein, R. S., 1980, Dephosphorylation of the 20,000-dalton light chain of myosin by two different phosphatases from smooth muscle, *J. Biol. Chem.* **255**:6535.

Pemrick, S. J., 1980, The phosphorylated L_2 light chain of skeletal myosin is a modifier of the actomyosin ATPase, *J. Biol. Chem.* **255**:8836.

Perrie, W. R., Smillie, L. B., and Perry, S. V., 1973, A phosphorylated light-chain component of myosin from skeletal muscle, *Biochem. J.* **135**:151.

Pires, E. M. V., and Perry, S. V., 1977, Purification and properties of myosin light-chain kinase from fast skeletal muscle, *Biochem. J.* **167**:137.

Rubenstein, P. A., and Spudich, J. A., 1977, Actin microheterogeneity in chick embryo fibroblasts, *Proc. Natl. Acad. Sci. U.S.A.* **74**:120.

Rüegg, J. C., Strassner, E., and Schirmer, R. H., 1965, Extraktion und Reinigung von Arterien-Actomyosin, Actin, und Extraglobulin, *Biochem. Z.* **343**:70.

Russell, W. E., 1973, Insolubilization and activation of arterial actomyosin by bivalent cations, *Eur. J. Biochem.* **33**:459.

Seidel, J. C., 1978, Chymotryptic heavy meromyosin from gizzard myosin: A proteolytic fragment with the regulatory properties of the intact myosin, *Biochem. Biophys. Res. Commun.* **85**:107.

Seidel, J. C., 1980, Fragmentation of gizzard myosin by α-chymotrypsin and papain, the effects on ATPase activity, and the interaction with actin, *J. Biol. Chem.* **255**:4355.

Sekine, R., Barnett, L. M., and Kielley, W. W., 1962, The active site of myosin adenosine triphosphatase. I. Localization of one of the sulfhydryl groups, *J. Biol. Chem.* **237**:2769.

Sherry, J. M. F., Gorecka, A., Aksoy, M. O., Dabrowska, R., and Hartshorne, D. J., 1978, Roles of calcium and phosphorylation in the regulation of the activity of gizzard myosin, *Biochemistry* **17**:4411.

Sheterline, P., 1980, Trifluoperazine can distingish between myosin light chain kinase-linked and troponin C-linked control of actomyosin interaction by Ca^{++}, *Biochem. Biophys. Res. Commun.* **93**:194.

Shibata, N., Yamagami, R., Yoneda, S., Akagami, H., Takeuchi, K., Tanaka, K., and Okamura, Y., 1973, Identification of myosin A, actin and native tropomyosin constitution of arterial contractile protein (myosin B) and their characteristics, *Jpn. Circ. J.* **37**:229.

Shoenberg, C. F., 1965, Contractile proteins of vertebrate smooth muscle, *Nature (London)* **206**:526.

Shoenberg, C. F., 1969, An electron microscope study of the influence of divalent ions on myosin filament formation in chicken gizzard extracts and homogenates, *Tissue Cell* **1**:83.

Shoenberg, C. F., and Haselgrove, J. C., 1974, Filaments and ribbons in vertebrate smooth muscle, *Nature (London)* **249**:152.

Siegman, M. J., Butler, T. M., Mooers, S. U., and Davies, R. E., 1976a, Calcium-dependent resistance to stretch and stress relaxation in resting smooth muscles, *Am. J. Physiol.* **231**:1501.

Siegman, M. J., Butler, T. M., Mooers, S. U., and Davies, R. E., 1976b, Cross-bridge attachment, resistance to stretch, and viscoelasticity in resting mammalian smooth muscle, *Science* **191**:383.

Silver, P. J., and DiSalvo, J., 1979, Adenosine 3′:5′-monophosphate-mediated inhibition of myosin light chain phosphorylation in bovine aortic actomyosin, *J. Biol. Chem.* **254**:9951.

Small, J. V., 1977, Studies on isolated smooth muscle cells: The contractile appartus, *J. Cell Sci.* **24**:327.

Small, J. V., and Sobieszek, A., 1977a, Studies on the function and composition of the 10-nm (100 Å) filaments of vertebrate smooth muscle, *J. Cell Sci.* **23**:243.

Small, J. V., and Sobieszek, A., 1977b, Ca-regulation of mammalian smooth muscle actomyosin via a kinase-phosphatase-dependent phosphorylation and dephosphorylation of the 20,000-M_r light chain of myosin, *Eur. J. Biochem.* **76**:521.

Small, J. V., and Squire, J. M., 1972, Structural basis of contraction in vertebrate smooth muscle, *J. Mol. Biol.* **67**:117.

Sobieszek, A., 1977a, Vertebrate smooth muscle myosin: Enzymatic and structural properties, in: *The Biochemistry of Smooth Muscle* (N. L. Stephens, ed.), pp. 413–443, University Park Press, Baltimore.

Sobieszek, A., 1977b, Ca-linked phosphorylation of a light chain of vertebrate smooth-muscle myosin, *Eur. J. Biochem.* **73**:477.

Sobieszek, A., and Bremel, R. D., 1975, Preparation and properties of vertebrate smooth-muscle myofibrils and actomyosin, *Eur. J. Biochem.* **55**:49.

Sobieszek, A., and Small, J. V., 1976, Myosin-linked calcium regulation in vertebrate smooth muscle, *J. Mol. Biol.* **102**:75.

Sobieszek, A., and Small, J. V., 1977, Regulation of the actin–myosin interaction in vertebrate smooth muscle: Activation via a myosin light-chain kinase and the effect of tropomyosin, *J. Mol. Biol.* **112**:559.

Somlyo, A. V., 1980, Ultrastructure of vascular smooth muscle, in: *Handbook of Physiology,* Section 2, *The Cardiovascular System,* Vol. II, *Vascular Smooth Muscle* (D. F. Bohr, A. P. Somlyo, and H. V. Sparks, eds.), pp. 33–67, American Physiology Society, Bethesda, Maryland.

Somlyo, A. P., Devine, C. E., Somlyo, A. V., and Rice, R. V., 1973, Filament organization in vertebrate smooth muscle, *Philos. Trans. R. Soc. London Ser. B* **265**:223.

Somlyo, A. V., Ashton, F. T., Lemanski, L. F., Vallières, J., and Somlyo, A. P., 1977, Filament organization and dense bodies in vertebrate smooth muscle, in: *The Biochemistry of Smooth Muscle* (N. L. Stephens, ed.), pp. 445–471, University Park Press, Baltimore.

Sparrow, M. P., and van Bockxmeer, F. M., 1972, Arterial tropomyosin and a relaxing protein fraction from vascular smooth muscle: Comparison with skeletal tropomyosin and troponin, *J. Biochem. (Tokyo)* **72**:1075.

Sparrow, M. P., Maxwell, L. C., Rüegg, J. C., and Bohr, D. R., 1970, Preparation and properties of a calcium ion-sensitive actomyosin from arteries, *Am. J. Physiol.* **219**:1366.

Stone, D., and Smillie, L. V., 1978, The amino acid sequence of rabbit skeletal α-tropomyosin: The NH₂-terminal half and complete sequence, *J. Biol. Chem.* **253**:1137.

Stone, D., Sodek, J., Johnson, P., and Smillie, L. B., 1974, Tropomyosin: Correlation of amino acid sequence and structure, *Proc. 9th FEBS Meet. (Budapest)* **31**:125.

Strasburg, G. M., and Greaser, M. L., 1976, The native subunit pattern of tropomyosin, *FEBS Lett.* **72**:11.

Stull, J. T., Blumenthal, D. K., deLanerolle, P., High, C. W., and Manning, D. R., 1978, Phosphorylation and regulation of contractile proteins, *Adv. Pharmacol. Ther.* **3**:171.

Suzuki, H., Onishi, H., Takahashi, K., and Watanabe, S., 1978, Structure and function of chicken gizzard myosin, *J. Biochem. (Tokyo)* **84**:1529.

Szent-Györgyi, A. G., Szentkiralyi, E. M., and Kendrick-Jones, J., 1973, The light chains of scallop myosin as regulatory subunits, *J. Mol. Biol.* **74**:179.

Takeuchi, K., and Tonomura, Y., 1977, Kinetic and regulatory properties of myosin adenosine-triphosphatase purified from arterial smooth muscle, *J. Biochem. (Tokyo)* **82**:813.

Vandekerckhove, J., and Weber, K., 1978a, Mammalian cytoplasmic actins are the products of at least two genes and differ in primary structure in at least 25 identified positions from skeletal muscle actins, *Proc. Natl. Acad. Sci. U.S.A.* **75**:1106.

Vandekerckhove, J., and Weber, K., 1978b, Actin amino-acid sequences: Comparison of actins from calf thymus, bovine brain, and SV 40-transformed mouse 3T3 cells with rabbit skeletal muscle actin, *Eur. J. Biochem.* **90**:451.

Vibert, P. J., Haselgrove, J. C., Lowy, J., and Poulsen, F. R., 1972, Structural changes in actin-containing filaments of muscle, *J. Mol. Biol.* **71**:757.

Wachsberger, P., and Kaldor, G., 1971, Studies on uterine myosin A and actomyosin, *Arch. Biochem. Biophys.* **143**:127.

Wachsberger, P. R., and Pepe, F. A., 1974, Purification of uterine myosin and synthetic filament formation, *J. Mol. Biol.* **88**:385.

Waisman, D. M., Singh, T. J., and Wang, J. H., 1978, The modulator-dependent protein kinase: A multifunctional protein kinase activatable by the Ca²⁺-dependent modulator protein of the cyclic nucleotide system, *J. Biol. Chem.* **253**:3387.

Walsh, M. P., Vallet, B., Autric, F., and Demaille, J. G., 1979, Purification and characterization of bovine cardiac calmodulin-dependent myosin light chain kinase, *J. Biol. Chem.* **254**:12136.

Walsh, M. P., Guilleux, J. C., and Demaille, J. G., 1981, Calcium- and cyclic AMP-dependent regulation of myofibrillar calmodulin-dependent myosin light chain kinases from cardiac and skeletal muscles, *Adv. Cyclic Nucleotide Res.* **14** (in press).

Watterson, J. G., Kohler, L., and Schaub, M. C., 1979, Evidence for two distinct affinities in the binding of divalent metal ions to myosin, *J. Biol. Chem.* **254**:6470.

Whalen, R. G., Butler-Browne, G. S., and Gros, F., 1976, Protein synthesis and actin heterogeneity in calf muscle cells in culture, *Proc. Natl. Acad. Sci. U.S.A.* **73**:2018.

Yamaguchi, M., Miyazawa, Y., and Sekine, T., 1970, Preparation and properties of smooth muscle myosin from horse esophagus, *Biochim. Biophys. Acta* **216**:411.

Yazawa, M., and Yagi, K., 1978, Purification of modulator-deficient myosin light-chain kinase by modulator protein–Sepharose affinity chromatography, *J. Biochem. (Tokyo)* **84**:1259.

Yazawa, M., Kuwayama, H., and Yagi, K., 1978, Modulator protein as a Ca^{2+}-dependent activator of rabbit skeletal myosin light-chain kinase: Purification and characterization, *J. Biochem. (Tokyo)* **84**:1253.

Yeaman, S. J., Cohen, P., Watson, D. C., and Dixon, G. H., 1977, The substrate specificity of adenosine 3′ : 5′-cyclic monophosphate-dependent protein kinase of rabbit skeletal muscle, *Biochem. J.* **162**:411.

Yerna, M.-J., Aksoy, M. O., Hartshorne, D. J., and Goldman, R. D., 1978, BHK 21 myosin: Isolation, biochemical characterization and intracellular localization, *J. Cell Sci.* **31**:411.

Yerna, M.-J., Dabrowska, R., Hartshorne, D. J., and Goldman, R. D., 1979, Calcium-sensitive regulation of actin–myosin interactions in baby hamster kidney (BHK-21) cells, *Proc. Natl. Acad. Sci. U.S.A.* **76**:184.

Zechel, K., 1979, Localization of the charge differences in the actins of rabbit skeletal muscle and chicken gizzard by two-dimensional gel electrophoretic analysis of tryptic fragments, *Hoppe-Seyler's Z. Physiol. Chem.* **360**:777.

15

Three-Dimensional Structure of Muscle Membranes Involved in the Regulation of Contraction in Skeletal-Muscle Fibers

Lee D. Peachey

1. Introduction

One area in cell biology in which considerable advance has been made over the last 25 years is the study of how internal membranes in skeletal-muscle cells [specifically the transverse tubular system (T-system) and sarcoplasmic reticulum (SR)] are involved in the control of the contractile state of the cell. This chapter will review this research and summarize the present state of our understanding in this area (for a broader review of muscle-cell biology over the last 25 years, see Franzini-Armstrong and Peachey, 1982). As will be seen, we now have a quite complete understanding of how the T-system of a variety of types of skeletal-muscle fibers is constructed and how it works. Furthermore, this knowledge from muscle cells, in which contractile and control mechanisms are highly elaborated and relatively easy to study, should turn out to be useful in studies of nonmuscle cells, where one might perhaps expect to find similar, though probably simpler, control systems.

2. Historical Background

At the time when biological electron microscopy was just getting started, in the early 1950s, a major outstanding question in muscle physiology was how depolarization of the surface of a muscle fiber by an action potential could

Lee D. Peachey • Department of Biology G7, University of Pennsylvania, Philadelphia, Pennsylvania 19104.

lead to contraction of the entire cross section of the fiber in the brief latent period between excitation and contraction. Hill (1948, 1949) had examined this coupling problem in two papers. In the paper published in 1948, he solved the differential equation for diffusion (Laplace's equation) in cylindrical coordinates and for the boundary conditions of a pulse of a diffusible substance released instantaneously at the surface of a cylindrical fiber and diffusing inward. This could, for example, represent a sudden entry of calcium ions during an action potential on a muscle-fiber surface. Hill's results showed that virtually no calcium would have reached the center of the fiber by this mechanism in the known latent period between electrical excitation and the beginning of the mechanical response. This was true both for a relatively slow-twitch muscle and for a fast-acting muscle. If diffusion of calcium from the fiber surface *were* the mechanism by which activation spread inward into the fiber, then only a part of the fiber cross section (that near the surface of the fiber) could be activated in the time available. Hill's second paper showed that virtually the entire cross section of a fiber is activated early in a contraction, so the calcium-diffusion mechanism could conclusively be ruled out, at least if the calcium originated at or near the fiber surface.

A number of other mechanisms were considered subsequently by other workers, but none of these seemed acceptable as an explanation for this coupling process. It was not until the late 1950s that two experimental results suggested the mechanism that is now widely accepted as the correct one. The first result came from the rapidly evolving field of biological electron microscopy, and the second was a physiological result on activation of a small part of the surface of a single muscle fiber.

In the early 1950s, electron microscopists observed in thin sections of striated muscle cells a delicate network of tubular and cisternal membrane systems that seemed to be the equivalent of the endoplasmic reticulum found in other types of cells (Bennett and Porter, 1953; Porter and Palade, 1957). The form of this reticulum was quite different in the muscle cell (where it was called the sarcoplasmic reticulum) than in cells of, for example, the exocrine pancreas. The special form of the reticulum in muscle cells, and its regular disposition in close association with the striations of the myofibrils, suggested that the reticulum might be expected to have a special role in the functioning of the muscle cell.

At about the same time, A. F. Huxley and Taylor (1958), in England, were performing experiments in which they contrived to depolarize only a small patch of the surface membrane of an isolated skeletal-muscle cell. They found that only regions of the sarcomere banding that corresponded to the location of the "triads" of the SR were responsive. This correlation held up in several kinds of muscle fibers, and the suggestion was clear that some component of the triad might be responsible for conveying activation into the fiber. Release of calcium from the SR throughout the fiber volume would reduce the diffusion distances to be covered to a micrometer or so, easily small enough for the time available (Peachey and Porter, 1959).

Since those studies were done, we have learned considerably more about the structure and function of these membrane systems. In this chapter, I will discuss only the membranes responsible for conducting the action potential into the depths of the fiber, and will concentrate on the more recent results, especially those obtained using high-voltage electron microscopy (HVEM). Attention will be focused on the central element of the triad, the transverse tubule, since this now is widely believed to be the element that is active in the inward spread of activation.

The T-system of a striated muscle cell is a predominantly transversely oriented network of interconnected tubules invaginating into the fiber from many points on its surface (Peachey and Schild, 1968). Initially, electron microscopists found the T-tubules in thin sections of fixed and embedded skeletal muscles. Soon after, it was realized that the T-tubular networks had been observed many years earlier in the light microscope by histologists such as Retzius (1881) and Veratti (1902). These keen observers had described delicate networks transversing muscle fibers stained by heavy-metal-impregnation methods. What electron microscopy initially added to these descriptions was a greater certainty that the networks were real (they had generally been pushed aside as artifacts by Retzius's and Veratti's contemporaries) and ultrastructural details of the membranous nature and structure of the tubules and their relationship to the SR. More recently, the T-system and SR have been studied by other electron-microscopic methods, including freeze–

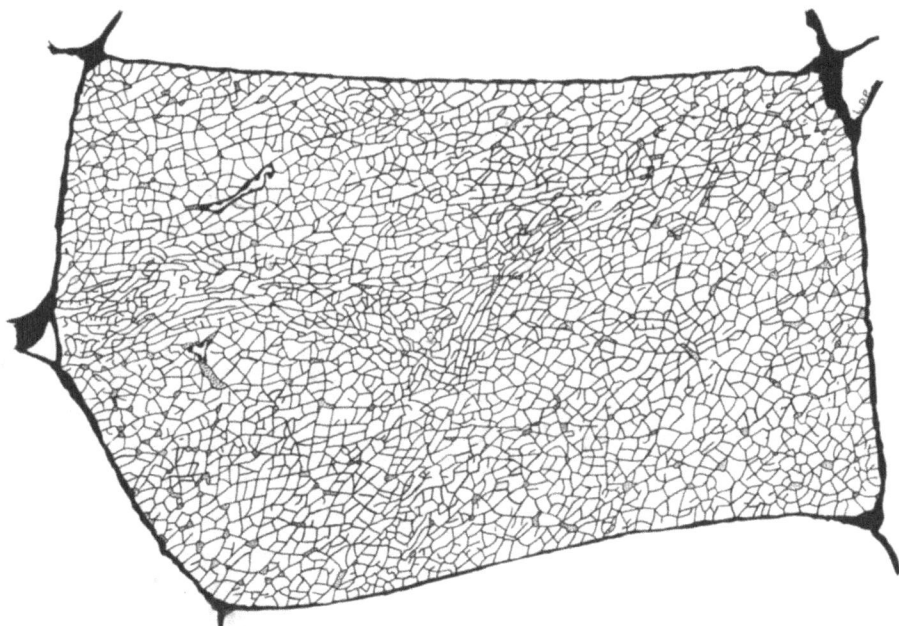

Figure 1. Complete reconstruction of the T-system network over the whole cross section of a frog twitch-muscle fiber. The reconstruction was made by tracing successive portions of the network in a set of high-voltage electron micrographs from serial slices of a muscle stained with the peroxidase method. From Peachey and Eisenberg (1978).

fracture and HVEM. These studies have improved our knowledge of the T-system as a network, including several aspects of its three-dimensional nature, and have better defined the nature of the critical junctional areas between T-system and SR. The remainder of this chapter will be devoted to a review of some of these recent results.

The early electron-microscopic studies provided a description of the T-system as a series of planar and continuous networks crossing the fiber transversely at regular intervals in registration with the sarcomere banding. The more recent morphological studies have revealed several ways in which the T-system deviates from this pattern, either by having a longitudinal component or with some form of interruption in its planar arrangement.

An early study of muscle-membrane systems using HVEM was done by Peachey and Eisenberg (1978), using the peroxidase tracer method to stain the extracellular space and T-system selectively. By following the T-tubule network through a set of serial sections 0.7 μm thick, it was possible to reconstruct a T-system network of a frog twitch fiber across the entire fiber cross section (Fig. 1). Such a reconstruction would have been extremely difficult using conventional thin-section techniques, because of the large number of serial sections that would be required and because of the difficulty of recognizing T-tubules in sections that are not selectively stained.

Some interesting points appeared in these reconstructions. Dead-end T-tubules were found fairly frequently. The interpretation of tubules with

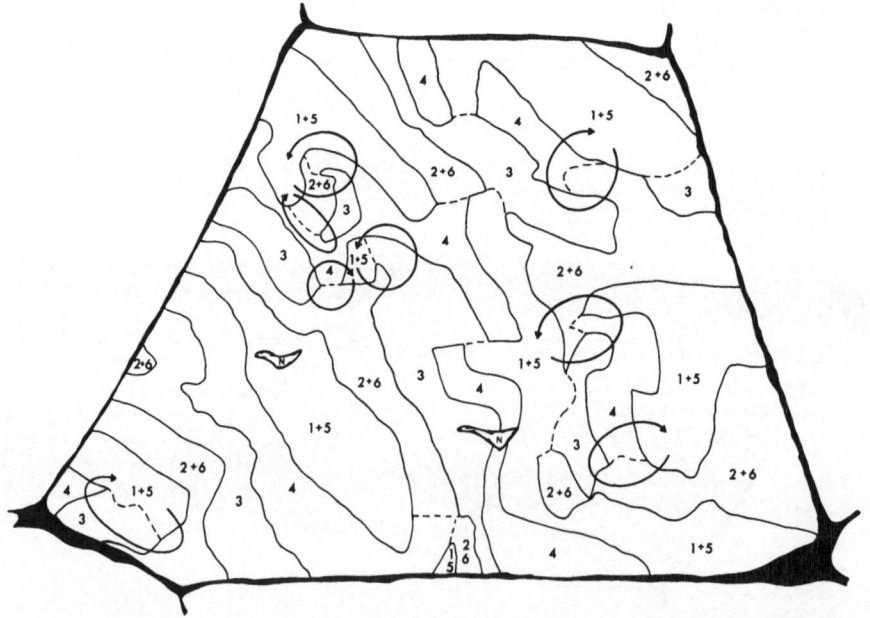

Figure 2. Results of a reconstruction similar to that in Fig. 1 laid out to show the slice numbers in the series and the areas where the T-system network appeared in each slice. The curved arrows show where helicoids are found. There is a total of eight helicoids in this reconstruction. From Peachey and Eisenberg (1978).

free ends had always been uncertain in thin sections, where there was the possibility that tubules, rather than ending, simply passed out of the plane of the section, but the ends can be clearly seen within the thickness of the section in the HVEM, especially when viewed stereoscopically. Relatively few of these free tubule ends were found, and our view of the T-system network was not changed very much quantitatively by this result. However, the result is an interesting one, in view of several possible functional correlates, including a possible role in myofibril splitting (Goldspink, 1971).

When, subsequently, the T-system was reconstructed in three dimensions over the entire width of a fiber and along the longitudinal axis for a distance of several sarcomeres, Peachey and Eisenberg (1978) found that longitudinally adjacent T-systems were often connected at one or more places in the fiber cross section by helicoidal ramps (Fig. 2). The number of helicoids found in any one fiber varied from zero to eight, in a total of 21 fibers reconstructed. Both right-handed and left-handed helicoids were found, even within the same fiber cross section.

The effect these helicoids might have on the physiology of the muscle fiber is not clear. The presence of helicoids does not require that the myofibrils have anything other than uniform sarcomere lengths, though helicoids are sometimes associated with vernier displacements of the striations, which *are* regions of the fiber with sarcomeres of different lengths.

At the center of each helicoid, there must be a dislocation in the sarco-

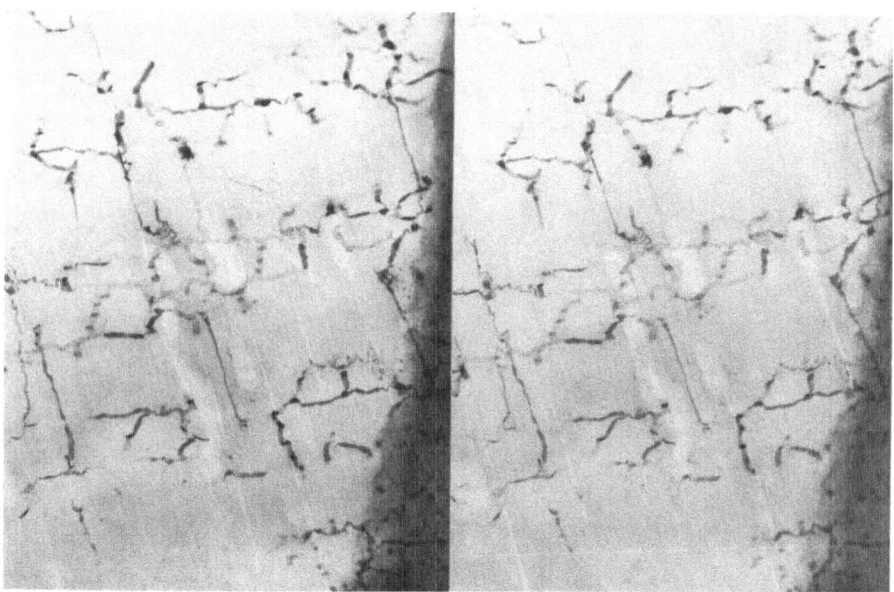

Figure 3. Stereo pair of high-voltage electron micrographs of a transverse slice of a frog muscle stained with the lanthanum method. A portion of the fiber surface is shown at the right. In addition to parts of several successive T-system networks, some longitudinal T-tubules can be seen. Thickness: 3 μm. 1000 kV. Total stereo tilt angle: 10° ×10,000.

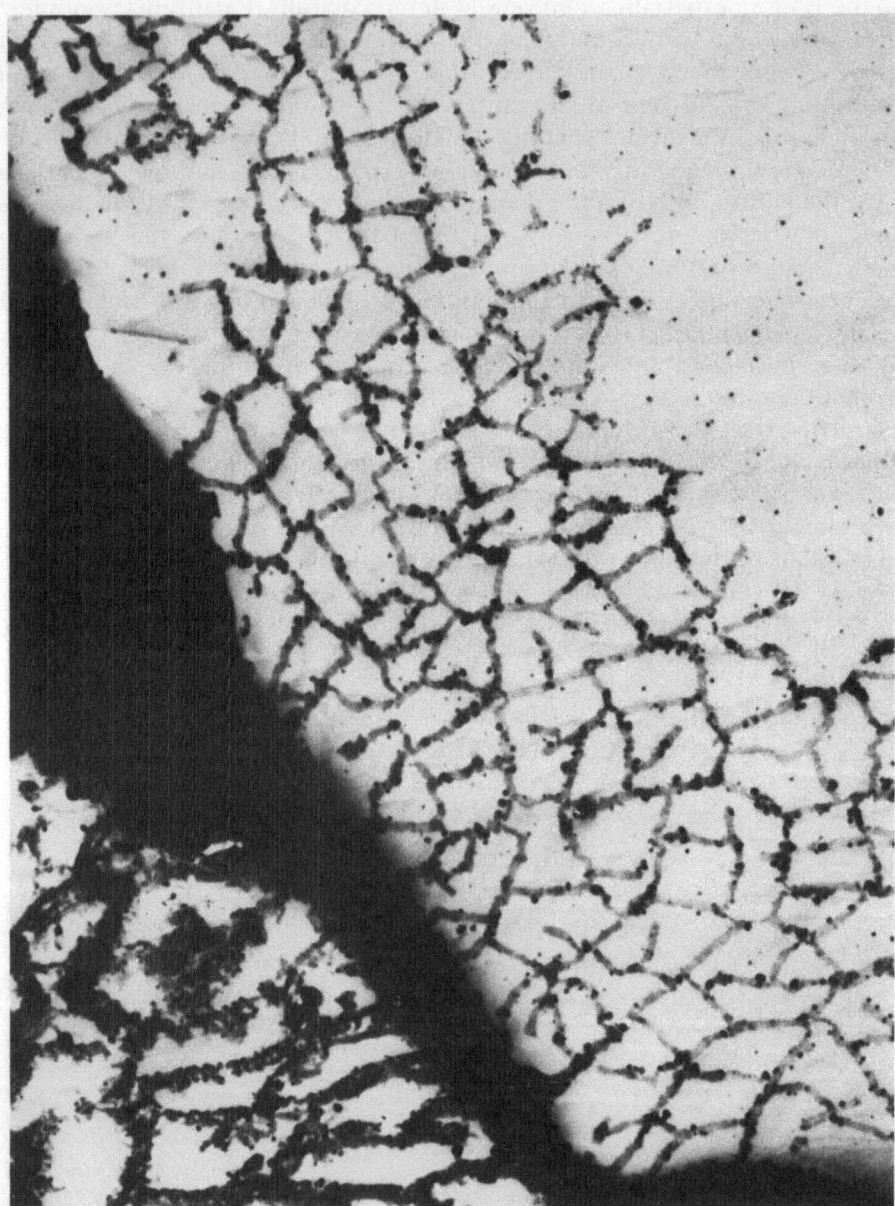

Figure 4. High-voltage electron micrograph of a frog muscle stained with the Golgi method. The dark band is the extracellular space between two muscle fibers, which appear in cross section. The fiber at the top is cut through the I-band, and the T-system is seen. The other fiber is cut through the A-band, and in this fiber, the SR is stained. Thickness: 1.5 μm. 1000 kV. ×5000.

mere banding, and thus of the planes in which the T-systems lie. One might expect there to be discontinuities in the T-system at those points, but another observation that has come out of the HVEM studies shows that that is not the case (Peachey, 1975). These studies have shown that longitudinal extensions of T-tubules are found commonly in the places where dislocations occur in the fibrillar banding pattern (Fig. 3).

The existence of longitudinal T-tubules was suspected from studies of thin sections, though rarely had a tubule been seen to extend all the way from one network longitudinally to the next in a vertebrate muscle (Jasper, 1967). Thus, the images provided by thin sections left a great deal of uncertainty about how frequently tubules with this orientation occurred and made it difficult in most cases to be sure about the identification of any particular tubule as a T-tubule. Selective stains and thick sections in the HVEM left no doubt, however, about the presence of longitudinally oriented T-tubules, and showed that they occur very commonly in association with dislocations in the band pattern of the adjacent myofibrils.

More recently, Dr. Clara Franzini-Armstrong and I have used the Golgi "black reaction" staining method used by Retzius and Veratti in the preparation of muscles from a variety of sources for HVEM. This procedure deposits a very dense precipitate in the T-system, making it especially useful for the

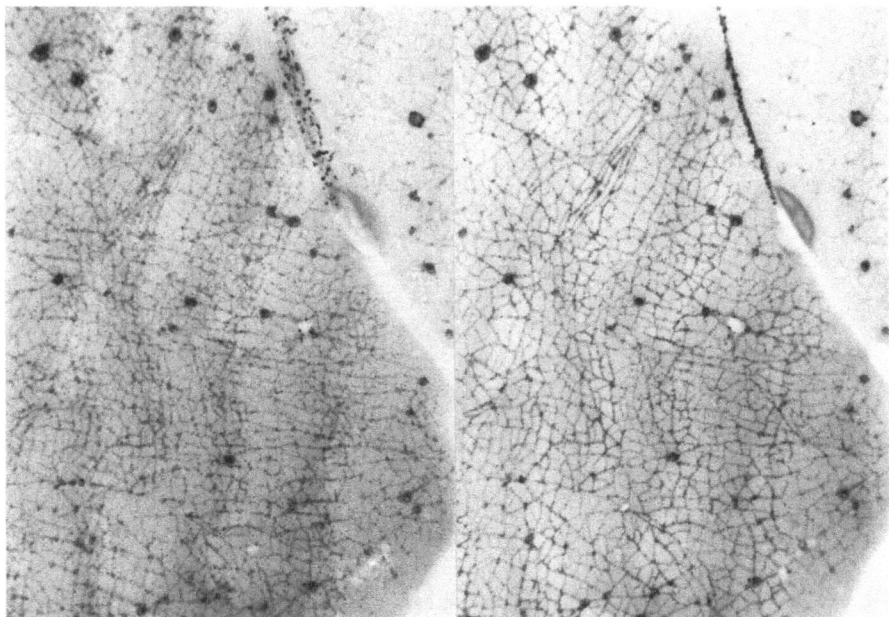

Figure 5. Stereo pair of high-voltage electron micrographs of a transverse slice of a Golgi-stained frog muscle. Complex undulations of the T-system networks are seen. Thickness: 5 μm. 1000 kV. Total stereo tilt angle: 24°. ×3000.

study of the three-dimensional arrangement of T-system membranes. Sometimes, rather irreproducibly, this method also stains the SR (Fig. 4).

The best views with this stain are the stereoscopic ones at low magnification. Such views of thick transverse slices of frog and rat skeletal-muscle fibers show that the planes occupied by the T-system networks often lie obliquely to the fiber axis and often undulate considerably in tbe longitudinal direction (Fig. 5). The number of layers of T-system seen within the thickness of the micrograph depends on the thickness of the slice and on the sarcomere length of the muscle fiber. Frequently, when the slice is thick enough to include more than one complete sarcomere, one can observe helicoids directly in these stereoscopic views. In frog muscle, single turns of the helicoid are seen for each sarcomere length included in the slice thickness. In both red and white rat muscles, a double helix is seen, with each of the two helices making one turn per sarcomere length (Peachey and Franzini-Armstrong, 1978). This, of course, is expected, because there are two T-systems in each sarcomere in these mammalian muscles. We have never observed a helix with a pitch of two sarcomeres, as was reported by Tiegs in the early part of this century based on his light-microscopic observations (see Tiegs, 1955).

For finer details of the structure of the T-tubules, the lanthanum stain is

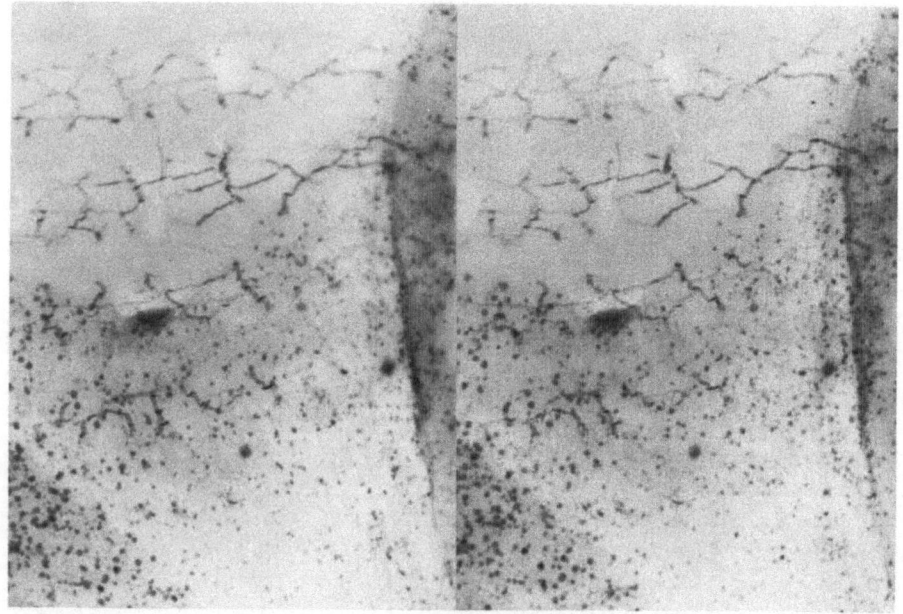

Figure 6. Stereo pair of high-voltage electron micrographs of an approximately longitudinal slice of a frog muscle stained with the lanthanum method. The dark dots are caveolae of the surface membrane of the fiber, which passes obliquely through the thickness of the slice. T-tubules connecting to the surface membrane can be seen. Thickness: 3 μm. 1000 kV. Total stereo tilt angle: 10° ×8000.

better than the Golgi stain (Peachey and Franzini-Armstrong, 1977), though we have found the lanthanum method to be relatively unreliable; in fact, most of the time it simply doesn't penetrate into the fiber, or if it does penetrate, it gets washed out later in the preparation procedure. However, when it does work, it gives some of the most striking pictures we have ever obtained. For example, this procedure is the only one so far that has provided three-dimensional views of the junctions of T-tubules with the surface membrane (Fig. 6).

Overall, HVEM has been kind to us in our quest for further knowledge of the three-dimensional arrangement of membranes in muscle cells. While it has been hard work to get the methods right, we ultimately have been able to see things we had no way of seeing before. It is to be hoped that this will continue and that it will be possible to apply these methods to other kinds of cells.

ACKNOWLEDGMENTS. This work was supported by grants from the Muscular Dystrophy Association (Henry M. Watts Center) and the National Institutes of Health (HL-15835, Pennsylvania Muscle Institute).

References

Bennett, H. S., and Porter, K. R., 1953, An electron microscope study of sectioned breast muscle of the domestic fowl, *Am. J. Anat.* **93**:61.

Franzini-Armstrong, C. F., and Peachey, L. D., 1982, Striated muscle: Contractile and control mechanisms, *J. Cell Biol.* **92**(1) Part 2 (in press).

Goldspink, G., 1971, Ultrastructure changes in striated muscle fibers during contraction and growth with particular reference to the mechanism of myofibril splitting, *J. Cell Sci.* **9**:123–138.

Hill, A. V., 1948, On the time required for diffusion and its relation to processes in muscle, *Proc. R. Soc. London Ser. B* **135**:446.

Hill, A. V., 1949, The abrupt transition from rest to activity in muscle, *Proc. R. Soc. London Ser. B* **136**:399.

Huxley, A. F., and Taylor, R. E., 1958, Local activation of striated muscle fibres, *J. Physiol.* **144**:426.

Jasper, D., 1967, Body muscles of the lamprey: Some structural features of the T system and sarcolemma, *J. Cell Biol.* **32**:219–227.

Peachey, L. D., 1975, Longitudinal elements of the T-system in the region of vernier band displacements in frog skeletal muscle, *J. Cell Biol.* **67**:327a.

Peachey, L. D., and Eisenberg, B. R., 1978, Helicoids in the T-system and striations of frog skeletal muscle fibers seen by high voltage electron microscopy, *Biophys. J.* **22**:145–154.

Peachey, L. D., and Franzini-Armstrong, C., 1977, Three-dimensional visualization of the T-system of frog muscle using high voltage electron microscopy and a lanthanum stain, in: *35th Annual Proceedings of the Electron Microscopy Society of America* (Boston) (G. W. Bailey, ed.), pp. 570–571, Claitor's, Baton Rouge.

Peachey, L. D., and Franzini-Armstrong, C., 1978, Observation of the T-system of rat skeletal muscle fibers in three dimensions using high voltage electron microscopy and the Golgi stain, *Biophys. J.* **21**:61a.

Peachey, L. D., and Porter, K. R., 1959, Intracellular impulse conduction in muscle cells, *Science* **129**:721–722.

Peachey, L. D., and Schild, R. F., 1968, The distribution of the T-system along the sarcomeres of frog and toad sartorius muscles, *J. Physiol.* **194**:249–258.

Porter, K. R., and Palade, G. E., 1957, Studies on the endoplasmic reticulum. III. Its form and distribution in striated muscle cells, *J. Biophys. Biochem. Cytol.* **3**:269.

Retzius, G., 1881, Zur Kenntnis der quergestreiften Muskelfaser, *Biol. Untersuch. Ser. I* **1**:1.

Tiegs, O. W., 1955, The flight muscles of insects—their anatomy and histology: With some observations on the structure of striated muscle in general, *Philos. Trans. R. Soc. Lond. Ser. B* **238**:221–348.

Veratti, E., 1902, Richerche sulla fine struttura della fibra muscolare striata, *Mem. Ist. Lomb. Classe Sci. Mat. Nat.* **19**:87–133.

16

Mechanism of the Apparent Inhibitory Effect of Mg^{2+} and Ca^{2+} on the Actomyosin MgATPase in Relation to the L_2 Light Chain

Suzanne M. Pemrick

1. Introduction

In the past, several laboratories had observed that the actomyosin MgATPase activity was inhibited by micromolar concentrations of either Mg^{2+} or Ca^{2+} (Bullard *et al.*, 1973; Murray, 1973; Lehman *et al.*, 1974; Bremel and Weber, 1975; Pemrick, 1976). Although the L_2 light chain was implicated (Pemrick, 1976), no mechanism was ever proposed for this cation sensitivity. The study reported in this chapter demonstrates that the "apparent" inhibitory effect of either Mg^{2+} or Ca^{2+} can be explained by an activating effect of free ATP, which requires a critical concentration of L_2-containing actomyosin complexes.

These findings are compatible with the hypothesis that L_2 of skeletal myosin is a modifier of the actomyosin ATPase. This hypothesis was based on the following observations: With regulated actin, partial removal of L_2 decreased the apparent Ca^{2+} affinity of troponin and maximal activation by decreasing the steady-state turnover of actomyosin–ADP–Pi complexes (Pemrick, 1977). With pure actin, phosphorylation of L_2 decreased the K_{app} of actin for myosin with no effect on the V_{max} (Pemrick, 1980).

Suzanne M. Pemrick • Department of Biochemistry, Downstate Medical Center, State University of New York, Brooklyn, New York 11203.

2. Methods

Myosin and globular (G)-actin were prepared from the back and leg muscles of rabbits as described previously (Pemrick, 1977). L_2-rich heavy meromyosin (HMM) was prepared by a modification (Pemrick, 1980) of the method of Weeds and Pope (1977). L_2-deficient myosin (1 mole L_2/mole) was prepared by 5,5'-dithiobis-(2-nitrobenzoic acid) (DTNB)–ethylenediamine tetraacetic acid (EDTA) treatment of myosin (Gazith *et al.*, 1970) as described by this laboratory (Pemrick, 1977). L_2-deficient and native myosin had identical K/EDTA-, Mg-, and Ca-ATPase activities.

Concentrated solutions of G-actin (8–12 mg/ml) were polymerized by the addition of 3 M KCl to a final concentration of either 100 mM KCl (for the actomyosin assays) or 30 mM KCl (for the acto-HMM assays). Actomyosin was prepared from filamentous (F)-actin and myosin by first mixing 0.5–4.0 mg/ml F-actin in 0.6 M KCl, followed by addition of soluble myosin (0.6 M KCl, 5 mM imidazole-Cl, pH 7 at 25°C) to a final concentration of 3 mg/ml. Acto-HMM was formed directly in the ATPase medium; no precautions were necessary. The actomyosin MgATPase activity was measured in a total volume of 2.0 ml in a medium of approximately 0.057 ionic strength: 20 mM imidazole-Cl (pH 7, 25°C); 30 mM KCl; 2 mM K_2-ethyleneglycol bis(β-aminoethyl ether)N,N'-tetraacetic acid (EGTA)–CaEGTA buffers (pH 7 at 25°C); 0.175–0.350 mg/ml actomyosin; 3 min incubation with 0.6–2 mM MgATP. The acto-HMM MgATPase activity was measured in 0.5–1.0 ml of a medium at 0.05 ionic strength: 5 mM imidazole-Cl (pH 7, 25°C); 1 mM K_2EGTA; 25 mM KCl (from F-actin); 2–25 μM F-actin; 0.45–80 μM HMM "heads"; 30 sec to 1 min incubation with 1.0 mM MgATP at various concentrations of Mg^{2+} (constant ionic strength). The MgATPase activity of HMM and myosin was determined as described above, but in the absence of actin. The remaining details of these assays have been described previously (Pemrick, 1977, 1980). The steady-state kinetic constants, K_{app} and V_{max}, were determined by a weighted fit of a linear-regression analysis (Wilkinson, 1961).

Calcium concentration was regulated by varying the K_2EGTA/CaEGTA ratio at a constant ionic strength assuming a K_d of 0.19 μM at pH 7 for Ca^{2+} and EGTA (Chaberek and Martell, 1959). The computations for determining Ca^{2+} were corrected for CaATP as described by Bremel and Weber (1975). Mg^{2+} concentration was regulated by assuming a K_d of 40 μM for Mg^{2+} and ATP at pH 7 (O'Sullivan and Perrin, 1964). If, as in these studies, the Mg^{2+} concentration is increased at a constant MgATP concentration, the ionic strength first decreases from 10 to 200 μM Mg^{2+}, due to a fall in the free ATP concentration, then remains constant (from 200 μM to 1.0 mM Mg^{2+}), and finally increases from 1 to 5 mM Mg^{2+}. At 1.0 mM MgATP, the total fluctuation in ionic strength is approximately 0.02, based on a ratio of 0.83 for ATP^{4-}/ATP^{3-} at pH 7 (O'Sullivan and Perrin, 1964). The actomyosin MgATPase system was insensitive to this variation in ionic strength (between 0.05 and 0.075). This was not true of the acto-HMM system. Therefore, in this instance, the ionic strength was held constant (at 0.05) from 20 μM to 5.0 mM Mg^{2+}.

3. Results and Discussion

When the actomyosin MgATPase activity was compared as a function of Ca^{2+} and Mg^{2+}, and at an actin concentration above K_{app}, several phenomena were apparent. First, Pi liberation was inhibited 50–60% by intermediate concentrations (100 μM) of either Ca^{2+} or Mg^{2+}. Second, the effect was not additive; that is Ca^{2+} plus Mg^{2+} was not more effective than either Ca^{2+} or Mg^{2+} alone. These results suggested that Ca^{2+} and Mg^{2+} were acting on the system by a common mechanism. Interestingly, if the same experiment was repeated with L$_2$-deficient actomyosin, there was no inhibitory response to either Mg^{2+} or Ca^{2+}. If the investigation had been concluded at this stage, these results could have been interpreted as a direct effect of L$_2$ (Pemrick, 1976). When the study was expanded, and the experiment repeated at actin concentrations below K_{app}, neither native nor L$_2$-deficient actomyosin was responsive to Mg^{2+} (20 μM to 5.0 mM) or to Ca^{2+} (0.01–100 μM). Therefore, this phenomenon was dependent on a critical concentration of actomyosin complexes, and was regulated by the steady-state kinetic constants, K_{app} and V_{max} (Table 1).

A casual inspection of the data in Table 1 indicates that the most distinctive characteristic was not the inhibitory effect of high Mg^{2+}, high Ca^{2+}, or partial removal of L$_2$, but rather the activating effect of low Mg^{2+} (20 μM, no Ca^{2+}) for native actomyosin. Under these conditions, both K_{app} and V_{max} were 2.5-fold greater than at either 100 μM Mg^{2+} (no Ca^{2+}) or 100 μM Ca^{2+} (20 μM Mg^{2+}). Substitution of 100 μM Mg^{2+} for 100 μM Ca^{2+} had no significant effect on either K_{app} or V_{max}. In the case of L$_2$-deficient actomyosin, the activating effect of low Mg^{2+} (no Ca^{2+}) was less dramatic (30% increase in K_{app}, 50% increase in V_{max}) but statistically significant. Furthermore, there was some

Table 1. *Apparent Inhibitory Effect of Ca^{2+} and Mg^{2+} on the Actomyosin MgATPase in Relation to the L$_2$ Light Chain*

Species	Conditions	K_{app} (μM actin)	V_{max}[a]
Native myosin	20 μM Mg^{2+} (2 mM K$_2$EGTA)	4.62 ± 0.12	9.20 ± 0.14
	100 μM Mg^{2+} (2 mM K$_2$EGTA)	1.98 ± 0.35[b]	3.66 ± 0.29[b]
	20 μM Mg^{2+} (100 μM Ca^{2+})[c]	2.46 ± 0.71	3.68 ± 0.51
L$_2$-deficient myosin	20 μM Mg^{2+} (2 mM K$_2$EGTA)	2.53 ± 0.27	3.37 ± 0.17
	100 μM Mg^{2+} (2 mM K$_2$EGTA)	1.96 ± 0.28[b]	2.21 ± 0.15[b]
	20 μM Mg^{2+} (100 μM Ca^{2+})[c]	4.52 ± 0.42	4.12 ± 0.22

[a] V_{max} = nmoles P$_i$ sec^{-1}·nmoles myosin "head"$^{-1}$.
[b] The steady-state kinetics were unaltered from 100 μM to 1.0 mM Mg^{2+} (no Ca^{2+}).
[c] At higher Ca^{2+} concentrations, the concentration of CaATP would be considerable, and therefore both CaATP and MgATP would compete for the hydrolytic site of myosin.

indication that L_2-deficient actomyosin could distinguish between high Ca^{2+} and high Mg^{2+}, producing a type of "Ca^{2+} sensitivity," due to an approximate 2-fold increase in both K_{app} and V_{max} at 100 μM Ca^{2+}.

Several conclusions could be drawn from the data in Table 1. With pure actin, the inhibitory effect of partial removal of L_2 could best be observed at low Mg^{2+}, and in the absence of Ca^{2+}. With one exception, the second-order rate constant (V_{max}/K_{app}) was unaffected by the parameters listed in Table 1. Therefore, these differences in K_{app} and V_{max} were due to changes in the rate constants controlling V_{max}. Low Mg^{2+}, in the absence of Ca^{2+}, activated the rate of P_i liberation from actomyosin complexes (V_{max}). Removal of 50% of L_2 decreased by 80% this activating effect of low Mg^{2+}.

In the experiments listed in Table 1, the Mg^{2+} and the Ca^{2+} concentration were varied at a constant MgATP concentration (1 mM). Therefore, as Mg^{2+} increased, free ATP decreased. To distinguish between an inhibitory effect of Mg^{2+} and an activating effect of free ATP, the actomyosin ATPase was analyzed as a function of Mg^{2+} (no Ca^{2+}) at two MgATP concentrations (Fig. 1). AT 2.0 mM MgATP, 500 μM Mg^{2+} was required for maximal inhibition,

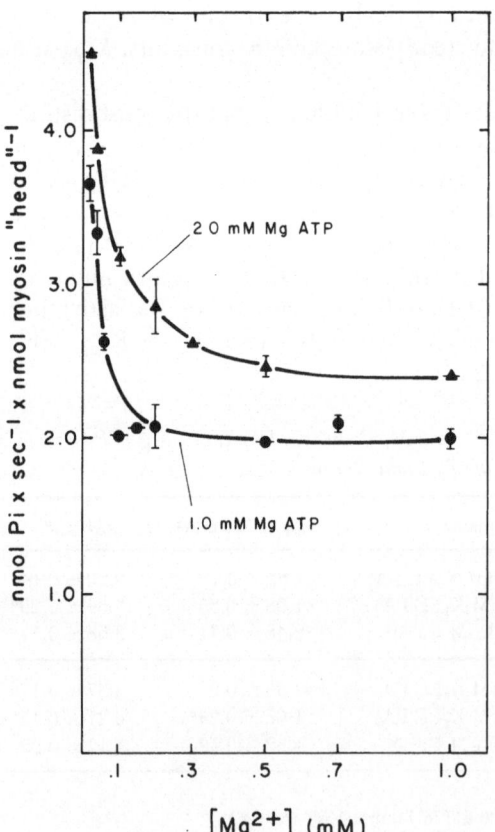

Figure 1. Native actomyosin (2 moles L_2/mole myosin) MgATPase activity in the absence of Ca^{2+}, as a function of the Mg^{2+} concentration: (▲) 2.0; (●) 1.0 mM MgATP. Data are expressed as means ± S.E. (N = 3). Note that the curves were not superimposable. At 2.0 mM MgATP, the Mg^{2+} concentration at 50% inhibition of the actomyosin MgATPase activity is 5-fold greater (500 μM Mg^{2+}) than that (100 μM Mg^{2+}) at 1.0 mM MgATP (3.6 μM actin, 0.6 μM myosin "heads").

whereas at 1.0 mM MgATP, 100 μM Mg^{2+} was sufficient. Therefore, this phenomenon, as characterized in Table 1, was not an Mg^{2+} effect. When the data were normalized to percentage maximal activity and compared as a function of the free ATP concentration, it became apparent that the data at three MgATP concentrations could be fitted by a single curve. Therefore, the 2.5-fold increase in K_{app} and V_{max} observed in Table 1 at low Mg^{2+} (native myosin) was due to an activating effect of free ATP (Fig. 2).

A similar effect could be demonstrated for the soluble acto-HMM system. It is generally accepted that the K_{app} of actin for HMM is much higher than that for myosin. For this reason, it was necessary to shift the concentration of HMM intermediates toward those that contained the products of ATP hydrolysis in order to observe the activating effect of free ATP. These were the conditions that Eisenberg and colleagues (Eisenberg and Kielley, 1972; Stein *et al.,* 1979) employed to make the empirical observation of a "refractory state." To illustrate, at high actin (25 μM) and low HMM (0.45 μM "heads") concentrations, free ATP had no effect on the acto-HMM activity (per nanomole HMM). However, at high HMM (80 μM "heads") and low actin (2 μM) concentrations, an activating effect of free ATP (rate per nanomole actin) was again observed (Fig. 3). If, at 2 μM actin, the HMM concentration was varied, the rate was linear and insensitive to free ATP until 30 μM HMM "heads," whereupon the rate leveled off at low free ATP (3 sec^{-1} · nmol $actin^{-1}$), but continued to rise at high free ATP, to a maximal activation of 6.3 sec^{-1} · nmol $actin^{-1}$. Similar to the situation for native actomyosin, free ATP (1.0 mM) increased both K_{app} (from 10 to 50 μM HMM "heads") and V_{max} (from 3.5 to 11 sec^{-1} · nmol $actin^{-1}$), with no significant effect on the second-order rate constant. Therefore, as for actomyosin, free ATP activated the rate of release of P_i from acto-HMM complexes.

Figure 2. Normalization of the actomyosin MgATPase activity as a function of the free ATP concentration (ATP^{4-} + ATP^{3-}) at 0.6 ($\times$), 1.0 ($\bullet$), and 2.0 mM ($\blacktriangle$) MgATP. The data have been normalized to the maximal activity at 4 mM free ATP (2.0 mM MgATP, 20 μM Mg^{2+}) of 4.5 nmol P_i·sec^{-1}·nmole myosin "heads"$^{-1}$. Within experimental error, these data were superimposable (3.6 μM actin, 0.6 μM myosin "heads").

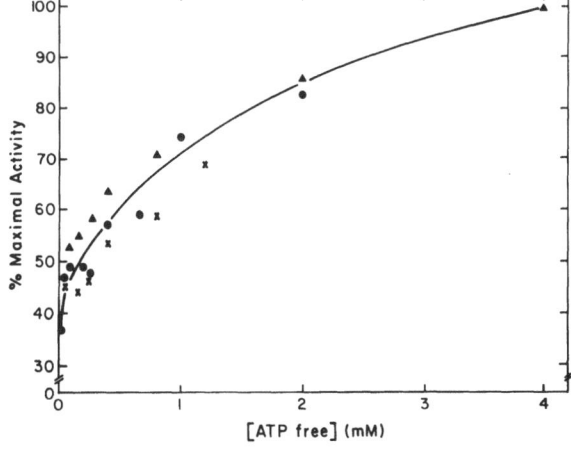

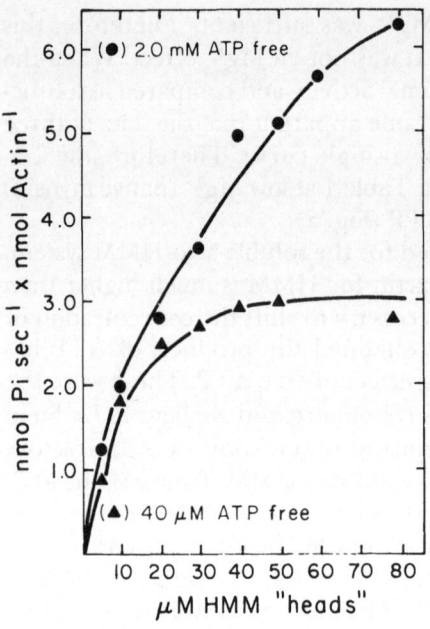

Figure 3. Acto-HMM MgATPase activity (no Ca^{2+}) at low (▲) and high (●) free ATP, as a function of the HMM concentration (1.0 mM MgATP, 2 μM actin, pH 7, 0.05 ionic strength throughout).

4. Conclusions

At this time, it is not possible to prove that the experimental conditions of high HMM and low actin were selecting for those HMM "heads" that contained L_2, although the implication is obvious. By analogy from the actomyosin system, this investigator tentatively concludes that at least 80% of the ability of free ATP to activate the system requires the full complement of L_2 (2 moles L_2/mole myosin or HMM).

Future studies on L_2 function, in relation to the actomyosin ATPase or to tension generation, must include a wide range of actin concentrations and, in the case of pure F-actin, must be limited to low Mg^{2+} and to low Ca^{2+} concentrations. Otherwise, the findings will be negative for the wrong reason (Srivastava *et al.*, 1980).

The physiological significance of this function for L_2 is unknown. It has been observed that at elevated concentrations of free ATP, higher concentrations of MgATP were required to inhibit the myofibrillar ATPase activity in the absence of calcium (relaxation) (Weber *et al.*, 1969). This parameter is the *in vitro* corollary of resting tension. This laboratory has demonstrated that loss of L_2 decreases (Pemrick, 1977) and phosphorylation of L_2 increases (Pemrick, 1980) the *in vitro* corollary of resting tension. The present results reported herein are the third indication that L_2 may be involved in regulating resting tension in skeletal muscle.

ACKNOWLEDGMENTS. This research was supported by Grant HL22401 from the United States Public Health Service and by a grant from the Muscular Dystrophy Association, Inc.

References

Bremel, R. D., and Weber, A., 1975, Calcium binding to rabbit skeletal myosin under physiological conditions, *Biochim. Biophys. Acta* **376**:366.

Bullard, B., Dabrowska, R., and Winkelman, L. B., 1973, The contractile and regulatory proteins of insect flight muscle, *Biochem. J.* **135**:277.

Chaberek, S., and Martell, A. E., 1959, Metal Buffers, *in: Organic Sequestering Agents*, pp. 174–215, Wiley and Sons, New York.

Eisenberg, E., and Kielley, W. W., 1972, Evidence for a refractory state of heavy meromyosin and subfragment-1 unable to bind to actin in the presence of ATP, *Cold Spr. Harbor Symp. Quant. Biol.* **37**:145.

Gazith, J., Himmelfarb, S., and Harrington, W. F., 1970, Studies on the subunit structure of myosin, *J. Biol. Chem.* **245**:15.

Lehman, W., Bullard, B., and Hammond, K., 1974, Calcium-dependent myosin from insect flight muscles, *J. Gen. Physiol.* **63**:553.

Murray, J. M., 1973, Cooperative alterations in the behavior of contractile proteins, Ph.D. thesis, University of Pennsylvania.

O'Sullivan, W. J., and Perrin, D. D., 1964, The stability constants of metal-adenine nucleotide complexes, *Biochemistry* **3**:18.

Pemrick, S. M., 1976, The effect of magnesium and calcium on the MgATPase activity of myosin in relation to the L_2 light chain, *Biophys. J.* **16**:70a.

Pemrick, S. M., 1977, Comparison of the calcium sensitivity of actomyosin from native and L_2-deficient myosin, *Biochemistry* **16**:4047.

Pemrick, S. M., 1980, The phosphorylated L_2 light chain of skeletal myosin is a modifier of the actomyosin ATPase, *J. Biol. Chem.* **255**:8836.

Srivastava, S., Cooke, R., and Wikman-Coffelt, J., 1980, Studies on the role of myosin light chain—LC_2 in tension generation, *Biochem. Biophys. Res. Commun.* **92**:1.

Stein, L. A., Schwarz, R. P., Jr., Chock, P. B., and Eisenberg, E., 1979, Mechanism of actomyosin adenosine triphosphatase. Evidence that adenosine 5'-triphosphate hydrolysis can occur without dissociation of the actomyosin complex, *Biochemistry* **18**:3895.

Weber, A., Herz, R., and Reiss, I., 1969, The role of magnesium in the relaxation of myofibrils, *Biochemistry* **8**:2266.

Weeds, A. G., and Pope, B., 1977, Studies on the chymotryptic digestion of myosin. Effects of divalent cations on proteolytic susceptibility, *J. Mol. Biol.* **111**:129.

Wilkinson, G. N., 1961, Statistical estimations in enzyme kinetics, *Biochem. J.* **80**:324.

17

Actin Activation of Phosphorylated Aortic Myosin

Dixie W. Frederiksen

1. Introduction

The interaction of actin and myosin in smooth muscle, like that in striated muscle, is controlled through changes in the intracellular calcium concentration. The calcium-sensitive mechanism responsible for the primary regulation in smooth muscle occurs through phosphorylation of myosin light chain (Barron *et al.*, 1980; Hathaway and Adelstein, 1979; Sherry *et al.*, 1978). The investigators cited have shown that myosin phosphorylation is prerequisite to actin activation of the MgATPase. A secondary modulation of myosin MgATPase by calcium was first reported by Chacko *et al.* (1977). This group showed that direct interaction of calcium with phosphorylated myosin from vas deferens increased the MgATPase activity to a level higher than that observed in the absence of calcium. Similar findings for phosphorylated myosin from vascular smooth muscle were reported by Rees and Frederiksen (1981).

The object of the study reported in this chapter has been to clarify further the nature of the action of calcium on phosphorylated myosin from porcine aortae. These experiments indicate that calcium serves to increase the affinity of phosphorylated myosin for actin.

2. Materials and Methods

Skeletal-muscle actin was isolated from the back and hind legs of 4- to 6-month-old New Zealand albino rabbits. Dehydrated muscle was stored at

Dixie W. Frederiksen • Department of Biochemistry, Vanderbilt University, Nashville, Tennessee 37232.

−20°C as an acetone powder (Noelken, 1962) before use. Actin was prepared from the powder by the method of Spudich and Watt (1971). The purified actin exhibited a single band on sodium dodecyl sulfate–polyacrylamide gel electrophoresis (SDS-PAGE) and contained no detectable contamination by tropomyosin or troponin.

Aortic actomyosin was purified from minced porcine aortic media as described by Frederiksen (1979). Briefly, the vessels were extracted at high ionic strength and in the presence of ATP; after an extended centrifugation to remove most of the actin and other contaminating proteins, the solution was brought to 30 mM in $MgSO_4$ and allowed to precipitate. The precipitated protein was used as actomyosin. The preparation contained both myosin-light-chain kinase (MLCK) and myosin-light-chain phosphatase (MLCP) in addition to actin, myosin, and tropomyosin. Phosphorylated and unphosphorylated actomyosins were prepared as described by Rees and Frederiksen (1981), except that the actomyosin concentration in the reaction mixture was reduced by 20%, to 3.5 mg/ml.

Phosphorylated aortic myosin was purified from phosphorylated actomyosin by agarose chromatography on Sepharose 4B by the method of Rees and Frederiksen (1981). The preparation was judged pure by a number of criteria. SDS-PAGE showed only three protein bands. These corresponded to the heavy and light polypeptide chain subunits of myosin. There was no band that migrated with the relative mobility of actin or tropomyosin. In addition, the purified phosphomyosin did not incorporate ^{32}P when incubated with $[\gamma\text{-}^{32}P]ATP$ and 0.1 mM $CaCl_2$ under conditions that activate MLCK. Purified ^{32}P-labeled phosphomyosin did not lose phosphate when incubated in 2.0 mM ethyleneglycol bis(β-aminoethyl ether)N',N'-tetraacetic acid (EGTA) under conditions optimal for MLCP activity.

Unphosphorylated aortic myosin was prepared by exactly the same procedure as phosphorylated myosin except that 2.0 mM EGTA was substituted for 0.1 mM $CaCl_2$ in all the buffers used. The unphosphorylated protein preparation contained no detectable contamination by actin, tropomyosin, kinase, or phosphatase.

Protein concentration was routinely determined by an automated (Technicon, Autoanalyzer) modification of the method of Lowry *et al.* (1951). Three to five separate dilutions of a particular protein stock solution were assayed to determine the stock-solution concentration. Bovine serum albumin standards were run concurrently with each assay.

SDS-PAGE was carried out by the method of Weber and Osborn (1969) with the modifications suggested by Frederiksen (1976). Protein was routinely electrophoresed at 5 mA/gel in 5-mm-diameter cylindrical gels consisting of 7.65% acrylamide–0.35% bisacrylamide in 3.5 mM SDS–50 mM NaP_i (pH 7.0).

Gels were stained with Coomassie brilliant blue and scanned at 560 nm with a Gilford spectrophotometer equipped with a scanning attachment.

ATPase assays. The actin activation of phosphorylated and unphosphorylated aortic myosin MgATPase was assayed as a function of skeletal-muscle actin concentration. Assays were performed at 35°C on solutions containing 35 mM KCl–4 mM $MgSO_4$–1 mM dithiothreitol–50 mM morphilinopropane-sulfonic acid at pH 7.0. P_i release was measured as a function of time. The phosphate was determined by the procedure of Martin and Doty (1949) as modified by Pollard and Korn (1973). Standard solutions of KH_2PO_4 were run with each assay.

3. Results

The MgATPase activity of phosphorylated and unphosphorylated aortic myosins were measured as a function of added skeletal-muscle actin in the presence and in the absence of calcium. Assays were performed as described in Section 2 in the low-ionic-strength buffer and in the presence (0.1 mM $CaCl_2$) or in the absence (2.0 mM EGTA) of calcium. The myosin concentration in these experiments was 0.2 mg/ml, or 0.44 μM, based on a myosin molecular weight of 452,000 (Frederiksen, 1979). Actin concentrations ranged from 0 to 8.85 μM. The data for phosphorylated myosin are shown in Fig. 1. The ATPase activity of myosin alone was subtracted from the activities measured in the presence of actin. The MgATPase exhibited a hyperbolic dependence on actin concentration in the presence and the absence of calcium. The activity in the presence of calcium, however, was consistently higher than that observed in the presence of EGTA. Unphosphorylated aortic myosin, on the other hand, showed no dependence on the calcium concentration.

Double-reciprocal plots of these data were linear (Fig. 2). The MgATPase activity at infinite actin concentration, V_{app}, was 8.13 μM P_i/μM phosphomyosin per min in both the presence and the absence of calcium. At 0.51 μM actin, the MgATPase activity in the presence of calcium was 50% of V_{app}. In the presence of EGTA, however, the activity of the enzyme was half of V_{app} at 1.18 μM actin. The concentration of actin required to produce 50% activation of the MgATPase is halved in the presence of calcium.

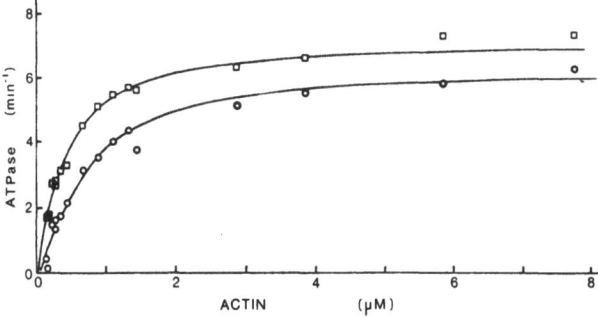

Figure 1. Actin activation of phosphorylated aortic myosin. The MgATPase activity of phosphomyosin was assayed as a function of added skeletal-muscle myosin (0.44 M) as described in the text at 35°C in 35 mM KCl–4 mM $MgSO_4$–1 mM dithiothreitol–50 mM morphilinopropane-sulfonic acid (pH 7.0) plus 0.1 mM $CaCl_2$ (□) or 2.0 mM EGTA (○).

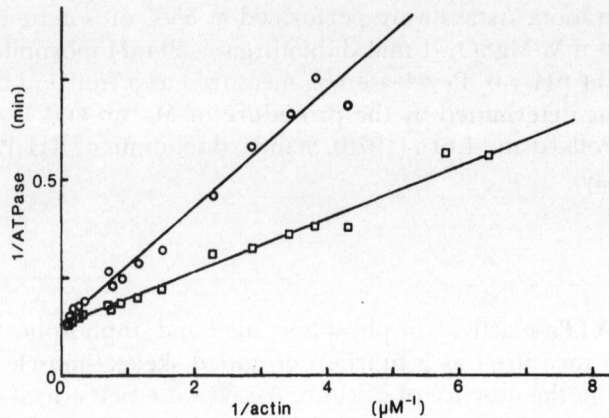

Figure 2. Actin activation of phosphomyosin in the presence and in the absence of calcium. The hyperbolic data of Fig. 1 are given in double-reciprocal form. Assays were performed in 0.1 mM CaCl$_2$ ($\square$) or 2.0 mM EGTA ($\bigcirc$).

4. Discussion

The results given here are the first to assign a molecular basis to the secondary modulation of phosphomyosin ATPase by calcium. The divalent cation seems to act by altering the affinity of phosphorylated myosin for actin. Chacko *et al.* (1977) and Rees and Frederiksen (1981) have shown that calcium influences the MgATPase of phosphorylated myosin in both vas deferens and arterial smooth muscle. This study has taken those observations a step further.

The effect of calcium on phosphorylated aortic myosin is not explained by contamination of the protein. Neither MLCK nor MLCP could be detected in the purified myosin preparations. Furthermore, the skeletal-muscle actin preparations contained no contamination by regulatory proteins. The action of calcium seems to be on myosin itself.

Because the affinity of phosphomyosin for actin approximately doubles in the presence of calcium, the cation may induce an interaction between the two heads of myosin.

ACKNOWLEDGMENTS. The author is grateful for the technical assistance of Phillip B. Flexon and Anna Chytil. She appreciates the generous gift of fresh porcine aortae from Kentucky Sausage Company, Nashville, Tennessee. This work was supported, in part, by N.I.H. Research Grants HL 18516 and NS 15077.

References

Barron, J. T., Bárány, M., Bárány, K., and Storti, R. V., 1980, Reversible phosphorylation and dephosphorylation of the 20,000-dalton light chain of myosin during the contraction–relaxation–contraction cycle of arterial smooth muscle, *J. Biol. Chem.* **255**:6238.

Chacko, S., Conti, M. A., and Adelstein, R. S., 1977, Effect of phosphorylation of smooth muscle myosin on actin-activation and Ca^{++}-regulation, *Proc. Natl. Acad. Sci. U.S.A.* **74**:129.

Frederiksen, D. W., 1976, Myosin-mediated Ca^{++}-regulation of actomyosin-adenosinetriphosphatase from porcine aorta, *Proc. Natl. Acad. Sci. U.S.A.* **73**:2706.

Frederiksen, D. W., 1979, Physical properties of myosin from aortic smooth muscle, *Biochemistry* **18**:1651.

Hathaway, D. R., and Adelstein, R. S., 1979, Human platelet myosin light chain kinase requires the Ca^{2+}-binding protein calmodulin for activity, *Proc. Natl. Acad. Sci. U.S.A.* **76**:1653.

Lowry, O. H., Rosebrough, N. J., Farr, A. L., and Randall, R. J., 1951, Protein measurement with the Folin reagent, *J. Biol. Chem.* **193**:265.

Martin, J. B., and Doty, D. M., 1949, Determination of inorganic phosphorus, *Anal. Chem.* **21**:965.

Noelken, M. E., 1962, Ph.D. thesis, The denaturation of paramyosin and tropomyosin and attempts to isolate actin, pp. 44–45, Washington University, St. Louis.

Pollard, T. D., and Korn, E. D., 1973, Acanthamoeba myosin. I. Isolation from *Acanthamoeba castellanii* of an enzyme similar to muscle myosin, *J. Biol. Chem.* **248**:4682.

Rees, D. D., and Frederiksen, D. W., 1981, Calcium regulation of porcine aortic myosin, *J. Biol. Chem.* **256**:357.

Sherry, J. M. F., Gorecka, A., Askoy, M. D., Dabrowska, R., and Hartshorne, D. J., 1978, Roles of calcium and phosphorylation in the regulation of the activity of gizzard myosin, *Biochemistry* **17**:4411.

Spudich, J., and Watt, S., 1971, The regulation of rabbit skeletal muscle contraction, *J. Biol. Chem.* **246**:4866.

Weber, K., and Osborn, M., 1969, The reliability of molecular weight determinations by dodecyl sulfate–polyacrylamide gel electrophoresis, *J. Biol. Chem.* **244**:4406.

18

The Regulation of Cardiac-Muscle Contraction by Troponin

James D. Potter, Michael J. Holroyde,
Steven P. Robertson, R. John Solaro, Evangelia G. Kranias,
and J. David Johnson

1. Introduction

The purpose of this review is to summarize the work we have done over the past several years regarding the regulation of cardiac muscle contraction by troponin (Tn). Studies have been carried out on the Ca^{2+}-binding properties of cardiac Tn (the $1:1:1$ molar-ratio complex of TnC, the Ca^{2+}-binding subunit; TnI, the inhibitory subunit; and TnT, the tropomyosin-binding subunit) and TnC, and the effects of Mg^{2+} and phosphorylation on these parameters have been measured. The relationship between these parameters and the activation of myofibrillar ATPase in cardiac and skeletal muscle will be discussed and related to the known physiological properties of cardiac muscle in different contractile states (e.g., β-adrenergic stimulation). Since the binding of Ca^{2+} to Tn is only the first step in the activation of muscle contraction, studies on Ca^{2+}-induced alterations in the interactions of Tn subunits, tropomyosin (Tm), and actin required to bring about muscle contraction or relaxation will also be discussed.

James D. Potter, Steven P. Robertson, Evangelia G. Kranias, and J. David Johnson • Section of Contractile Proteins, Department of Pharmacology and Cell Biophysics, University of Cincinnati College of Medicine, Cincinnati, Ohio 45267. *Michael J. Holroyde* • Department of Physiology, University of Cincinnati College of Medicine, Cincinnati, Ohio 45267. *R. John Solaro* • Department of Physiology and Department of Pharmacology and Cell Biophysics, University of Cincinnati College of Medicine, Cincinnati, Ohio 45267.

2. Ca²⁺ Binding to Cardiac Troponin (CTn) and the Regulation of Contraction

It is well established that the regulation of cardiac and skeletal muscle depends on the Ca^{2+} concentration in the myofilament space (Ebashi *et al.*, 1969; Weber and Murray, 1973). The Ca^{2+} receptor in both systems has been shown to be Tn (Ebashi *et al.*, 1969), although the properties of Tn from these two muscle types vary. For example, the cardiac subunits of Tn differ from their counterparts in skeletal muscle with respect to molecular weight, amino acid composition (Perry, 1979; Brekke and Greaser, 1976; Potter *et al.*, 1977), immunochemical reactivity (Hirabayashi and Perry, 1973), and metal-binding parameters (Ebashi *et al.*, 1969; Potter *et al.*, 1977; Leavis and Kraft, 1978).

The Ca^{2+}-binding properties of skeletal TnC (STnC) and skeletal Tn (STn) have been well characterized (Potter and Gergely, 1975). STnC contains two sites that have a high affinity for Ca^{2+} ($K_{Ca} = 2 \times 10^7 M^{-1}$) and also bind Mg^{2+} competitively ($K_{Mg} = 4 \times 10^3 M^{-1}$) and have been called the Ca^{2+}–Mg^{2+} sites (Potter and Gergely, 1975). STnC also contains two sites of slightly lower Ca^{2+} affinity ($K_{Ca} = 3.2 \times 10^5 M^{-1}$) that are specific for Ca^{2+} (Potter and Gergely, 1975) and have been called the Ca^{2+}-specific sites. Numerous studies have indicated that the Ca^{2+}-specific sites are the regulatory sites of muscle contraction (Potter and Gergely, 1975; Potter *et al.*, 1977; Johnson *et al.*, 1979, 1981; Robertson *et al.*, 1981b).

In contrast to those of STnC and STn, the Ca^{2+}-binding properties of cardiac TnC (CTnC) and CTn, until recently, have not been well understood. A comparison between the sequence studies of van Eerd and Takahashi (1976) on CTnC and the sequences of parvalbumin and STnC (Collins *et al.*, 1977) suggested that Ca^{2+}-binding region I of CTnC would not bind Ca^{2+} due to the substitution of a leucine and an alanine for two key Ca^{2+}-coordinating aspartic acid residues. Direct Ca^{2+}-binding measurements of CTnC (Potter *et al.*, 1977) have shown that it indeed contains only three Ca^{2+}-binding sites, two Ca^{2+}–Mg^{2+} sites like those on STnC, and one Ca^{2+}-specific site. These direct binding studies have been confirmed by indirect measures of Ca^{2+} binding (Leavis and Kraft, 1978). Since the sequence of region I in CTnC appeared defective, it became obvious from the binding measurements on CTnC that region I is the location of one of the two Ca^{2+}-specific sites in STnC (Potter *et al.*, 1977). Previous studies have shown (Potter *et al.*, 1976) that region III contained one of the two Ca^{2+}–Mg^{2+} sites of STnC. Furthermore, by comparing the sequences and Ca^{2+}-binding properties of a variety of Ca^{2+}-binding proteins (e.g., CTnC, STnC, calmodulin, parvalbumin, DTNB light chain), it was possible to show that region I and II of STnC and CTnC contain the Ca^{2+}-specific sites and that regions III and IV contain the two Ca^{2+}–Mg^{2+} sites. Leavis *et al.* (1978) and Leavis and Kraft (1978) have come to a similar conclusion based on fragment and binding studies.

Although the studies cited above support the original proposal of van Eerd and Takahasi (1976), there are several conflicting reports with regard to the Ca^{2+}- and Mg^{2+}-binding properties of CTnC and CTn. Burtnick and Kay

(1977) have reported that CTnC contains three sites of equal affinity and that they are all Ca^{2+}–Mg^{2+}-type sites. Although Stull and Buss (1978), using bovine CTn, and Kohama (1979), using chicken CTn, both report that CTn binds 1.5–2.0 moles of Ca^{2+} per mole, there was disagreement as to the number of high- and low-affinity binding sites.

2.1. Ca^{2+} Binding to Cardiac Troponin, Troponin C, and Troponin C–Troponin I Complex

To clarify the discrepancies discussed above, we have carried out a detailed study of the Ca^{2+}-binding properties of CTn, CTnC, and CTnC–CTnI complex. We have used direct Ca^{2+}-binding techniques to study the Ca^{2+}–Mg^{2+} and Ca^{2+}-specific sites of these proteins (Potter and Gergely, 1975; Holroyde *et al.*, 1980) and indirect-fluorescence techniques to study the Ca^{2+}-specific sites (Johnson *et al.*, 1980). Both techniques agree very well (Table 1) and indicate that CTnC contains two Ca^{2+}–Mg^{2+} sites of affinity similar to that of the two Ca^{2+}–Mg^{2+} sites in STnC and a single Ca^{2+}-specific site of affinity similar to that of the two Ca^{2+}-specific sites in STnC (Table 1). One notable difference, however, is the presence of positive cooperativity in the binding of Ca^{2+} to Ca^{2+}–Mg^{2+} sites in CTnC (Potter *et al.*, 1977; Holroyde *et al.*, 1980) not seen in Ca^{2+} binding to the Ca^{2+}–Mg^{2+} sites of STnC. Interestingly, this cooperativity is lost in CTn or in the CTnC–CTnI complex (Holroyde *et al.*, 1980). The affinity of the four sites on STnC and the three sites on CTnC increases by an order of magnitude on formation of the TnC–TnI complex or in Tn (see Table 1).

Table 1. Comparison of the Ca^{2+}-Binding Properties of Cardiac- and Skeletal-Muscle TnC, TnC–TnI Complex, and Tn

Protein[a]	MgCl$_2$ concentration (M)	n_1 (mol/mol)	K_1 (M^{-1})	n_2 (mol/mol)	K_2 (M^{-1})
STnC[b]	—	2	2.1×10^7	2	3.2×10^5
	2×10^{-3}	2	2.8×10^6	2	1.1×10^5
CTnC[c]	—	2	1.4×10^7	1	2.5×10^{5d}
	4×10^{-3}	2	3.6×10^6	1	2.5×10^{5d}
STnC–STnI[b]	—	2	2.2×10^8	2	3.5×10^6
	2×10^{-3}	2	4.3×10^6	2	4.3×10^6
CTnC–CTnI[c]	—	2	3.2×10^8	1	1×10^{6d}
	4×10^{-3}	2	3.2×10^7	1	1×10^{6d}
STn[b]	—	2	5.3×10^8	2	4.9×10^6
	2×10^{-3}	2	5.4×10^6	2	5.4×10^6
CTn[c]	—	2	3.7×10^8	1	2.5×10^{6d}
	4×10^3	2	2.4×10^7	1	2.4×10^{6d}

[a] Skeletal and (C) cardiac troponin (Tn) Ca^{2+}-binding (TnC) and inhibitory (TnI) subunits.
[b] from Potter and Gergely (1975).
[c] from Holroyde *et al.* (1980).
[d] Similar results were obtained from fluorescence measurements (Johnson *et al.*, 1980).

Thus, the cardiac proteins appear to be very similar in their Ca^{2+}-binding properties to those in skeletal muscle with the exception that one of the Ca^{2+}-specific sites is absent in CTnC.

2.2. Relationship of Troponin Ca^{2+} Binding to the Activation of Myofibrillar ATPase

Previous studies have suggested that only the Ca^{2+}-specific sites of the skeletal and cardiac Tn's are involved in the activation of myofibrillar ATPase. This finding has been based on static as well as kinetic measurements.

The initial studies (Potter and Gergely, 1975) were carried out by varying the free Mg^{2+} concentration and then examining the Ca^{2+} dependence of myofibrillar ATPase at these different Mg^{2+} concentrations. Since no shift was observed in the Ca^{2+} dependence of myofibrillar ATPase, which would have been expected if the $Ca^{2+}-Mg^{2+}$ sites were involved, it was concluded that only the Ca^{2+}-specific sites were involved in regulation. It should be pointed out that there is still some question about the effect of Mg^{2+} in these systems, and we are currently investigating this. Recent studies on cardiac myofibrillar ATPase (Holroyde *et al.*, 1980) as a function of pCa showed that activation of cardiac myofibrillar ATPase occurred over the same Ca^{2+} concentration range as binding to the single Ca^{2+}-specific site in CTn. Thus, in both the cardiac and the skeletal system, it appeared that regulation occurred only with Ca^{2+} binding to the Ca^{2+}-specific regulatory sites and not with Ca^{2+} binding to the $Ca^{2+}-Mg^{2+}$-type sites.

Additional evidence comes from studying the kinetics of Ca^{2+} and Mg^{2+} exchange with the various Ca^{2+}-binding sites on Tn. Our initial studies (Johnson *et al.*, 1979) showed that the rate of dissociation of Ca^{2+} from the $Ca^{2+}-Mg^{2+}$ sites of STnC was much slower ($\approx 1-5$ sec^{-1}) than that from the Ca^{2+}-specific sites (≈ 300 sec^{-1}), again suggesting that only the Ca^{2+}-specific sites are involved in regulation. Recent work on Tn (Johnson *et al.*, 1981; Potter *et al.*, 1980) in contrast to TnC has shown that the Ca^{2+} off rate from the $Ca^{2+}-Mg^{2+}$ sites is even slower than in TnC (0.6 sec^{-1}), while the off rate of Ca^{2+} from the Ca^{2+}-specific sites remains quite rapid (≈ 23 sec^{-1}). In addition, the off rate for Mg^{2+} is also very slow (2 sec^{-1}) (Potter *et al.*, 1980). Since the free Mg^{2+} in muscle is in the millimolar range, the $Ca^{2+}-Mg^{2+}$ sites would be partially saturated with Mg^{2+} when the muscle is relaxed. This Mg^{2+} would have to dissociate before Ca^{2+} could completely fill these sites, thus slowing the on rate of Ca^{2+} (which is diffusion-limited to both classes of sites) to these sites to essentially the off rate of Mg^{2+}. Thus, not only would Ca^{2+} dissociate slowly from the $Ca^{2+}-Mg^{2+}$ sites, but also the association of Ca^{2+} would be very slow, making it unlikely for these sites to play a regulatory role (on a rapid time scale) in muscle contraction (Potter *et al.*, 1980; Robertson *et al.*, 1981b; Johnson *et al.*, 1979, 1981). In contrast, the Ca^{2+}-specific sites bind and dissociate Ca^{2+} rapidly without interference from Mg^{2+} and thus are able to exchange Ca^{2+} rapidly enough to be involved in regulation. We have recently shown that this also holds true for the single Ca^{2+}-specific site of CTn (Robertson *et al.*, 1981a).

Thus, it appears that in both the cardiac and the skeletal system, the Ca^{2+}-specific sites are clearly the "regulatory" sites and that the Ca^{2+}–Mg^{2+} sites probably play a structural role, since they always have either Ca^{2+} or Mg^{2+} bound to them depending on recent past muscle activity.

2.3. Role of Cyclic-AMP-Dependent Phosphorylation of CTnI in the Regulation of Muscle Contraction

Another unique feature about CTn is its ability to be phosphorylated at serine 20 of CTnI in the CTn complex by cyclic AMP (cAMP)-dependent protein kinase (Moir *et al.*, 1980; Stull and Buss, 1977). In contrast, STnI phosphorylation does not occur in the STn complex. CTnI has an additional 26 residues on its N terminus that STnI does not contain (Wilkinson and Grand, 1978), which probably accounts for the difference in the ability of cAMP-dependent protein kinase to phosphorylate CTnI and not STnI in the respective Tn complexes.

Much interest has been focused on this finding since, in beating hearts, it has been possible to correlate the degree of positive inotrophy brought about by β-adrenergic agonists and the degree of CTnI phosphorylation (England, 1975; Solaro *et al.*, 1976; Ezrailson *et al.*, 1977). The mechanism that may be responsible for this relationship is not known, but several investigators have noted a shift (to lower pCa) in the Ca^{2+} dependence of cardiac myofibrillar ATPase and force development on phosphorylation of CTnI (Bailin, 1979; Holroyde *et al.*, 1979; Mope *et al.*, 1980). This may be brought about by a lowering of the affinity of CTn for Ca^{2+}, although Stull and Buss (1978) were unable to demonstrate any change in affinity.

We have recently been able to show a reduction in the affinity of the single Ca^{2+}-specific site on CTn with phosphorylation of CTnI by cAMP-dependent protein kinase (Robertson *et al.*, 1981b; Solaro *et al.*, 1981). To measure the effect CTnI phosphorylation has on the Ca^{2+}-binding properties of the CTn complex, we have reconstituted CTn from CTnI, CTnT, and CTnC. To selectively follow Ca^{2+} binding at the Ca^{2+}-specific site of TnC, this subunit was labeled with the fluorescence probe IAANS before the troponin complex, CTn_{IA}, was reconstituted (Johnson *et al.*, 1980). The Ca^{2+}-dependent changes in the fluorescence of this complex provide an accurate measure of the Ca^{2+} affinity of this site in CTn (Johnson *et al.*, 1980). Two different methods were used to obtain phosphorylated CTn_{IA}. The first was to form CTn_{IA} using nonphosphorylated CTnI (0.1 mol P/mole) and then to phosphorylate CTn_{IA} with cAMP-dependent protein kinase and ATP. Control samples were treated similarly, but without added ATP. This procedure catalyzed the incorporation of 0.9 ± 0.3 mol P/mole CTn_{IA}. Autoradiograms of 12.5% polyacrylamide slab gels run in the presence of sodium dodecyl sulfate, indicated that more than 98% of the ^{32}P incorporated into CTn_{IA} was associated with CTnI. The second method used was to first phosphorylate the CTnI subunit with the same enzymes described above to a level of 1.7 ± 0.3 mole P/mole CTnI. Control nonphosphorylated CTnI samples were treated under identical conditions, but without added ATP, and contained 0.1 mole P/mole CTnI. These two samples of CTnI were then used to form phosphorylated and nonphosphorylated CTn_{IA} preparations. Independent of the method used to obtain the phosphorylated CTn_{IA} complexes, a decrease in the pCa required to achieve 50% of the maximal Ca^{2+}-induced fluorescence change in CTn_{IA} was observed, thus clearly demonstrating a decrease in the affinity of the single Ca^{2+}-specific site for Ca^{2+}. Another indication of a lowered Ca^{2+} affinity was that Ca^{2+} dissociation from CTn_{IA} was also found to increase on CTnI phosphorylation. The measured rate constants for Ca^{2+} removal and the $pCa_{50\%}$ values for nonphosphorylated and phosphorylated CTn_{IA} were used to model the time course of Ca^{2+} binding to the single Ca^{2+}-specific site of Tn in response to Ca^{2+} transients similar to *in vivo* cardiac Ca^{2+} transients. Tension development would be expected to have a time-course similar to that of Ca^{2+} binding, with some lag due to other steps in the process subsequent to binding (e.g., conformational changes).

The results of these calculations are shown in Fig. 1. In Fig. 1A, the Ca^{2+}-binding properties of the Ca^{2+}-specific site are illustrated for the dephospho- state. In Fig. 1B, the off rate of Ca^{2+} has been increased to the value we have measured for phosphorylated CTn. In addition, the amplitude and time–course of the Ca^{2+} transient have been increased to simulate the known (Blinks *et al.*, 1980) changes in Ca^{2+} release and uptake from the sarcoplasmic reticulum (SR) in response to β-adrenergic stimulation (which may be brought about by cAMP-dependent protein kinase phosphorylation of the SR). The end result is that in the phosphorylated state, the rate and extent of Ca^{2+} binding are increased, as is the rate of Ca^{2+} dissociation. This follows the well-known change in cardiac force development, which also rises to a greater extent more rapidly and relaxes faster in the inotropic state produced by catecholamines.

Thus, two cAMP-dependent phosphorylations are probably at work in producing the inotropic state, phosphorylation of the SR, which allows more rapid release and uptake of Ca^{2+}, and the phosphorylation of CTnI in CTn, which allows CTn to exchange its "regulatory" Ca^{2+} more rapidly to work synergistically with the increased pumping capability of the SR.

3. Interactions of Cardiac-Thin-Filament Proteins in the Regulation of Muscle Contraction

Another area of prime interest is in the interaction of all the cardiac-thin-filament proteins (actin, Tm, Tn) and the influence of Ca^{2+} on these interactions. The skeletal-muscle thin-filament system has been studied very thoroughly (Potter and Gergely, 1974; Hitchcock *et al.*, 1973), and our recent results on the cardiac system will be compared with that system (Table 2).

These interactions are, of course, of considerable interest, since Ca^{2+} binding to Tn is only the first step in the activation of muscle contraction and

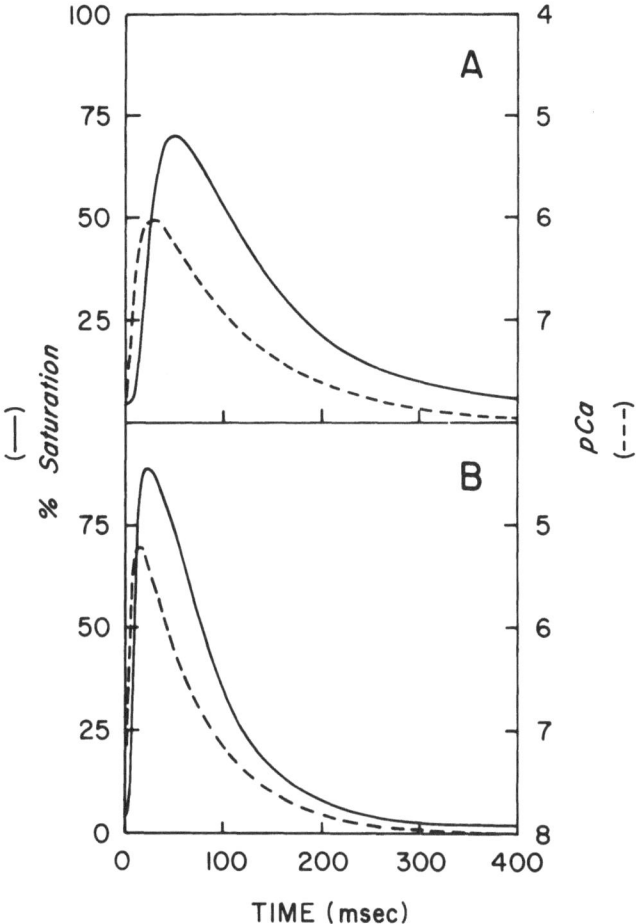

Figure 1. Time–course of Ca^{2+} binding to the single Ca^{2+}-specific site of CTn as a function of C-TnI phosphorylation. The results were calculated using the following set of equations:

$$pCa = 8 - a(e^{-Ft} - e^{-Rt})$$

$$\frac{dx}{dt} = k_{off}[10^{(pCa_{50\%} - pCa)}(100 - x) - x]$$

where t is in seconds, k_{off} is the rate of Ca^{2+} dissociation from CTn, pCa$_{50\%}$ is the pCa at half saturation of this site, and x is the percentage saturation with Ca^{2+} of the single Ca^{2+}-specific site of CTn. (A) Dephosphorylated case: a = 2.87, F = 10, R = 100, k_{off} = 14.5, pCa$_{50\%}$ = 6.63. (B) Phosphorylated case: a = 3.73, F = 15, R = 200, k_{off} = 21.0, pCa$_{50\%}$ = 6.31.

results in an alteration of these interactions that then allows myosin and actin to interact.

Starting with CTnT, one of its primary interactions is with CTm, which can be demonstrated by either viscosity or sedimentation measurements to have a stoichiometry of 1:1 (Table 2). The other primary interaction of CTnT is with CTnI, which occurs only when the two sulfhydryls on CTnI are re-

Table 2. Comparison of Cardiac and Skeletal Troponin-Subunit Interactions

	Interaction[a]			Cardiac	Skeletal
1. TnT + Tm	$\longrightarrow$		TnT–Tm	+	+
2. TnT + TnI	$\longrightarrow$		TnT–TnI	+	+
3. Tm–TnT+TnI	$\longrightarrow$		Tm–TnT–TnI	+	+
4. TnI.+ A	$\longrightarrow$		TnI–A	+	+
5. TnI + A–Tm	$\longrightarrow$		TnI–A–Tm	+	+
6. TnI + TnC		$+/-Ca^{2+}$	TnC–TnI	+	+
7. TnC + TnT		$+Ca^{2+}$ $\overrightarrow{}$ $\underset{-Ca^{2+}}{\overleftarrow{}}$	TnC–TnT	+	+
8. TnC–TnI	$\overset{X}{\longrightarrow}$			–	–
9. TnC–TnI		$\underset{+Ca^{2+}}{\overset{-Ca^{2+}}{\rightleftarrows}}$	TnC–TnI–A–Tn	+	+
10. TnI + Tm	$\overset{X}{\longrightarrow}$			–	–
11. TnC–TnI	$\overset{X}{\longrightarrow}$			–	–

[a] (A) Actin; (CI) TnC–TnI; (CT) TnC–TnT.

duced (Table 2). Similar findings have been reported in the skeletal system (Horwitz *et al.*, 1979). Again, the stoichiometry is $1:1$. This has been measured by gel-exclusion chromatography. CTnT is essentially excluded on Sephadex G-200 and CTnI is included, thus any interaction between the two proteins can be detected by the presence of CTnI in the excluded CTnT peak. This is also observed when CTnT is complexed with CTm (Table 2). These measurements are made at fairly high ionic strengths due to the low solubility of both these proteins in the absence of CTnC. Again, there is clearly a strong stoichiometric interaction between these two proteins, which suggests that this is a vital interaction, not previously thought to be of any importance in the interaction of the Tn subunits. That is because previously it was thought that the primary interaction was between TnT and TnC. Our recent experiments suggest that the interactions between CTnT and CTnC are weak and rather nonspecific and occur only in the presence of Ca^{2+} at low ionic strengths (they are abolished even in the presence of Ca^{2+} above 0.3 M KCl). The reaction (Table 2) appears to be nonspecific, because even at high molar ratios of added CTnC to CTnT $(6:1)$, a stoichiometry of only about 0.5 mole CTnC bound per CTnT could be obtained. Our current feeling, then, is that CTnC probably does not interact with CTnT in the CTn complex (Fig. 2B), although it is still possible that this may occur in the presence of all three subunits and Ca^{2+}, and we have included that possibility in our current working model of Tn, Tm, and actin interactions (Fig. 2C). The primary interactions of CTnT, then, are with CTm and with CTnI. Another important point about CTnT and STnT is their highly asymmetric shape, which has been determined from gel-exclusion chromatography (Stokes radius = 7.8 nm for CTnT and 4.1 nm for STnT) and from fluorescence and sedimentation methods (Prendergast and Potter, 1979). The fluorescence and sedimentation studies indicate that the length of CTnT is approximately 10 nm. This strongly suggests that since

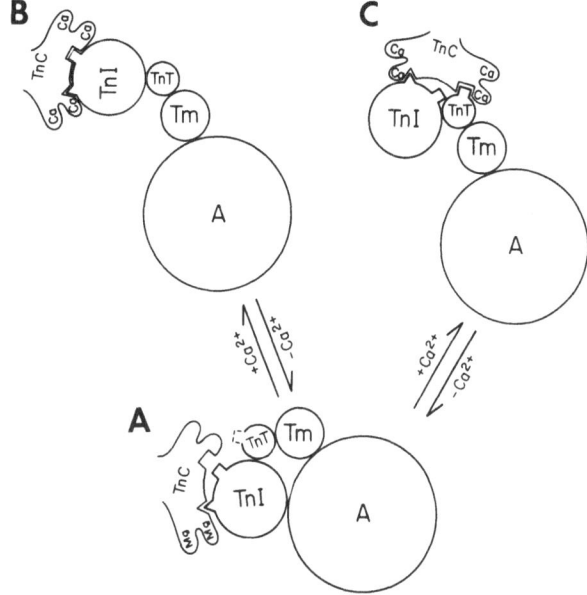

Figure 2. Model for the regulation of cardiac- and skeletal-muscle contraction. In the absence of Ca²⁺ [relaxed state (A)], the two Ca^{2+}-Mg^{2+} sites of TnC are shown to contain Mg^{2+} and the two Ca^{2+}-specific sites (one in the case of CTnC) to contain no metal. TnI is bound to actin (inhibitory) in this state (A). Note that in all cases (A–C), TnT interacts with Tm and TnI. On going from the relaxed state (A) to the contractile state (increasing Ca^{2+} concentration), two possibilities are illustrated (B, C). The currently favored case (B) is that on the binding of Ca^{2+} to TnC, it interacts more strongly with TnI (B) and brings about the dissociation of TnI from actin (activation). Note that the changes in the interaction between TnC and TnI and the dissociation of TnI from actin are brought by Ca^{2+} binding to the Ca^{2+}-specific site(s). Although Ca^{2+} is drawn in the Ca^{2+}-Mg^{2+} sites (B, C), it is unlikely from kinetic arguments made in the text that these sites will bind very much Ca^{2+} in a normal contraction, whereas the Ca^{2+}-specific sites can readily exchange Ca^{2+} during the time-course of contraction. The less-favored possibility that TnC can interact with TnT in the presence of Ca^{2+} is also illustrated (C)

TnT covers a considerable portion of Tm, it plays an important role in affecting the structure of Tm (Fig. 2).

The other interaction of CTnI is with CTnC, where a stoichiometric complex forms and is stable in the presence or absence of Ca^{2+} in nondenaturing solvents (Table 2). Thus, in contrast to earlier ideas, the Tn complex is probably put together via TnT-TnI-TnC (Fig. 2), rather than via TnT-TnC-TnI. CTnI will bind to C-actin (and actin–Tm complex), but not to Tm (Table 2), and accounts for the well-known inhibition of actomyosin ATPase by TnI. The most important of all these interactions is that between the CTnC–CTnI complex and the actin–Tm complex in the absence of Ca^{2+} (see Table 2), since this probably represents the key, if not the only, Ca^{2+}-dependent event in these interactions. For it is known that under these conditions, the TnC–TnI complex can inhibit actomyosin ATPase when present in a molar ratio with actin (Perry *et al.*, 1972). Note that the TnC–TnI complex does not bind to Tm or to actin alone regardless of the Ca^{2+} concentration. In the case of the TnC–TnI complex, dissociation from the actin–Tm complex occurs as the free Ca^{2+} concentration is increased (Table 2) and probably corresponds to the Ca^{2+}-dependent event that results in the relaxation of muscle contraction in the complete system as shown in Fig. 2.

ACKNOWLEDGMENTS. This work was supported by grants from the National Institutes of Health (HL 22619-3 A, 3 B, 3 E, and HL 22231), the American Heart Association (78 1167, 79 1001), the Muscular Dystrophy Association, the Southwestern Ohio Heart Association, and postdoctoral training grants HL 07382-03 and AM 05998-02. R.J.S. and E.G.K. are holders of Research Career Development Awards.

References

Bailin, G., 1979, Phosphorylation of a bovine cardiac actin complex, *Am. J. Physiol.* **236**:C41.

Blinks, J. R., Lee, N. K. M., and Morgan, J. P., 1980, Ca^{2+} transients in mammalian heart muscle: Effects of inotropic agents on aequorin signals, *Fed. Proc. Fed. Am. Soc. Exp. Biol.* **39**:854.

Brekke, C. J., and Greaser, M. L., 1976, Separation and characterization of the troponin components from bovine cardiac muscle, *J. Biol. Chem.* **251**:866.

Burtnick, L. D., and Kay, C. M., 1977, The calcium-binding properties of bovine cardiac troponin C, *FEBS Lett.* **75**:105.

Collins, J. H., Greaser, M. L., Potter, J. D., and Horn, M. J., 1977, Determination of the amino acid sequence of troponin C from rabbit skeletal muscle, *J. Biol. Chem.* **252**:6356.

Ebashi, E., Endo, M., and Ohtsuki, I., 1969, Control of muscle contraction, *Q. Rev. Biophys.* **2**:351.

England, P. J., 1975, Correlation between contraction and phosphorylation of the inhibitory subunit of troponin in perfused rat heart. *FEBS Lett.* **50**:57.

Ezrailson, E. G., Potter, J. D., Michael, L., and Schwartz, A., 1977, Positive inotropy induced by ouabain, by increased frequency, by calcium, by RO2-2985 (X537A) and by isoproterenol: The lack of correlation with phosphorylation of TnI, *J. Mol. Cell. Cardiol.* **9**:693.

Hirabayashi, T., and Perry, S. V., 1973, An immunochemical study of the calcium ion-binding protein (troponin C) and inhibitory protein (troponin-I) of the troponin complex and their interaction, *Biochim. Biophys. Acta* **351**:273.

Hitchcock, S. E., Huxley, H. E., and Szent-Györgyi, A. G., 1973, Calcium sensitive binding of troponin to actin-tropomyosin: A two-site model for troponin action, *J. Mol. Biol.* **80**:825.

Holroyde, M. J., Howe, E., and Solaro, R. J., 1979, Modification of calcium requirements for activation of cardiac myofibrillar ATPase by cyclic AMP dependent phosphorylation, *Biochim. Biophys. Acta* **586**:63.

Holroyde, M. J., Robertson, S. P., Johnson, J. D., Solaro, R. J., and Potter, J. D., 1980, The Ca^{2+} and Mg^{2+} binding sites on cardiac troponin and their role in the regulation of myofibrillar adenosine triphosphatase, *J. Biol. Chem.* **255**:11688.

Horwitz, J., Bullard, B., and Mercola, D., 1979, Interaction of troponin subunits, *J. Biol. Chem.* **254**:350.

Johnson, J. D., Charlton, S. C., and Potter, J. D., 1979, A fluorescence stopped flow analysis of Ca^{2+} exchange with troponin C, *J. Biol. Chem.* **254**:3497.

Johnson, J. D., Collins, J. H., Robertson, S. P., and Potter, J. D., 1980, A fluorescent probe study of Ca^{2+} binding to the Ca^{2+}-specific sites of cardiac troponin and troponin C, *J. Biol. Chem.* **255**:9635.

Johnson, J. D., Robertson, S. P., Schwartz, A., and Potter, J. D., 1981, Ca^{2+} exchange with troponin and the regulation of muscle contraction, in: *The Regulation of Muscle Contraction: Excitation-Contraction Coupling* (A. D. Grinnell and M. A. B. Brazier, eds.), p. 241, Academic Press, New York.

Kohama, K., 1979, Divalent cation binding properties of slow skeletal muscle troponin in comparison with those of cardiac and fast skeletal muscle troponins, *J. Biochem. (Tokyo)* **86**:811.

Leavis, P. C., and Kraft, E. L., 1978, Calcium binding to cardiac troponin C[1,2], *Arch. Biochem. Biophys.* **186**:411.

Leavis, P., Rosenfeld, S., Gergely, J., Grabarek, Z., and Drabikowski, W., 1978, Proteolytic fragments of troponin C, *J. Biol. Chem.* **253**:5452.

Moir, A. J. G., Solaro, R. J., and Perry, S. V., 1980, The site of phosphorylation of troponin I in the perfused rabbit heart, *Biochem. J.* **185**:505.

Mope, L., McClellan, G. B., and Winegrad, S., 1980, Calcium sensitivity of the contractile system and phosphorylation of troponin in hyperpermeable cardiac cells, *J. Gen. Physiol.* **75**:271.

Perry, S. V., 1979, The regulation of contractile activity in muscle, *Biochem. Soc. Trans.* **7**:593.

Perry, S. V., Cole, H. A., Head, J. F., and Wilson, J. F., 1972, Localization and mode of action of the inhibitory protein component of the troponin complex, *Cold Spring Harbor Symp. Quant. Biol.* **37**:251.

Potter, J. D., and Gergely, J., 1974, Troponin, tropomyosin and actin interactions in the Ca^{2+} regulation of muscle contraction, *Biochemistry* **13**:2697.

Potter, J. D., and Gergely, J., 1975, The calcium and magnesium binding sites on troponin and their role in the regulation of myofibrillar adenosine triphosphatase, *J. Biol. Chem.* **250**:4628.

Potter, J. D., Seidel, J. C., Leavis, P., Lehrer, S. S., and Gergely, J., 1976, Effect of Ca^{2+} binding on troponin C, *J. Biol. Chem.* **251**:7551.

Potter, J. D., Johnson, J. D., Dedman, J. R., Schreiber, W. E., Mandel, F., Jackson, R. L., and Means, A. R., 1977, Calcium-binding proteins: Relationship of binding, structure, conformation and biological function, in: *Calcium Binding Proteins and Calcium Function* (R. H. Wasserman, R. A. Corradino, E. Carafoli, R. H. Kretsinger, D. H., MacLennan, and F. L. Siegel, eds.), p. 239, Elsevier, New York, Amsterdam, Oxford.

Potter, J. D., Robertson, S. P., Collins, J. H., and Johnson, J. D., 1980, The role of the Ca^{2+} and Mg^{2+} binding sites on troponin and other myofibrillar proteins in the regulation of muscle contraction, in: *Calcium Binding Proteins: Structure and Function* (Siegel, F. L., Carafoli, E., Kretsinger, R. H., MacLennan, D. H., and Wassenman, R. H., eds.), p. 279, Elsevier/North-Holland, New York.

Prendergast, F. G., and Potter, J. D., 1979, Solution conformation and hydrodynamic properties of rabbit skeletal TnT, *Biophys. Soc. Abstr.* **25**:250a.

Robertson, S. P., Johnson, J. D., Holroyde, M. J., Kranias, E. G., Potter, J. D., and Solaro, R. J., 1981a, The effect of troponin I phosphorylation on the Ca^{2+}-binding properties of the Ca^{2+}-regulatory site of bovine cardiac troponin, *J. Biol. Chem.* (in press).

Robertson, S. P., Johnson, J. D., and Potter, J. D., 1981b, The time course of Ca^{2+} exchange with calmodulin, troponin, parvalbumin and myosin in response to transient increases in Ca^{2+}, *Biophys. J.* **34**:559.

Solaro, R. J., Moir, A. J. G., and Perry, S. V., 1976, Phosphorylation of troponin I and the inotropic effect of adrenaline in the perfused rabbit heart, *Nature (London)* **262**:615.

Solaro, R. J., Holroyde, M. J., Robertson, S. P., Johnson, J. D., and Potter, J. D., 1981, Troponin I phosphorylation: A unique regulator of the amounts of calcium required to activate cardiac myofibrils, *Cold Spring Harbor Symp. Protein Phosphorylation* **8**:901–913.

Stull, J. T., and Buss, J. E., 1977, Phosphorylation of cardiac troponin by cyclic adenosine 3'-5'-monophosphate-dependent protein kinase, *J. Biol. Chem.* **252**:851.

Stull, J. T., and Buss, J. E., 1978, Calcium binding properties of beef cardiac troponin, *J. Biol. Chem.* **253**:5932.

Van Eerd, J. P., and Takahashi, K., 1976, Determination of the complete amino acid sequence of bovine cardiac troponin C, *Biochemistry* **15**:1171.

Weber, A., and Murray, J. M., 1973, Molecular control mechanism in muscle contraction, *Physiol. Rev.* **53**:612.

Wilkinson, J. M., and Grand, R. J. A., 1978, Comparison of amino acid sequence of troponin I from different striated muscles, *Nature (London)* **271**:31.

19

Fluorescent-Probe Studies of Contractile Proteins

Robert A. Mendelson

1. Introduction

The present knowledge of the mechanism of muscle contraction and cell movement at the molecular level comes from an accumulation of experimental evidence obtained using a wide variety of biochemical and biophysical techniques. In relatively recent times, the use of intrinsic and extrinsic fluorescence probes has provided useful information about the kinetic intermediates, mobility, binding, orientation, intramolecular distances, and site environment of the globular head region of the myosin molecule, both free in solution and as a part of the intact muscle "cross-bridges" that are thought to be the impellers of biological movement.

The use of tryptophan as an intrinsic probe has the experimental advantage of allowing one to study the protein (or its complexes) without chemical modification that could have possible deleterious effects on its enzymology. On the other hand, there are many kinds of experiments that could not be done without the introduction of extrinsic probes that have special physical characteristics (such as long lifetimes) and that allow labeling of the protein at a specific site. Many of the experiments to be described rely on the presence of a fast-reacting cysteine group on the myosin head.

A unique advantage of fluorescence techniques is that with readily available light sources and detectors, they have a very high degree of sensitivity;

Robert A. Mendelson • Department of Biochemistry and Biophysics and The Cardiovascular Research Institute, University of California, San Francisco, California 94143.

thus, experiments at low protein concentrations may be performed and dynamic experiments are frequently possible.

Before reviewing some of the experiments that have been performed to study the cross-bridge mechanism, I will discuss the underlying physical principles of the fluorescence techniques that have been used.

2. Physical Principles of Fluorescence Techniques

2.1. Background

The physical basis of fluorescence has been reviewed extensively elsewhere (Weber, 1972; Yguerabide, 1972); consequently, only those points necessary for later development will be reviewed.

An energy-level diagram of a typical fluorophore is shown in Fig. 1. Excitation of the first excited (singlet) state occurs by absorption of a photon of energy $h\nu$ into a vibrational level near the singlet ground electronic state (S_0). Since the vibrational levels have a Boltzmann distribution and their number is usually large for organic fluorophores, a broad (10–50 nm) absorption spectrum is typically observed. The excited vibrational level rapidly decays (lifetime $\simeq 10^{-11}$ sec) by nonradiative transitions to the lowest vibrational level of the first excited singlet state (S_1). This level may then decay to the ground state by several routes: the liberation of a photon from S_1 (fluorescence) occurs usually within 10^{-7} sec; a photon may also arise from the "forbidden" triplet to singlet transition (T_0 to S_0), which usually occurs within 10^{-3} sec; or, finally, heat may be liberated to the solvent by collisional interactions. In fact, the nature of the dye environment is crucial in determining the relative amounts of these processes, and thus the quantum yield and fluorescence lifetime (τ) are dependent on the dye site environment.

Additionally, the fluorophores possess transition moments (Kauzmann, 1957) for absorption (in the direction of unit vector $\hat{\alpha}$) and emission (in the direction along a unit vector $\hat{\varepsilon}$) along which plane-polarized light is optimally

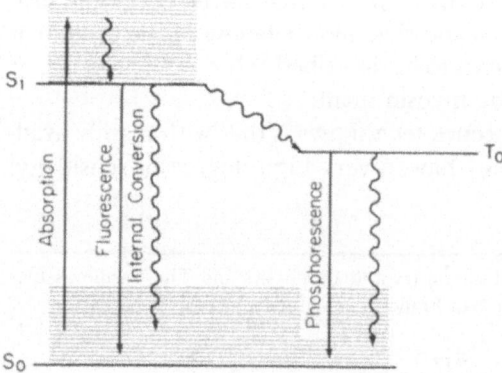

Figure 1. Energy-level diagram of a typical fluorophore. Decays that emit a photon are shown by solid lines.

absorbed and emitted, respectively. The probability of photon absorption is proportional to a geometric factor

$$P_{\text{abs}} \propto [\hat{\alpha} \cdot \hat{E}_{\text{ex}}]^2 \tag{1a}$$

and the probability of observing a fluorescence photon is proportion to

$$P_{\text{obs}} \propto [\hat{\epsilon} \cdot \hat{E}_{\text{em}}]^2 \tag{1b}$$

where $\hat{E}_{\text{ex}}$ and $\hat{E}_{\text{em}}$ are the unit vectors along the orientations of the excitation polarizer and emission analyzer, respectively. In an "ideal" fluorophore, having only radiative decay, the transition moments coincide ($\hat{\alpha} = \hat{\epsilon}$). In real fluorophores, the angle between the moments (λ) typically varies between 10 and 30°. These directional properties and the availability of fluorophores having lifetimes up to 100 nsec, a time that is comparable to that which it takes biological macromolecules to undergo significant rotational Brownian movements, allows the measurement of rotational diffusion coefficients of these molecules (or moieties) if the probe is rigidly attached. Selective labeling of a protein moiety, in this case the head region of myosin, allows information about the segmental flexibility of that moiety to be obtained using the time-resolved fluorescence anisotrophy decay (TRFAD) method.

If the macromolecule to which the probe is attached is occasionally bound to a much larger macromolecule or is not tumbling freely because of steric constraints, the evolution in time of the polarized intensities will be affected; thus, measurements of affinity constants (cf. Mendelson *et al.*, 1975; Highsmith *et al.*, 1976) and information about the constraints (see below) may be obtained.

In cases where there is little or no rotational Brownian movement (on a time scale of the fluorescence lifetime), the fluorophores embedded within the structure report their orientation via the spatial anisotropy of emitted polarized light. As will be discussed below, this has been applied to the study of cross-bridge orientation changes in muscle fibers (Aranson and Morales, 1969; dos Remedios *et al.*, 1972; Nihei *et al.*, 1974a; Borejdo and Putnam, 1977), and more recently a search has been made for fluctuations in the average polarization during contraction (Borejdo *et al.*, 1979). Such fluctuations give information about the cross-bridge cycling frequency.

The use of energy transfer to measure distances between labeling sites has been applied to muscle proteins by many workers. This subject is reviewed in Cooke (1981).

2.2. Time-Resolved Fluorescence Anisotrophy Decay

2.2.1. Theory

In the TRFAD technique, the sample is illuminated by a nanosecond burst of plane polarized light. This generates a $\cos^2 \theta_\alpha$ distribution [where θ_α is the angle between $\hat{\alpha}$ and E_{ex} in equation (1a)] of fluorophores at time $t = 0$. As time progresses and the angular orientation of the macromolecule bearing the (rigidly attached) dye is randomized by rotational Brownian motion, the initial distribution disappears at a rate determined by rotational diffusion coefficients of the macromolecule. Experimentally, this randomization is measured by observing the polarization anisotropy $r(t)$ as a function of time:

$$r(t) \equiv \frac{I_{\parallel}(t) - I_{\perp}(t)}{I_{\parallel}(t) + 2 I_{\perp}(t)} \tag{2}$$

where $I_{\parallel}(t)$ and $I_{\perp}(t)$ are orthogonal polarized intensities measured at 90° to the excitation direction (see Fig. 2). For spherical molecules

$$r(t) = r(0)e^{-t/\phi_{sp}} \tag{3}$$

where ϕ_{sp} is the relaxation time, which may be determined from

$$\phi_{sp} = \frac{V\eta}{kT} \tag{4}$$

Here η is the viscosity, V is the hydrated molecular volume and k is Boltzman's constant.

If the macromolecule is described as a general ellipsoid having three different semiaxes, the theoretical expression for $r(t)$ is (Belford *et al.*, 1972; Ehrenberg and Rigler, 1972)

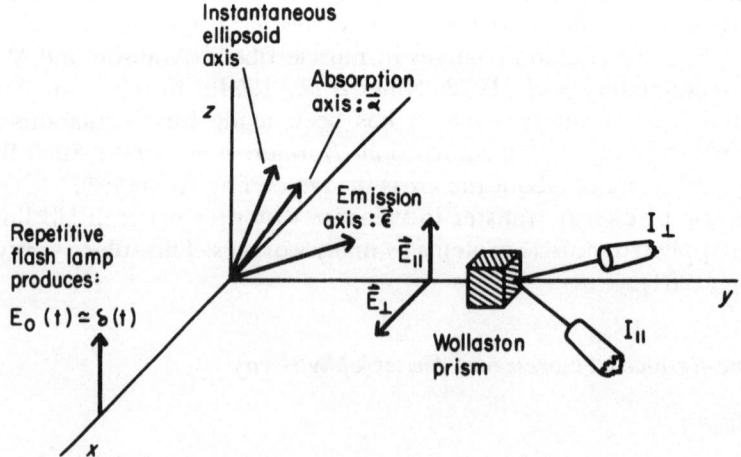

Figure 2. Geometry of the TRFAD method. The cuvette containing labeled macromolecules is located at the origin.

$$r(t) = \sum_{i=1}^{5} a_i \ (\alpha_1, \alpha_2, \alpha_3, \ \epsilon_1, \epsilon_2, \epsilon_3) \ e^{-t/\phi_i} \tag{5}$$

where the α_k and ϵ_k are direction cosines of the kth absorption and emission dipoles and where ϕ_i is a function of rotational diffusion constants about the semiaxes. In particular, for ellipsoids of revolution, the expression reduces to

$$r(t) = a_1 \ (\alpha_1, \epsilon_1) e^{-D_\perp t} + a_2(\alpha_1, \alpha_3, \epsilon_1, \epsilon_3) e^{-(5D_\perp + D_\parallel)t}$$

$$+ a_3 \ (\alpha_1, \alpha_3, \epsilon_1, \epsilon_3) e^{-(2D_\perp + 4D_\parallel)t} \tag{6}$$

where $D_\parallel$ and $D_\perp$ are the rotational diffusion constants about the major and minor axes. The values for $D_\parallel$ and $D_\perp$ may be calculated from the molecular weight and axial ratios using the equations of Perrin (1936).

In addition to information about rotational diffusion, this technique *simultaneously* provides information about the polarity of the dye environment. The denominator in equation (2) is proportional to the total intensity

$$I(t) = I(o)e^{-t/\tau} \propto [I_\parallel(t) + 2I_\perp(t)] \tag{7}$$

where τ is the excited-state lifetime. Thus, measurements of τ with different ligands or solvents present in a solution of fluorescent-labeled macromolecules yield information about perturbations of the site.

An important special case of fluorescence anisotropy decay occurs when the body bearing the fluorophore can undergo only restricted rotary Brownian movement. The fluorescence anisotropy decay will then be altered as $t \rightarrow \infty$ because each excited fluorophore cannot reach all points in space as it approaches equilibrium. Although, in general, the boundaries can be of any size and shape, and the moiety to which the fluorophore is attached can experience anisotropic three-dimensional Hookean restoring forces, the problem of calculating the anisotropy decay has been attempted only for isotropic Hookean forces (Harvey and Cheung, 1980) and freely diffusing particles in cone-shaped boundaries; Kinesota *et al.* (1977) have analytically treated the problem of an ideal fluorophore lying along the major axis of an ellipsoid tumbling within a cone, while Mendelson and Cheung (1978) have numerically simulated the same problem for many dye orientations. Unless ancillary information is available, the data obtained in TRFAD experiments can probably be adequately described by such simple models.

Numerical modeling was accomplished by forming the products of equation (1a) (at $t = 0$), $\sin \theta_\alpha$ (0), and equation (1b) (at time t). Cylindrical averaging yielded equations that were used to calculate polarized intensities emitted by psuedomolecules at various times during their random rotational trajectories. These trajectories were generated using the Einstein–Perrin (see Perrin, 1934) equations.

$$\delta\omega_i = \pm \ \sqrt{2D_i \ \delta t} \tag{8}$$

where D_i are the ellipsoidal diffusion constants about the ellipsoid's semiaxes and the $\delta\omega_i$ are the angular increments about these axes in a time interval δt. The signs were chosen by a random-number generator. Results from these calculations will be discussed in later sections.

To compare the observed data with any theoretical predictions, one must deconvolute the intensities to account for the instrumental time resolution. The observed polarized intensities $I_k^0(t)$ are a convolution of the true time intensities, $I_k(t - T)$, with the total instrumental "prompt" response function $L(t)$:

$$I_k^0(t) = \int_0^t L(t)\, I_k\,(t - T)\, d\,T \qquad (9)$$

For single exponential intensities (which arise in TRFAD of spheres, for example) this can be inverted by a simple least-squares procedure (see Yguerabide, 1972). In many cases, this is adequate to describe the data and the system response. The more complicated general case involving several exponentials has also recently received much attention (see Wahl, 1979).

2.2.2. Instrumentation

An apparatus for measuring TRFAD is shown in Fig. 3 (Mendelson, *et al.*, 1975). Briefly, its operation is described as follows: a weak light flash of 1- to 4-nsec duration is filtered, polarized, and focused onto a 1-cm-square rectangular cuvette. In addition to exciting a small fraction of the fluorophores, this flash illuminates a photomultiplier that starts a time-to-amplitude converter (TAC). The arrival of a single, polarized, fluorescent photon of appropriate wavelength in either of the photomultipliers (situated normal to the direction of excitation) causes the TAC to stop. Subsequently, the TAC output amplitude is digitized and stored in the computer memory.

The design of the apparatus is similar to that used by Yguerabide *et al.* (1970) for measuring immunoglobulin G (IgG) flexibility except that it is optimized for following the relatively slow motions of muscle proteins. In this regard, the use of two photomultipliers rather than one, both a timing method and a pulse-height method of pile-up rejection, and a high-power air-filled flash lamp has allowed total true counting rates of up to 20 kHz at lamp flash rates of up to 100 kHz to be used. By the use of fluorophores with lifetimes of 20 nsec, sufficient data can be obtained within an hour to allow determination of effective relaxation time of 100–1000 nsec to within 10% accuracy.

The electronics necessary for these improvements were assembled from standard modules except for the analog to digital converter–computer (PDP 8/e) interface. This was designed (Mendelson and Ferrin, unpublished) with the necessary control electronics to allow high-speed storage of events by directly accessing the computer's memory without interfering with its instruction cycles.

An extensive machine-language program was written to allow control of

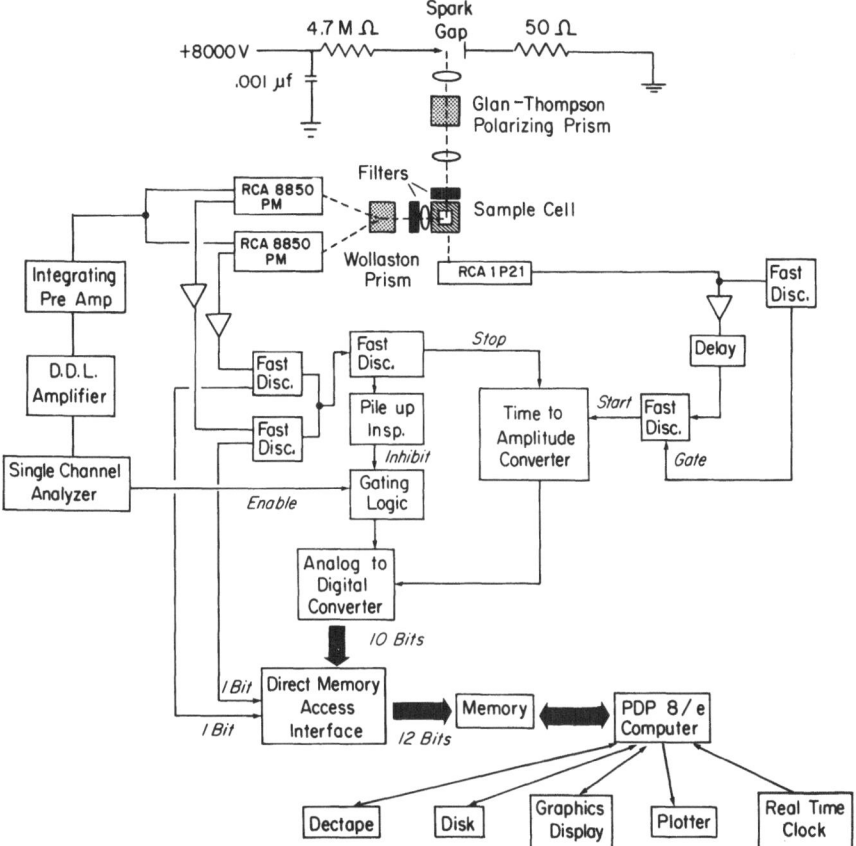

Figure 3. Electronics of a TRFAD apparatus optimized to study slow motions in muscle proteins. From Mendelson *et al.* (1975).

data-taking, display, manipulation, and tape storage under computer or by operator control. Displays of the polarized intensities, or polarization anisotropy, and least-squares fits can be obtained during data accumulation without interruption. After the experiments, the stored data can be plotted and more extensive data analysis can be performed.

2.3. Polarized Fluorescence from Fibers

2.3.1. Theory

If the dye lifetime is short compared to any motions of the dye (caused by the dye moving on the macromolecule or by motion of the macromolecule itself, or both), the polarization anisotropy is independent of time after excitation by a flash. In such a case, a measurement of polarization anisotropy with a constant-intensity light source is sufficient. Such a measurement will then be sensitive only to an appropriate average of dye orientations. As will be seen below, the polarization is a function only of the angle the dye makes with the fiber axis in axially symmetric systems having ordered states.

Historically, the "polarization function" P was defined by

$$P \equiv \frac{I_{\parallel} - I_{\perp}}{I_{\parallel} + I_{\perp}} \tag{10}$$

rather than r [equation (2)]. For a completely randomized immobile set of fluorophores, values of P approaching 0.5 can be obtained as the pure dipole situation ($\hat{\alpha} = \hat{\epsilon}$) is approached.

The problem of fluorophores decorating a helix was first considered theoretically by Tregear and Mendelson (1975). They noted that the absorption ($\hat{\alpha}$) and emission ($\hat{\epsilon}$) dipoles may be translated to a common origin. The problem may then be thought of as being equivalent to calculating the polarization from two concentric cones formed by the coupled dipole pairs. Thus, the observed intensities are proportional to the circular averages of the products of equations (1a) and (1b). For example, if a helix is illuminated by rays of light normal to the helix axis, with polarization parallel to the helix axis ($\parallel$) (Fig. 4), and the emission light is observed coaxially ($\psi = 0$) with the excitation light through a polarizer oriented normal ($\perp$) to the helix axis, the observed intensity will be proportional to a geometric factor

$$_{\parallel}I_{\perp} \propto \overline{[\hat{\alpha} \cdot \hat{E}_{ex}]^2 [\hat{\epsilon} \cdot \hat{E}_{o:}]^2} = {}^1\!/_2 \cos^2\theta_\alpha \sin^2\theta_\epsilon \tag{11}$$

where bar denotes averaging around the helix axis.

In muscle fibers, an additional complication arises if $\hat{\alpha} \neq \hat{\epsilon}$ and if observation is not made coaxially with the exciting light. In muscle sarcomeres, one has helices of the same handedness symmetrically oriented with respect to the M-line of each sarcomere (C_2 symmetry). The general expressions for intensities of observation in the x–y plane at an observation angle of x have been derived by Mendelson and Morales (1977) for the pure random and pure helical cases as a function of ψ (see Fig. 4):

Helical arrays in muscle

$$_{\parallel}I^h_{\parallel} \propto \cos^2\theta_\alpha \cos^2\theta_\epsilon \tag{12}$$

$$_{\perp}I^h_{\parallel} \propto {}^1\!/_2 \sin^2\theta_\alpha \cos^2\theta_\epsilon$$

$$_{\parallel}I^h_{\perp} \propto {}^1\!/_2 \cos^2\theta_\alpha \sin^2\theta_\epsilon$$

$$_{\perp}I^h_{\perp} \propto {}^1\!/_8 \sin^2\theta_\alpha \sin^2\theta_\epsilon [1 + 2\cos^2\beta + 2(1 - 2\cos^2\beta \sin^2\psi)]$$

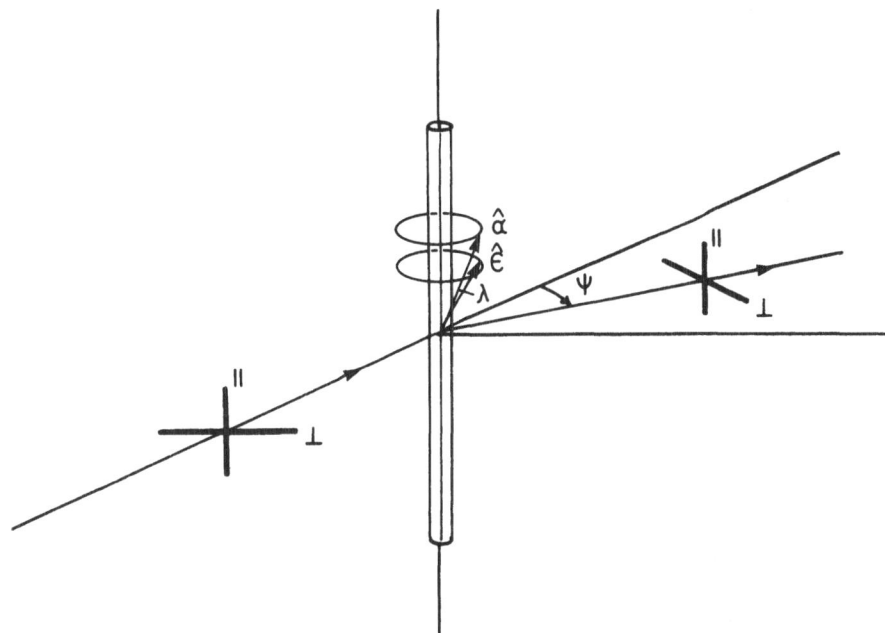

Figure 4. Geometry of fiber experiments using the static fluorescence polarization. The absorption dipole (α) and the emission dipole (ϵ) have declinations θ_α and θ_ϵ from the fiber axis. Equations for the polarizations are computed by averaging absorption and emission contributions around this axis.

Random

$$_\parallel I^r_\parallel \propto \frac{1}{15} \ (1 + 2 \cos^2\lambda)$$

$$_\perp I^r_\parallel \propto \frac{1}{15} \ (2 - \cos^2\lambda)$$

$$_\parallel I^r_\perp \propto \frac{1}{15} \ (2 - \cos^2\lambda)$$

$$_\perp I^r_\perp \propto \frac{1}{15} \ (1 + 2 \cos^2\lambda) + \frac{1}{15} \ (1 - 3 \cos^2\lambda) \sin^2\psi$$

where

$$\beta \equiv \phi_\alpha - \phi_\epsilon = \cos^{-1} \left[\ \frac{\cos \lambda - \cos \theta_\alpha \cos \theta_\epsilon}{\sin \theta_\alpha \sin \theta_\epsilon} \ \right]$$

and

$$\cos \lambda = \hat{\alpha} \cdot \hat{\epsilon}$$

In a realistic case, the labeling may not be completely specific, or the moieties to which dye is attached could be disordered, or both, so that the modeling of polarized intensities must usually consider a linear combination of at least two kinds of disorder. Mendelson and Wilson (1981) have recently found additional analytical expressions for certain other kinds of moiety disorder. Their expressions (and those above) are most meaningfully applied if the absorption and emission dipoles are coincident; consequently, future physiological experiments will be aimed at achieving this experimental situation.

2.3.2. Instrumentation

Instrumentation for such measurements is relatively simple for static measurements. A fine-focus Xe or Hg–Xe lamp is focused onto a species held in a microscope stage. Observations are usually made using two photomultipliers and a Wollaston prism, a setup allows simultaneous measurement of both intensities and hence is not subject to amplitude fluctuations of the source. Excitation and emission solid angles should be suitably small and optical components should be of high quality so that the measured polarized intensities are close to true. Signal-averaging or photon-counting techniques are useful for cases where the fluorescence intensity is weak.

3. Experiments on Myosin and Its Complexes

3.1. Background

In the sliding-filament model of muscle contraction (H. E. Huxley and Hanson, 1954; A. F. Huxley and Niedergerke, 1954), it is postulated that contraction is caused by cyclic interaction of individual force generators acting between interdigitating, hexagonally arranged, filaments. These two filaments are classified as the "thick filaments," which contain primarily myosin, and the "thin filaments," which contain actin and the control proteins, troponin and tropomyosin. In early electron microscopy of vertebrate muscle (see H. E. Huxley, 1957, 1960), radial protrusions from the thick to thin filaments were clearly seen in the region where the two filaments longitudinally overlapped. Subsequent electron-microscopy studies (Reedy et al., 1965) and X-ray diffraction (H. E. Huxley and Brown, 1967) on glycerinated insect muscle in relaxed and rigor states indicated that in projection, the cross-bridges are about normal and at about 45° to the thick filament, respectively. These results and quick-release tension measurements on living fibers (A. F. Huxley and Simmons, 1971) suggested that contraction occurs by a change in cross-bridge declination between the extremes of 90 and 45° states. It is postulated that this change causes a spring in the cross-bridges to be stretched as the S-1 fragment rocks between different orientations on actin.

The structure of myosin, of course, must be integrated with the notions of cross-bridges obtained in these experiments. Myosin is a Y-shaped molecule with a molecular weight of about 450,000. Historically, the myosin subfragment nomenclature arose from its ability to be split into two large proteolytic fragments: a soluble portion called heavy meromyosin (HMM) and an insoluble, rod-shaped portion called light meromyosin (LMM). HMM can be further proteolyzed (see Lowey *et al.*, 1969) into two largely globular portions called subfragment-1 (S-1) and the remaining helical rod portion, subfragment-2 (S-2). The LMM portion is responsible for thick-filament formation under low-salt conditions. S-1 (molecular weight ≈ 130,000), the arms of the "Y," contains two light chains, the ATPase- and actin-binding sites, and a highly reactive thiol called SH_1 (Sekine and Kielley, 1964). The S-2 portion is thought to transmit force to the thick filament as S-1 develops torque by rotation at the acto–S-1 interface. In such a scheme (see H. E. Huxley, 1969), there should be a flexible joint between S-1 and S-2 and a bendable joint within the rod. A simplified view of the solution kinetics (see Inoué *et al.*, 1979; Taylor, 1979; Trentham *et al.*, 1976) of the ATPase is as follows:

$$
\begin{array}{c}
\text{M} \cdot \text{ATP} \longrightarrow \text{M} \cdot {}^{**}\text{ADP} \cdot \text{P}_i \\
\text{ATP} \nearrow \quad \uparrow \qquad\qquad \downarrow \qquad \searrow \text{A} \\
\text{A} \cdot \text{M} \leftarrow\!\!\!\rightleftharpoons \text{A} \cdot \text{M} \cdot \text{ADP} \cdot \text{P}_i \\
\underset{\text{work}}{}
\end{array}
$$

MgATP is hydrolyzed, and after many kinetic intermediates, the product state $\text{M} \cdot {}^{**}\text{ADP} \cdot \text{P}_i$ is formed. In the absence of actin, the release of products is very slow. Thus, in the presence of actin in solution or in an activated muscle, the binding occurs primarily with this myosin intermediate. After attachment to actin, the work-producing intermediates occur, with the final state presumably being the A · M state, which is thought to have the rigor orientation. The addition of MgATP rapidly breaks the A · M link, and the cycle is started anew.

3.2. Segmental Flexibility of Subfragment-1

In 1970, when our fluorescence work was started, little was known about the relationship of the cross-bridges seen in electron microscopy to the structure of myosin. The work of Lowey *et al.* (1969) showed that a rather specific proteolytic region existed that defined S-1; this suggested itself as a possible "hinge" site that would allow rotation of S-1 relative to S-2.

To test this hypothesis, we (Mendelson *et al.*, 1973) adopted a strategy similar to that used by Yguerabide *et al.* (1970) to study the segmental flexibility of the F_{ab} moieties of IgG antibodies. The single, fast-reacting thiol on each myosin head (SH_1) provided a labeling site in much the same manner as did the antigen site on IgG. This thiol had been previously labeled by Seidel *et al.* (1970) with a nitroxide spin label having an ioadacetimide reactivity. The timely synthesis of the dye *N*-iodoacetylamino-1-naphthylamine-1-sulfonic acid (1,5-IAEDANS) (Hudson and Weber, 1973) having this iodoacetamide reactivity and a relatively long lifetime (21 nsec) made a fluorescence-depolarization experiment possible. It was found (Mendelson *et al.*, 1973;

Takashi *et al.,* 1976) that the Ca^{2+}-ATPase saturated (at about 5–7 times the unmodified activity) when 2 moles of dye were added per myosin. This behavior is characteristic of SH_1 labeling. Takashi *et al.* (1976) found radioactive 1,5-IAEDANS largely in a single spot on a peptide map of trypsin-digested labeled myosin. Mendelson and Cheung (1978) observed that the ratio of fluorescence from rod to that from the parent myosin was about 1:120, so that the fluorescence can effectively be considered as coming entirely from S-1 moieties.

The fluorescence-depolarization curves of S-1, HMM, and HMM + excess actin, myosin, and aggregated myosin are shown in Fig. 5. As can be seen, there was an observable decay of $r(t)$ for S-1, HMM, and soluble myosin. However, myosin aggregated in low-ionic-strength buffer or bound to actin showed no observable decay on this time scale. Since the two independent methods of immobilizing S-1 moieties into macromolecular complexes gave no significant polarization decay, it was likely that this decay came from mo-

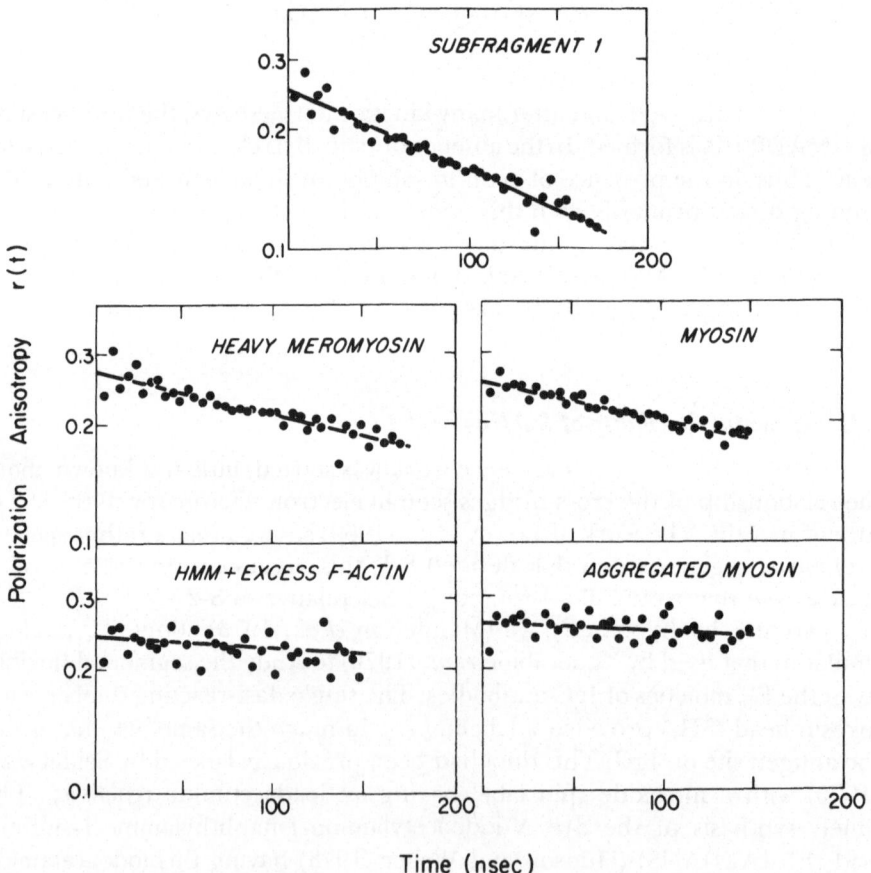

Figure 5. Fluorescence-depolarization curves from myosin and its subfragments. From Mendelson *et al.* (1973).

tion of the entire moiety rather than motion of the dye on the moeity. In fact, a search of possible solutions to equation (6) showed that if the shape was taken as a prolate ellipsoid of revolution, its axial ratio was at least 3.5. Even this minimum axial ratio gave an S-1 that was large in comparison with the interfilament spacing, that again it was unlikely that there was either significant motion of the dye relative to S-1 or internal flexing within S-1 that made a contribution to the data.

The data were characterized by a single exponential decay time ϕ. For S-1, HMM, and myosin, the ϕ values were 220, 400, and 450 nsec, respectively, at 5°C. Thus, even though the molecular weight quadrupled and the particle size increased by a factor of about 10 on incorporation of S-1 into myosin, the relaxation time only doubled. Since the dipoles were approximately along an equivalent major axis of S-1 and thus sensing motion about the minor axes (Mendelson *et al.*, 1973), rigid configurations in which the S-1 major axes (or dye axes) were approximately parallel to the myosin rod portion were ruled out because the predicted ϕ value, which is approximately proportional to the cube of the length, would be far too large. Many other rigid configurations, in which motion *about* the rod axis could be giving rise to the observed decay, could also be ruled out. For example, if the heads were in a "T" configuration, the molecular asymmetry and molecular weight (compared to S-1 alone) would cause increased hydrodynamic drag, and the ratio S-1 HMM would be far greater than observed. Other configurations such as a Y-shaped myosin could not be ruled out unambiguously. For this reason, Mendelson and Cheung (1978) measured the relative relaxation times of single-headed myosin and myosin in soluble conditions.

Single-headed myosin was purified by exploiting the fact that the affinity of the two-headed species for actin is much greater than is that of the single-headed species. Thus, actomyosin could be selectively removed by centrifugation of a myosin digest in a solution containing substiochometric amounts of actin under high-salt conditions. This method produced a mixture containing less than 5% myosin but having about 25% rod. However, the fluorescence from the rod was shown to be negligible. Figure 6 shows the results from two different control myosins and single-headed myosin. The ϕ value for single-headed myosin was within 10% of that of myosin. This confirmed that flexibility existed within HMM because the removal of one of the arms of the "Y" would be expected to have a large effect on the rotary diffusion about the rod axis. Other, less plausible, situations are still possible. However, by combining certain information about S-1 inferred from other experiments, it is possible to virtually exclude many of these cases (Mendelson and Cheung, 1978).

In an attempt to quantitate the degree of flexibility, the two myosin heads were modeled as two halves of a once-broken rod. Yu and Stockmeyer (1967) showed that the diffusion constants for each of these halves pivoting freely around the "break point" was twice that of the unbroken rod. By calculating this diffusion constant and by assuming that the dye dipoles lie approximately along or straddle the the S-1 rod long axis [the first term in equation (6) is dominant], the ratio of relaxation times of S-1 to myosin could be calculated:

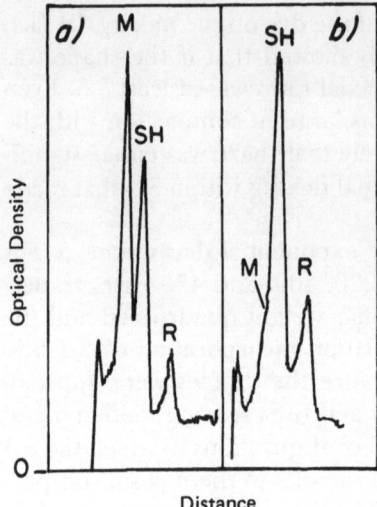
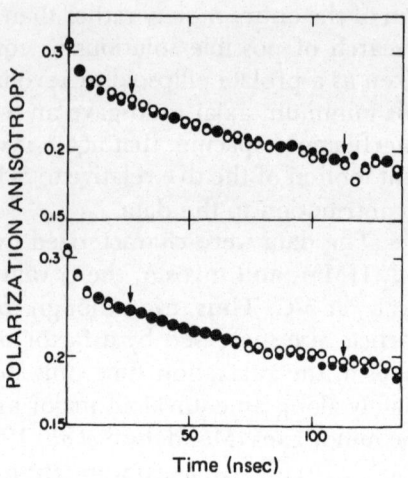

Figure 6. Single-headed-myosin experiments comparing the fluorescence depolarization of myosin and single-headed myosin. (a, b) Nondenaturing gels of the papain digest of myosin (M) and the single-headed-myosin preparation (SH). (R) Rod. Top right: Depolarization of the myosin digest (○) and single-headed myosin (●); bottom right: single-headed myosin (●) and its parent myosin (○). In the myosin digest, 75% of the fluorescence was from myosin. From Mendelson and Cheung (1978).

$$\frac{\phi_{\text{S-1}}}{\phi_{\text{myo}}} \simeq \frac{D_{\text{myo}}}{D_{\text{S-1}}} = \frac{2\,D_{\text{S-1}}\,(2l_{\text{S-1}})}{D_{\text{S-1}}\,(l_{\text{S-1}})} = 0.4 \tag{13}$$

where $l_{\text{S-1}}$ is the S-1 rod length. (This calculated ratio was not a strong function of $l_{\text{S-1}}$.) Experimentally, this ratio is approximately 0.5. Thus, it was concluded that the data are in agreement with a model of an asymmetric S-1 that can pivot by a universal joint between S-1 and S-2.

The small difference in ϕ values between myosin and HMM is consistent with this picture. The S-2 may not be sufficiently massive and asymmetric in comparison with the two S-1 moieties to fix the S-1–S-2 joint; the motion of the center of mass of the entire HMM complex may contribute to the observed depolarization. However, the additional length and mass of the LMM moiety in myosin should decrease the center-of-mass motion by approximately 10-fold and thus effectively fix the S-1–S-2 joint on this time scale. The observed small increase in ϕ on HMM incorporation into myosin is consistent with this "freezing out" of a small amount of center-of-mass motion.

The shape of S-1 is now known from X-ray scattering results to be more complex than a simple object of revolution (Mendelson and Kretzschmar, 1980), so that such a model should be thought of only as an approximation. However, the long relaxation time for S-1 indicates that the dyes are sensing some slow motion of the long axis of an equivalent general ellipsoid. The S-1 moiety of myosin could well exhibit varying degrees of torsional movement. In our more recent experiments, with improved instrumentation, the polarization decay of myosin and single-headed myosin is somewhat biphasic, indicating the possibility that such motion contributes to the observed data. (To minimize torsional contributions to our results, the least-squares fits were taken over a region excluding the first 25 nsec.) Given the number of possible modes of motion, the agreement with the simple model considered here is noteworthy.

Saturation-transfer electron paramagnetic resonance (STEPR) experiments by Thomas *et al.* (1975) also measured the relaxation times (τ_2) of free S-1 and the S-1 moiety. The good agreement found with the fluorescence work using this method, electron microscopy (Elliott and Offer, 1978) and other methods may be taken as verification that the S-1 moiety has considerable freedom at the S-1–S-2 junction (see Chapter 20).

It is interesting to note that this flexibility and S-1 shape were not measurably altered by physiologically relevant metals or nucleotides (Mendelson *et al.*, 1975; Mendelson and Cheung, 1978). Difference studies using MgAMP-P-N-P to stimulate the M · ATP binding state and MgATP to generate the rate-limiting state (M · **ADP · P_i) showed no measurable change in ϕ; if the S-1 shape was approximated by a prolate ellipsoid, the change in length of the long axis was less than 1.0 nm (Mendelson *et al.*, 1975). This is consistent with the hydrodynamic and circular dichroism work of Gratzer and Lowey (1969). Thus, it appears that S-1 (and HMM) do not reach out to attach to actin. This result is independent of the absence or presence of Ca^{2+} when physiological concentrations of Mg^{2+} are present.

3.3. Lack of Effect of Ca²⁺ on the Thick Filaments

A question that has received considerable attention in the recent past is whether, in vertebrate striated muscle, there exists a Ca^{2+}-sensitive switch on the thick filament whose purpose would be to radially move the cross-bridges toward the thin filament or otherwise cause them to change orientation in the absence of actin. Haselgrove (1975), measuring X-ray diffraction in highly stretched living muscle, found that the thick-filament-layer lines had essentially the same intensities as in overlapped muscle when the muscle was contracting or when forced into rigor by poisoning. Further, Morimoto and Harrington (1974) found from viscosity and sedimentation experiments that when the Ca^{2+} concentration was raised from 10^{-8} to 10^{-4} M, a change occurred in the hydrodynamic parameters of native and synthetic thick filaments. They suggested that these changes were caused by a radial fanning-out of the S-1 moieties.

We performed two similar experiments using fluorescence-polarization techniques. In the first experiment, glycerinated fibers were labeled with the 1,5-IAEDANS fluorophore (Nihei *et al.*, 1974a), which showed selectivity toward the myosin head. The polarization P was measured as a function of sarcomere length (Fig. 7) and used as a phenomenological indicator of cross-bridge orientation (Nihei *et al.*, 1974b). At average sarcomere lengths greater than about 4 μm (where no overlap of sarcomeres exists), there were no indications of differences in $P_\perp$ between relaxing, rigor, and contracting mus-

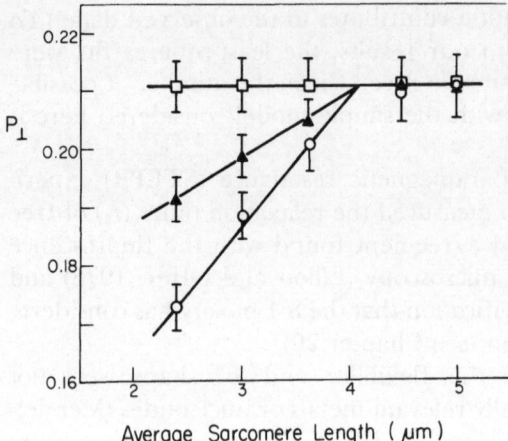

Average Sarcomere Length (μm)

Figure 7. Fluorescence polarization generated by relaxed ($\square$), contracting ($\blacktriangle$), and rigor ($\bigcirc$) 1,5-IAEDANS-labeled glycerinated fibers as a function of sarcomere length. From Nihei *et al.* (1974a).

cle. This result was taken as evidence that the change in orientation or spatial dispersion that S-1 undergoes when the cross-bridges are adjacent to actin is not seen in the absence of actin.

The second experiment used fluorescence depolarization to investigate whether increasing the amount of Ca^{2+} induced an increased rotational Brownian motion (Mendelson and Cheung, 1976) of myosin heads. A removal of the inhibition of rotational motion would be expected if there were a net radial displacement of S-1 away from the thick-filament lattice [as changes in the ratios of equatorial X-ray reflections suggest (H. E. Huxley and Brown, 1967)].

Experiments were performed using the same preparation used by Morimoto and Harrington (1974). Filament formation was monitored using electron microscopy and analytical ultracentrifugation. Figure 8 shows representative data taken under relaxed conditions (pCa = 8.5) and contracting conditions (pCa = 4). Even though the filament order was very sensitive to pH and KCl concentration changes (Table 1) in agreement with electron microscopy (Kaminer and Bell, 1966), no changes were induced by Ca^{2+} addition.

A quantitative estimate of the expected induced change was made by assuming S-1 to be a 13.0-nm-long rod that moved out 4.0 nm from the filament in the presence of Ca^{2+}. The model assumed that when Ca^{2+} was present, all the S-1 moieties were diffusing freely except for steric constraints imposed by the thick filament. Using the measured relaxation time ($\phi_0 \simeq 1000$ nsec) for the case in which no calcium was present, a change of 100 [($\phi_{Ca} - \phi_0)/\phi_0$] $\simeq -50\%$ was expected on Ca^{2+} addition. This model depends only weakly on the shape of S-1; a 4.0-nm movement allows the head to tumble through $7\pi/2$ steradians, and thus the ϕ_{Ca} value should be virtually the same as for soluble myosin. Thus, we conclude that Ca^{2+} does not induce the rotational motion that would be expected if the cross-bridges moved radially away from the thick filaments.

These experiments were not done in a living muscle, so that one could argue that something was altered in the glycerination of the fibers, reconstitu-

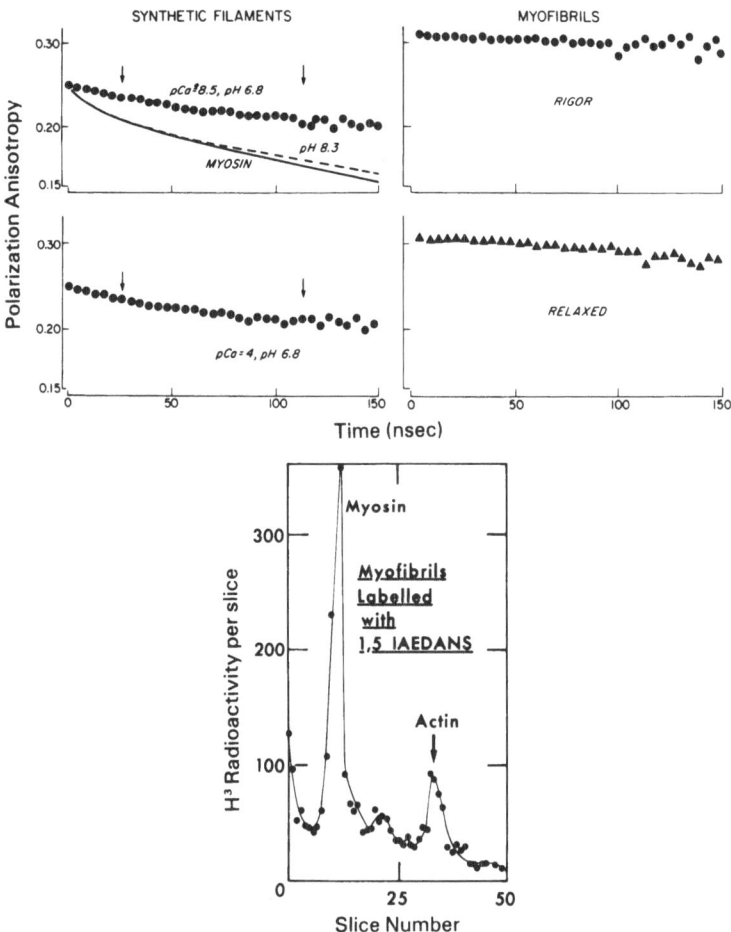

Figure 8. TRFAD experiments (top) on synthetic filaments with and without Ca²⁺ and labeled myofibrils in relaxing and rigor solutions. Radioactive-1,5-IAEDANS-labeled fibers were tested for their ability to relax by stiffness measurements. After homogenization, the TRFAD experiments were performed and sodium dodecyl sulfate gels were made. Subtraction of a blank gel gave the result in the bottom panel. In part from Mendelson and Cheung (1978).

tion of the filaments or by labeling. However, subsequent to our experiments, the X-ray fiber experiments were redone by Yagi and Matsubara (1980) and H. E. Huxley *et al.* (1980), and synthetic-filament results were reexamined by detailed cross-linking experiments (Sutoh and Harrington, 1977). All experimental results now agree that there is no significant radial movement of the cross-bridges in the absence of actin on Ca²⁺ addition. Recent X-ray scattering experiments (Mendelson and Kretzschmar, 1980) find S-1 to have a longest chord of 12 ± 1 nm, so that it could nearly span the interfilament spacing (about 14 nm); attachment might then be accomplished by only a small amount of radial diffusion.

Table 1. Effect of Ca²⁺, pH, and KCl on the Mobility of the S-1 Moiety in Synthetic Filaments

Change in conditions[a]	$100\Delta\phi/\phi^b$ (%)	Trials[c]	Preparations
In pH 6.8 solution, pCa 8.5 → pCa 4			
Filaments made from unchromatographed myosin	0.1 ± 2	25	—
Filaments made from myosin chromatographed on DEAE-50	−2 ± 3	9	—
Filaments made from unchromatographed myosin + 5 mM MgATP	3 ± 5	11	—
Solution pH 8.3 → pH 6.8	50–70		5
Solution pH 8.3 → pH 6.5	250		1
In pH 6.8 solution, 90 mM KCl → 110 mM KCl	−15		2
Solution with 0.6 M KCl (free myosin) changed to pH 6.8 solution	80–160		5

[a] The pH 6.8 solution had a protein content of 1 mg/ml, 90 mM KCl, 0.3 mM MgCl₂, and 40 mM imidazole ($\mu = 0.12$). The pH 8.3 solution had a protein content of 1 mg/ml, 135 mM KCl, and 2 mM tris ($\mu = 0.137$).
[b] All experiments were at 5°C.
[c] Ca²⁺ addition trials spanned five preparations; each half of the difference trial lasted 1 hr.

3.4. Mobility and Orientation of Subfragment-1 Moieties in Myofibrils and Muscle

Fluorescence-depolarization studies in relaxed glycerinated myofibrils were performed to examine the mobility of the heads in a native filament (see Fig. 8) (Mendelson and Cheung, 1976). Fibers were labeled following the methods of Duke *et al.* (1976). A small difference in mobility was observed between the rigor and the relaxed states; however, after correction for a small fraction of slow motion seen in myosin-extracted myofibrils, the difference became immeasurably small. Using the TRFAD method, we could set a lower limit of only $\phi \geq 3000$ nsec.* Based on the observation of highly developed thick-filament layer lines in X-ray diffraction, we reasoned that all S-1 moieties would be at the same average orientation with respect to the fiber axis. Using the method described in Section 2.2.1., we found that we could set a lower limit of 12.5° for the half angle through which S-1 could freely diffuse.

Recently, Thomas *et al.* (1980) reported STEPR experiments on myofibrils and synthetic filaments. They also found that a large immobilization occurred on lattice formation; an effective relaxation time of 10 μsec was determined by calibrating their spectra against that expected from an isotropically rotating molecule. Taken alone, this result agrees with the model that we considered; however, another experiment using oriented, spin-labeled, glycerinated rabbit fibers (Thomas and Cooke, 1980) indicates that the myosin heads are almost completely randomly dispersed in space. Further, results on oriented, fluorescent-labeled, glycerinated fibers (Borejdo and

---

*In Mendelson and Cheung (1978), the value was incorrectly stated as $\phi \simeq 3000$ nsec, rather than $\phi \geq 3000$ nsec. This was a typographical error and does not affect the conclusions in any way.

Putnam, 1977) could also be indicating a great deal of randomness (Mendelson and Wilson, 1981). Thus, with the caveat that X-ray diffraction work by Rome (1972) indicates that there is increased disorder in relaxed glycerinated fibers and hence that work on this preparation should be treated with caution, the immobilization seen in relaxed fibers could be due to a slow diffusion over a large angular range, rather than a rapid and small diffusion of heads about a single equilibrium orientation. One possibility that agrees with both the orientation and the mobility measurements is that the heads are completely disordered and freely diffusing (through $< 12.5°$) about random equilibrium positions. Another possibility is that the oriented-fiber-experiment results do not have a unique interpretation. Mendelson and Wilson (1981) have developed methods for calculating expected EPR spectra and fluorescence-polarization values for various kinds of three-dimensional disordering of the cross-bridges; work is under way to apply these to the available data. Still another possibility is that a mixture of various states of S-1 moieties exists. For example, if we treat the fluorescence-depolarization results as arising from a completely free fraction and a completely oriented, immobile fraction, then less than 25% of the heads are in the former category.

No disagreement exists in the interpretation of rigor-state data; both oriented-fluorescence and EPR data indicate a rather definite orientation of the labels, suggesting a single population of cross-bridge orientations in the rigor state. Recent modeling of X-ray diffraction from rigor insect muscle (Offer and Elliott, 1978; Holmes *et al.*, 1980) predicts that only about one half of the cross-bridges are bound to actin. Whether the EPR and fluorescence-polarization data and this model can be reconciled is unclear at this time.

4. Summary

Figure 9 summarizes schematically how fluorescence results might harmonize with the rotating-cross-bridge model of force generation (H. E. Huxley, 1969; A. F. Huxley and Simmons, 1971). The asymmetric detached heads are immobile and taken to be in a rather definite orientation. However, they occasionally undergo a small amount of radial diffusion that allows them to bind to actin when Ca^{2+} has switched on the thin-filament regulation mechanism. Binding may be facilitated by increased rotational diffusion of the S-1 moiety. After attachment, the flexible joint allows the head to undergo significant rotation, which draws its center of mass toward the thin filament. Work is done against a spring with longitudinal (not shown) and transverse compliance. This transverse compliance probably arises from the S-1–S-2 joint's sustaining a bending moment (a leaf spring) that causes the heads to be radially restored to near the thick-filament backbone and immobilized (indicated by a notch in Fig. 9) after the power stroke is completed. If the muscle is shortening, this restoration positions the head near a new actin-binding site. Ultimate validation of this model, of course, awaits further experiments.

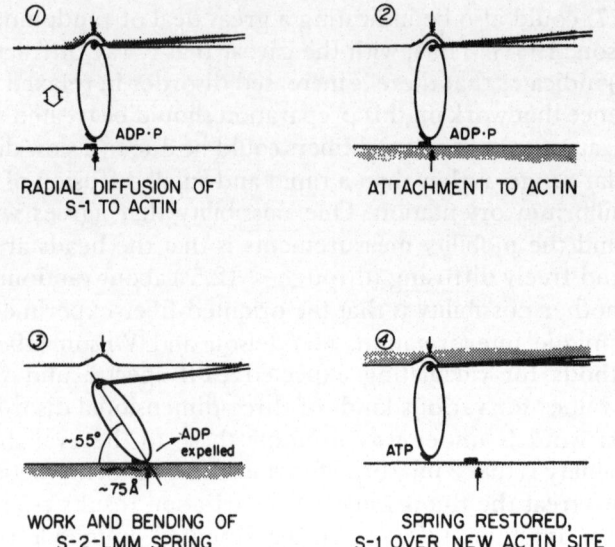

Figure 9. Schematic diagram of a model of the events in the cross-bridge cycle. The immobilization of S-1 seen in the relaxed state is indicated by a notch in the thick filament. For simplicity, a crossbridge in an isotonically contracting muscle is shown.

ACKNOWLEDGMENTS. The author gratefully acknowledges his colleagues, who are the true authors of this work. In particular, he thanks Professor Manuel Morales, without whose encouragement, support, and help this work would not have been possible. This work was supported by NSF grant PCM-75-22698 and U.S.P.H.S. Grants HL06285 and HL 11683 and AHA Grant CI8.

References

Aranson, J., and Morales, M. F., 1969, Polarization of tryptophan fluorescence in muscle, *Biochemistry* **8**:4512.

Belford, G. G., Belford, R. L., and Weber, G., 1972, Dynamics of fluorescence polarization in macromolecules, *Proc. Natl. Acad. Sci. U.S.A.* **69**:1392.

Borejdo, J., and Putnam, S., 1977, Polarization of fluorescence from single skinned glycerinated rabbit psoas fibers in rigor and relaxation, *Biochim. Biophys. Acta* **459**:578.

Borejdo, J., Putnam, S., and Morales, M. F., 1979, Fluctuations in polarized fluorescence: Evidence that muscle cross bridges rotate repetitively during contraction, *Proc. Natl. Acad. Sci. U.S.A.* **76**:6346.

Cooke, R., 1981, Fluorescence as a probe of contractile systems, *Methods Enzymol.* (in press).

Dos Remedios, C. G., Millikan, R. G. C., and Morales, M. F., 1972, Polarization of tryptophane fluorescence from single striated muscle fibers, *J. Gen. Physiol.* **59**:103.

Duke, J., Takashi, R., Ue, K., and Morales, M. F., 1976, Reciprocal reactivities of specific thiols when actin binds to myosin, *Proc. Natl. Acad. Sci. U.S.A.* **73**:302.

Ehrenberg, M., and Rigler, R., 1972, Polarized fluorescence and rotational Brownian motion, *Chem. Phys. Lett.* **14**:539.

Elliott, A., and Offer, G., 1978, Shape and flexibility of the myosin molecule, *J. Mol. Biol.* **123**:505.

Gratzer, W. B., and Lowey, S., 1969, Effect of substrate on conformation of myosin, *J. Biol. Chem.* **244**:22.

Harvey, S. C., and Cheung, H. C., 1980, Transport properties of particles with segmental flexibility. II. Decay of fluorescence polarization anisotropy from hinged macromolecules, *Biopolymers* **19**:913.

Haselgrove, J. C., 1975, X-ray evidence for conformational changes in the myosin filaments of vertebrate striated muscles, *J. Mol. Biol.* **92**:113.

Highsmith, S., Mendelson, R. A., and Morales, M. F., 1976, Affinity of myosin S-1 for F-actin, measured by time-resolved fluorescence anisotropy, *Proc. Natl. Acad. Sci. U.S.A.* **73**:133.

Holmes, K. C., Tregear, R. T., and Barrington Leigh, J., 1980, Interpretation of low angle x-ray diffraction from insect flight muscle in rigor, *Proc. R. Soc. London Ser. B* **207**:13.

Hudson, E., and Weber, G., 1973, The synthesis and characterization of two fluorescent sulfhydryl reagents, *Biochemistry* **12**:4154.

Huxley, A. F., and Niedergerke, 1954, Structural changes in muscle during contraction: Interference microscopy of living muscle fibers, *Nature (London)* **173**:971.

Huxley, A. F., and Simmons, R. M., 1971, Proposed mechanism of force generation in striated muscle, *Nature (London)* **233**:533.

Huxley, H. E., 1957, The double array of filaments in cross-striated muscle, *J. Biophys. Biochem. Cytol.* **3**:631.

Huxley, H. E., 1960, Muscle cells, in: *The Cell*, Vol. 4 (J. Brachet and A. Mirsky, eds.), pp. 365–911, Academic Press, New York.

Huxley, H. E., 1969, The mechanism of muscular contraction, *Science* **164**:1356.

Huxley, H. E., and Brown, W., 1967, The low-angle x-ray diagram of vertebrate striated muscle and its behavior during contraction and rigor, *J. Mol. Biol.* **30**:383.

Huxley, H. E., and Hanson, J., 1954, Changes in cross-striations of muscle during contraction and stretch and their structural interpretation, *Nature (London)* **173**:973.

Huxley, H. E., Faruqi, A. R., Bordas, J., Koch, M. H. J., and Milch, J. R., 1980, The use of synchrotron radiation in time-resolved X-ray studies of myosin layer-line reflections during muscle contraction, *Nature (London)* **284**:140.

Inoué, H., Takenaka, T., and Tonomura, Y., 1979, Functional implications of the two-headed structure of myosin, *Adv. Biophys.* **13**:1.

Kaminer, B., and Bell, A. L., 1966, Myosin filamentogenesis—Effects of pH and ionic concentration, *J. Mol. Biol.* **20**:391.

Kauzmann, W., 1957, *Quantum Chemistry*, Chapter 15, Academic Press, New York.

Kinesota, K., Jr., Kawato, S., and Ikegami, S., 1977, A theory of fluorescence polarization decay in membranes, *Biophys. J.* **20**:289.

Lowey, S., and Slayter, H. S., Weeds, A. G., and Baker, H., 1969, Substructure of the myosin molecule. I. Subfragments of myosin by enzymic degradation, *J. Mol. Biol.* **42**:1.

Mendelson, R. A., and Cheung, P., 1976, Muscle crossbridges: Absence of direct effect of calcium on movement away from the thick filaments, *Science* **194**:190.

Mendelson, R. A., and Cheung, P., 1978, Intrinsic segmental flexibility of the S-1 moiety of myosin using single-headed myosin, *Biochemistry* **17**:2139.

Mendelson, R. A., and Kretzschmar, K. M., 1980, Structure of subfragment 1 from low-angle x-ray scattering, *Biochemistry* **19**:4103.

Mendelson, R. A., and Morales, M. F., 1977, The theory of fluorescence polarization from fluorescent labelled muscle fibers, *Biochim. Biophys. Acta* **459**:590.

Mendelson, R. A., and Wilson, M., 1981, Three dimensional disorder in helical systems: Application to dipolar ESR and fluorescent probes on muscle cross-bridges, *Biophys. J. Abstr.* **33**:82.

Mendelson, R. A., Morales, M. F., and Botts, J., 1973, Segmental flexibility of the S-1 moiety of myosin, *Biochemistry* **12**:2250.

Mendelson, R. A., Putnam, S., and Morales, M. F., 1975, Time dependent fluorescence depolarization and lifetime studies of myosin subfragment-one in the presence of nucleotide and actin, *J. Supramol. Struct.* **3**:162.

Morimoto, K., and Harrington, W. F., 1974, Evidence for structural changes in vertebrate thick filaments induced by calcium, *J. Mol. Biol.* **88**:693.

Nihei, T., Mendelson, R. A., and Botts, J., 1974a, The site of force generation in muscle contraction as deduced from fluorescence polarization studies, *Proc. Natl. Acad. Sci. U.S.A.* **71**:274.

Nihei, T., Mendelson, R. A., and Botts, J., 1974b, Use of fluorescence polarization to observe changes in attitude of S-1 moieties in muscle fibers, *Biophys. J.* **14**:236–242.

Offer, G., and Elliott, A., 1978, Can a myosin molecule bind to two actin filaments?, *Nature (London)* **271**:325.

Perrin, F., 1934, Mouvement Brownien d'un ellipsoide. I. Dispersion diélectrique pour des molécules ellipsoidales, *J. Phys. Radium VII* **5**:497.

Perrin, F., 1936, Mouvement Brownien d'un ellipsoide. II. Rotation libre et dépolarisation des fluorescences: Translation et diffusion de molécules ellipsoidales, *J. Phys. Radium VII* **7**:1.

Reedy, M., Holmes, K. C., and Tregear, R. T., 1965, Induced changes in orientation of cross-bridges, *Nature (London)* **207**:1276.

Rome, E., 1972, Relaxation of glycerinated muscle: Low-angle x-ray diffraction studies, *J. Mol. Biol.* **65**:331.

Seidel, J. E., Chap, M., and Gergely, J., 1970, Effect of nucleotides on spin labels bound to S_1 thiol groups of myosin, *Biochemistry* **9**:3265.

Sekine, T., and Kielley, W. W., 1964, The enzymmatic properties of N-ethyl maleimide modified myosin, *Biochim. Biophys. Acta* **81**:336.

Sutoh, K., and Harrington, W. F., 1977, Cross-linking of myosin thick filaments under activating and rigor conditions: A study of radial disposition of the cross-bridges, *Biochemistry* **16**:2441.

Takashi, R., Duke, J., Ue, K, and Morales, M. F., 1976, Defining the "fast-reacting" thiols of myosin by reaction with 1,5 IAEDANS, *Arch. Biochem. Biophys.* **175**:279.

Taylor, E., 1979, Mechanism of actomyosin ATPase and the problem of muscle contraction, *CRC Crit. Rev. Biochem.* **6**:103.

Thomas, D. D., and Cooke, R., 1980, Orientation of spin-labelled myosin heads in glycerinated muscle fibers, *Biophys. J.* **32**:891.

Thomas, D. D., Seidel, J. C., Hyde, J. S., and Gergely, J., 1975, Motion of subfragment-1 in myosin and its supramolecular complexes: Saturation transfer electron paramagnetic resonance, *Proc. Natl. Acad. Sci. U.S.A.* **72**:1729.

Thomas, D. D., Ishiwata, S., Seidel, J., and Gergeley, J., 1980, Submillisecond rotational dynamics of spin-labelled myosin heads in myofibrils, *Biophys. J.* **32**:873.

Tregear, R. T., and Mendelson, R. A., 1975, Polarization from a helix of fluorophores and its relation to that obtained from muscle, *Biophys. J.* **15**:455.

Trentham, D. R., Eccleston, J. F., and Bagshaw, C. R., 1976, Kinetic analysis of ATPase mechanisms, *Q. Rev. Biophys.* **9**:2.

Wahl, P., 1979, Analysis of fluorescence anisotropy decays by a least squares method, *Biophys. Chem.* **10**:91.

Weber, G., 1972, Uses of fluorescence in biophysics and some recent developments, *Annu. Rev. Biophys.* **1**:553.

Yagi, N., and Matsubara, I., 1980, Myosin heads do not move on activation in highly stretched vertebrate striated muscle, *Science* **207**:307.

Yguerabide, J., 1972, Nanosecond fluorescence spectroscopy of macromolecules, *Methods Enzymol.* **26**:498.

Yguerabide, J., Epstein, H. F., and Stryer, L., 1970, Segmental flexibility in an antibody molecule, *J. Mol. Biol.* **51**:573.

Yu, H., and Stockmeyer, W. H., 1967, Intrinsic viscosity of a once broken rod, *J. Chem. Phys.* **47**:1369.

20

Myosin Flexibility

Stephen C. Harvey and Herbert C. Cheung

1. Introduction

The description of the molecular events that produce tension in a contracting muscle is often called *mechanochemistry,* a term that well describes the synthesis of the two key elements in the contractile cycle. One of those elements is mechanical: we need to be able to describe in detail, with resolution down to the atomic level, the sequence of conformational changes that cause the thick and thin filaments to slide past one another. The other element is chemical: we need to determine the sequence of biochemical reactions in the contractile cycle and to measure the energetic and kinetic parameters of each of those reactions. Finally, the problem requires synthesis: we need to explain the coupling of the mechanical and chemical events. Only then will we understand how the chemical energy released by the hydrolysis of ATP is converted to the mechanical work performed by the contracting muscle.

At present, our understanding of the mechanical side of the problem is somewhat more limited than our biochemical understanding. This is partly because many chemical studies are inherently dynamic, whereas most of the physical studies provide only static information (electron microscopy, for example) or give information only on average structures, viewed over time scales much longer than one cycle of contractile events (X-ray fiber diffraction, for instance). Further, the physical studies are still hampered by limited resolution, particularly because X-ray crystallographers have so far had no success with myosin or its subfragments, much less with the actomyosin complex.

But progress is being made on the structural questions. This review ex-

Stephen C. Harvey and Herbert C. Cheung • Biophysics Section, Department of Biomathematics, University of Alabama in Birmingham, Birmingham, Alabama 35294.

amines flexibility within myosin, an important question because of the obvious role that such flexibility must play in the conformational changes that cause contraction. Other reviews are available on the higher-level organization of the contractile and regulatory proteins in the thick and thin filaments (Weber and Murray, 1973; Squire, 1975; Mannherz and Goody, 1976; Murphy, 1979) (also see Chapters 5, 11, and 18). The current state of our understanding of how the chemical and mechanical events are coupled has been reviewed by Marston *et al.* (1979) and by Taylor (1979).

1.1. Sliding-Filament Model and Myosin Flexibility

The foundation of research into the molecular basis of muscle contraction is the sliding-filament model, set forth when electron microscopy revealed that the thick and thin filaments remain of constant length while the sarcomere shortens (H. E. Huxley, 1957). Subsequent refinements of this model first postulated that force generation results from the interaction of actin with the myosin heads, since the latter reach across from the thick to the thin filament (H. E. Huxley, 1957), and later suggested that the interaction requires the myosin heads to rotate as the filaments move past one another (H. E. Huxley, 1969, 1971). Thus, even without direct evidence for the rotation of the myosin heads, the first myosin hinge was introduced into the model as a joint between subfragment-1 (S-1) and the rod. In addition, the problem of how each cross-bridge could generate the same force regardless of interfilament spacing was solved by proposing a second hinge within the rod, probably near the proteolytically sensitive region where heavy meromyosin (HMM) and light meromyosin (LMM) are joined (H. E. Huxley, 1969, 1971). The locations of these hinges are shown schematically in Fig. 1.

There have been a number of later refinements and modifications of the sliding filament–rotating cross-bridge model originally put forward by H. E. Huxley (for example, Marston *et al.*, 1979; Morales and Botts, 1979; Eisenberg and Hill, 1978; Eisenberg *et al.*, 1980; Schoenberg, 1980a,b), and all of them regard the putative hinge between S-1 and the rod as essential. Flexibility within the rod is generally regarded as useful in these models, if not an absolute requirement.

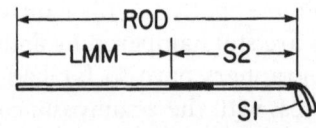

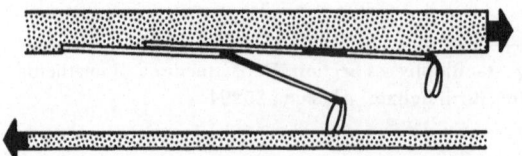

Figure 1. Schematic diagram of the geometry of the myosin molecule in solution (*top*) and aggregated into the thick filament in muscle (*bottom*). The black regions represent the areas of proteolytic sensitivity where flexibility has often been suggested to occur.

A second kind of model, supposing a more active role for subfragment-2 (S-2) was first set forth by A. F. Huxley and Simmons (1971). Although the myosin head is still believed to rotate, the most important aspect of rod flexibility in their model is not the lateral bending, allowing the head to reach over to the thin filament, but the elastic stretching of S-2, acting to transmit the tensile force to the thick filament.

An even more active role for rod flexibility was proposed by Harrington (1971). He suggested that the local change in pH arising from the protons released by ATP hydrolysis on one myosin head would cause a portion of S-2 in a neighboring myosin, attached to the thin filament, to undergo a helix–coil transition, thereby shortening that S-2 and generating tension. A comparison of Harrington's model with the Huxley–Simmons model is shown in Fig. 2.

Thus, the basic mechanochemical models of muscle contraction are built on two flexible regions within myosin: first, S-1 is joined to the myosin rod by a fairly free hinge or swivel; second, the trypsin-sensitive region within the rod allows or causes motion between LMM and S-2, either bending, or stretching, or both. Section 2 of this chapter reviews the evidence for the hinge between S-1 and the rod, while the studies on flexibility within the rod are reviewed in Section 3.

2. Hinge Between Subfragment-1 and the Rod

There is an ample body of evidence for the existence of segmental flexibility about the joint where the myosin heads join the rod, and there is virtually no controversy about that flexibility. Since all the evidence has not previously been collected in a single article, this section is intended as a brief but comprehensive review of that evidence. The studies are divided into two groups, solution studies on single molecules (Section 2.1) and studies on supramolecular aggregates, including both synthetic filaments and whole muscle (Section 2.2).

Figure 2. Comparison of two models of contractility. (A) Model of A. F. Huxley and Simmons. (1) Resting state; (2) attachment of S-1 to F-actin; (3) S-1 rotates while attached to actin, and the elastic component of S-2 is stretched; (4) power stroke, resulting from contraction of elastic component; (5) return of cross-bridge to resting state. (B) Model of Harrington. (1) Resting state; (2) S-1 swivels and attaches to actin; (3) S-2 is released from the surface of the thick filament; (4) power stroke, resulting from helix–coil transition within S-2; (5) return of cross-bridge to resting state. Reproduced by permission from Tsong *et al.* (1979).

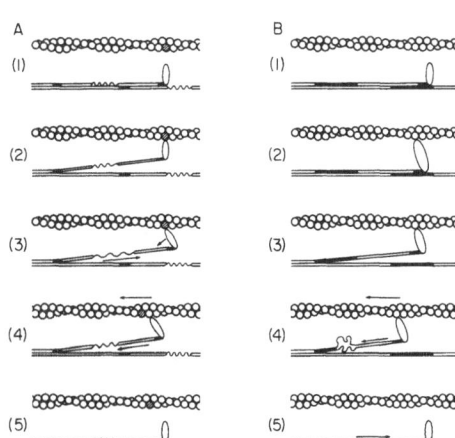

2.1. Studies on Single Molecules

2.1.1. Fluorescence Depolarization

The first direct evidence for flexibility within myosin came from fluorescence-depolarization studies (Mendelson *et al.*, 1972, 1973), following the lead of Yguerabide *et al.* (1970), who had demonstrated segmental flexibility in an immunoglobulin using the single photon-counting method (for a review of the method, see Yguerabide, 1972; Wahl, 1975). Mendelson and co-workers observed that the rate of decay of the polarization anisotropy of fluorescently labeled S-1 was consistent with the fluorophore's being rigidly attached to an elongated ellipsoidal molecule with an axial ratio of at least 3.5. When they labeled HMM or whole myosin, the rate of decay went down, but it was more rapid than would have been expected for a rigid macromolecule of that size. From this, they concluded that the myosin head has considerable rotational mobility about a swivel or hinge where it is attached to the rod. The effects of divalent cations on the extent of rotational mobility as detected by fluorescence depolarization were examined by Highsmith (1978). Since all the fluorescence work is reviewed in Chapter 19, it will not be discussed in greater detail here.

2.1.2. Saturation-Transfer Electron Paramagnetic Resonance

One shortcoming of fluorescence depolarization for the investigation of intramolecular motions in very large macromolecules or supramolecular aggregates is that fluorescence lifetimes are typically much shorter than the time scale of the motions being studied. Saturation-transfer electron paramagnetic resonance (STEPR) was developed to study these large-scale motions in the microsecond and submillisecond range. The method has been reviewed by Dalton (1976), by Thomas *et al.* (1976), and by Hyde (1978).

Myosin flexibility has been examined by STEPR by Thomas and co-workers (Thomas *et al.*, 1975a,b; Thomas, 1978). To facilitate a comparison with the fluorescence-depolarization studies (Mendelson *et al.*, 1972, 1973), in which an iodoacetamide fluorophore had been used, Thomas and co-workers used an iodoacetamide spin label to selectively label S-1 at either the SH-1 or SH-2 sulfhydryls. Effective rotational correlation times were determined by comparing the experimental spectra with computer-simulated reference spectra where the only parameter varied was τ_2, the rotational correlation time for the reorientation of the molecule-fixed hyperfine and g tensors of the spin label, and with the spectra of hemoglobin in glycerol–water mixtures.

The STEPR studies essentially confirmed the fluorescence-depolarization results. First, the rotational correlation time for S-1 is about 4 times as long as would be expected for a spherical molecule of the molecular weight of S-1, but it is approximately what would be expected for a rigid prolate ellipsoid of axial ratio 4–5 if the principal axis of the label is nearly aligned with the long axis of the ellipsoid. Thus, S-1 is an elongated molecule. Second, the rotational correlation time for S-1 in HMM and in whole myosin is increased over that for free S-1 by a factor of 2 or less. This is considerably less than would be expected if HMM and myosin were rigid, so the STEPR measurements confirm the earlier finding of a hinge between the head and tail portions of myosin.

2.1.3. Electric Birefringence

Kobayashi and Totsuka (1975) examined the S-1–S-2 hinge using electric birefringence. They found that HMM, S-1, and S-2 all show a positive birefringence and that HMM and S-2 have large intrinsic Kerr constants.

The birefringence of S-1 was found to decay with a single exponential time constant of 250 nsec, consistent with that for a rigid ellipsoidal molecule with an axial ratio of 7, again indicating that S-1 is elongated. Although the Kerr constant is sensitive to pH, the rotational correlation time was found to be unchanged over the pH range from 6.8 to 9.0, dropping to 170 nsec at pH 11.5.

The birefringence decay of HMM is not a single exponential, but it can be well fitted to a two-exponential decay with time constants of 700 nsec and 5.9 μsec. Since the electric birefringence experiments must be performed at low salt concentrations, Kobayashi and Totsuka carefully examined their solutions by Sephadex chromatography and by sedimentation, to see whether the fast component could be due to a small molecule contaminating their solutions. No such contaminant was found chromatographically. The sedimentation pattern in 0.06 M KCl gave a single sedimentation constant, but a minor component with a smaller sedimentation coefficient was observed at low salt. The total amount of that minor component was too small to account for the amplitude of the fast component of the birefringence decay, however. Thus, these experiments also indicate that there is bending motion within HMM, probably at a hinge between S-1 and S-2.

2.1.4. Nuclear Magnetic Resonance

High resolution [¹H]-NMR studies by Highsmith *et al.* (1979) identified the fraction and location of highly mobile amino acid side chains in S-1, HMM, and myosin. The spectra show strong, narrow signals, quite different from the spectrum for LMM. The peaks are unexpectedly narrow for such large molecules, and the sharpness must be a consequence of some intramolecular motion besides the anticipated rotation of the methyl groups responsible for the signal. Since the narrow line widths are similar to those for the denatured proteins, the motions in the native proteins must be comparable to those in a random coil.

The amount of mobile structure, 22% in S-1, is significant. By showing that this mobility does not reflect an unraveled structure or loose ends of the peptide chain, and that it is not due to motion at the nucleotide binding site, Highsmith and co-workers were able to argue that the mobility could be either at the actin binding site, or at the joint between S-1 and S-2, or both. Because there is so much mobility involved, the flexibility of the S-1–S-2 hinge appears well substantiated by the NMR measurements.

2.1.5. Thermodynamics of Binding to Actin

Peller (1975) noted that a comparison of the binding of S-1 to actin with that of HMM to actin should provide information on the extent of intramolecular motion in myosin, since S-1 and HMM are one- and two-headed subfragments. He calculated theoretical binding isotherms with a parameter to represent the effective local concentration of the second head once the first head had attached itself to the actin filament. That parameter was then calculated using the dissociation constants for S-1–actin and HMM–actin reported by Margossian and Lowey (1973). It was found that the effective local concentration was so high that the second head must have considerable freedom of rotational motion relative to the first, so that it can seek out binding sites on the F-actin filament.

2.1.6. Hydrodynamic Properties

There have been many hydrodynamic studies on myosin and its subfragments, but their interpretation in terms of specific molecular models has been difficult, because of the absence of theoretical formalisms for predicting the hydrodynamic properties for models of irregular shape and possessing internal flexibility. One such formalism has recently been developed (García de la Torre and Bloomfield, 1978; García Bernal and García de la Torre, 1980) and applied to the investigation of intramolecular flexibility in myosin (García de la Torre and Bloomfield, 1980).

The study was carried out in three stages. First, from a collation of all the available data on the physical properties of S-1, a pearlike shape was selected for each of the myosin heads, with final dimensions including the swelling necessary to account for hydration. Second, a similar procedure provided model dimensions for S-2 and for the rod; the rod model included a possible hinge between S-2 and LMM (see Section 3). Third, these elements were combined into the model for whole myosin, allowing for possible bending between the heads and the rod. The algorithm allows the calculation of the translational and rotational diffusion coefficients and the intrinsic viscosities for rigid and flexible models.

The accuracy of the experimental data does not permit the unambiguous answer to the question of whether or not S-1 is attached to the rod by a hinge or swivel, because the heads do not make a very large contribution to the hydrodynamic properties of either HMM or myosin. Nonetheless, García de la Torre and Bloomfield (1980) did conclude that "the experimental properties are compatible with flexibility at the joint of S1 and S2."

2.1.7. Electron Microscopy

The visualization of myosin in the electron microscope has the potential for providing the most direct information on the location and extent of intramolecular flexibility. Elliott and Offer (1978) have carried out an analysis of the geometry of myosin using stereographic photography and the shadow-casting technique. Their preparatory methods include the use of a cryoprotectant and drying at low temperatures to improve the preservation of the molecular shape. The negative staining method was used by Takahashi (1978) for a similar study.

The angle that the head makes with the tail can be determined to about ±10°, and the distribution of angles was analyzed for a large number of molecules in each study. At the level of resolution of these investigations, the heads behave as nearly rigid bodies attached to the rod by a hinge. Elliott and Offer observed no correlations in the angles the two heads make with the tail, indicating that each one can independently explore all orientations. By contrast, Takahashi reported that in nearly half the molecules observed, one head made an angle of about 45° with the tail; in 52% of the molecules, there was an angle of approximately 90° between the heads. Thus, he found preferred orientations. Nonetheless, the wide variety of observed conformations and the fact that the largest angle that was observed between the heads was 225° indicate that the heads pivot quite freely about the S-1–S-2 hinge.

2.2. Studies on Synthetic Filaments and Whole Muscle

2.2.1. Fluorescence Depolarization

When fluorescently labeled myosin is aggregated into synthetic filaments, the motion of the heads is restricted, producing higher steady-state polarizations and longer rotational correlation times (Mendelson *et al.,* 1973; Mendelson and Cheung, 1976, 1978). But even in these filaments, there is some residual depolarization, indicating that S-1 can still rotate about the S-1–S-2 hinge. By comparing the observed decays with model decay curves from a Monte Carlo simulation for the restricted motion of a fluorophore constrained to diffuse within a cone, Mendelson and Cheung (1978) estimated that in synthetic filaments and native thick filaments, the myosin head can move freely within a cone of half angle less than 12.5°.

A particularly important study on glycerinated fibers is that of Borejdo *et al.* (1979) in which fluctuations in the polarization of fluorescence were monitored. They used an iodoacetamide derivative of rhodamine to label sulfhydryl groups on the myosin head. The fluorescence was resolved into orthogonally polarized components, and the fluctuations in polarization were measured by a photon-counting system with long time resolution. No fluctuations were observed in relaxation or rigor or when the label was on actin, but there were fluctuations from the labeled heads during steady-state contraction, dominated by a low-frequency mode near 2 Hz. This is near the measured ATPase frequency, and since the fluctuations are produced by changes in fluorophore orientation but not in fluorophore number, the experiments provide evidence of repetitive rotation of the heads during contraction.

2.2.2. Saturation-Transfer Electron Paramagnetic Resonance

As indicated in Section 2.1.2, STEPR was developed specifically for the examination of rotational motions in the microsecond-to-submillisecond range, and the method has been applied to study the motions of S-1 in synthetic filaments, both in the absence and in the presence of F-actin (Thomas *et al.*, 1975a,b; Thomas, 1978). As in the solution studies, a slower but still observable rotational motion of S-1 is observed on the formation of filaments. Thomas (1978) was careful to point out that the ability to analyze spectra in terms of specific geometric details of the motion is still limited, so that the spectral changes could be due to the restriction on the permitted range of motion rather than an actual slowing of the rotation. Because this is consistent with the model of Mendelson and Cheung (1978) mentioned in Section 2.2.1, the STEPR measurements again serve to confirm the fluorescence-depolarization results.

One important STEPR experiment showed that it is possible to immobilize the myosin heads nearly completely when spin-labeled myosin monomers are added to F-actin to form the rigor complex. This result is significant, since it indicates that the motion being observed in STEPR (and probably in fluorescence depolarization) does indeed arise from the motion of the entire head, and not from some smaller region undergoing independent local motion within S-1.

2.2.3. X-Ray Diffraction

Striated muscle is characterized by a rich and potentially very informative low-angle X-ray diffraction pattern, and changes in that pattern as the muscle changes from relaxation to rigor or contraction, either isometric or isotonic, have been studied for many years. When the X-ray data are combined with information from other studies, particularly electron microscopy, it is possible to interpret the structural changes accompanying different physiological states without having to treat or alter the muscle in any way.

H. E. Huxley and Brown (1967) showed that the spacings of all reflections remain approximately constant when a resting muscle is stimulated to contract isometrically, but the relative intensities of the various reflections do change. In particular, there is a marked change in the ratio of the two principal equatorial reflections, $I_{1,0}/I_{1,1}$. To account for this change, it has been shown that the myosin head must move both azimuthally (Lymn, 1975) and radially (Haselgrove and Huxley, 1973; Haselgrove *et al.*, 1976). This motion would clearly be facilitated by a hinge at the base of the head, so the X-ray data are further evidence of flexibility at that point.

3. Flexibility within the Rod

While the hinge between S-1 and the rod is, as far as we know, universally accepted, the question of flexibility within the rod is not so easily answered. There have been a number of studies aimed at determining whether the rod is flexible and, if so, how large the flexible region is, and what kinds of motion (stretching, bending, or both) occur. Different investigators, using different approaches and different experimental conditions, have reached different conclusions. This section reviews those studies, attempting to correlate the various results into a consistent picture.

3.1. Evidence for Flexibility

3.1.1. Electric Birefringence

One of the simplest tests for flexibility in a rod-shaped molecule is to measure the rotational diffusion coefficient for the end-over-end tumbling motion. The sensitivity of this test rests on the fact that for a very elongated particle, the rotational diffusion coefficient is roughly proportional to the inverse third power of the length. If the observed rate of rotational diffusion is much faster than this, then the molecule is probably flexible.

The myosin rod is known to be a coiled coil of alpha helices (Lowey *et al.*, 1969; Burke *et al.*, 1973), so its dimension can be calculated from the known molecular weight, and the rotational diffusion coefficient can be calculated from the equation given by Broersma (1960) for a rigid rod. The actual diffusion coefficient can be determined experimentally by observing the rate of decay of electric birefringence. This procedure was used by Highsmith *et al.* (1977) to examine the flexibility of the myosin rod.

In those experiments, a strong uniform electric field is applied to a solution of rods, and the rods become oriented because of the interaction of their electric dipoles (permanent or induced) with the field. The solution thus acquires optical anisotropy, becoming birefringent. When the field is turned off, the rate of decay of the birefringence gives the rotational diffusion coefficient. Highsmith and co-workers argued that the coiled-coil conformation of the rod should give it circular symmetry, so the only detectable motion in these experiments should be the sought-after tumbling of the rod axis. They observed relaxations that were well described by single-exponential decay times, confirming their assumption that only a single mode of molecular rotation contributes to the relaxation of birefringence.

The observed decay time for LMM was 13.1 μsec, very close to the value expected for a rigid cylindrical molecule of molecular weight 140,000. In the case of S-2, the decay time of 6.0 μsec was shorter than that expected of a cylinder, with the difference being statistically significant at the level $P <$ 0.001. Similarly, the rod gave a decay time of 41.2 μsec, again statistically significantly below the expected value of 57.3 μsec ($P < 0.001$). Highsmith *et al.* (1977) concluded from these results that LMM is rigid, that S-2 is flexible, and that the rod is flexible. Their data could not determine whether S-2 is flexible along its entire length or only within a restricted region. Their conclusion that rod flexibility occurs within S-2, probably near the S-2–LMM junction, was confirmed by hydrodynamic modeling of the rotational diffusion of rigid and hinged rods (García de la Torre and Bloomfield, 1980).

To make comparisons between these results and others that will be presented below, it should be noted that the electric birefringence measurements were made on samples at 3°C in a low-salt buffer (2 mM $Na_4P_2O_7$, pH 9.3). The low salt concentration was necessary so that solution conductivity did not interfere with the measurements; the high pH requirement then follows because of the tendency to aggregate at low salt, which is characteristic of myosin, rods, and LMM. The authors themselves pointed out that the observed flexibility could be modified under physiological conditions (Highsmith *et al.*, 1977).

3.1.2. Electron Microscopy

The electron-microscopy studies of Elliott and Offer (1978) and Takahasi (1978) described in Section 2.1.7 included analyses of bending within the myosin tail. Although different preparatory conditions were used (shadowcasting by Elliott and Offer; negative staining by Takahashi), both investigations reported that many of the molecules showed some curvature or bending within the rod. Sharp bends occurred frequently enough in the same region that the location of a hinge within the rod could be identified with fair accuracy. Even though there are some differences in the results of those two studies (discussed in more detail in Sections 3.2.4 and 3.3.3), the electron micrographs do provide evidence for flexibility within the rod.

3.1.3. Intrinsic Viscosity

One physical property of elongated rigid molecules that is quite sensitive to changes in shape is the intrinsic viscosity. If significant flexibility is introduced into a rod, the intrinsic viscosity can be expected to drop markedly. Burke *et al.* (1973) examined the dependence of the viscosity of solutions of myosin rods and LMM on temperature, pH, and protein concentration. They also measured the content of alpha helix of these molecules using optical rotatory dispersion (ORD). A portion of their results is presented in Fig. 3 and 4.

The gradual loss of helical structure as the rod is melted is shown in Fig. 3. A plot of the derivative curve shows two distinct cooperative melting transitions with melting temperatures of 44 and 55°C. Similar results have been reported by Samejima *et al.* (1976).

Figure 4 shows the temperature dependence of the reduced viscosity of LMM and rods. The intrinsic viscosity is the infinite dilution limit of the reduced viscosity, and these two properties should show similar responses to

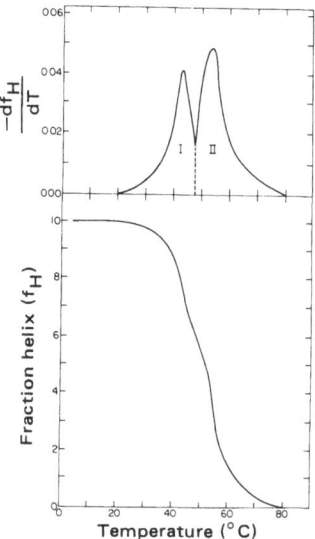

Figure 3. Thermal denaturation of myosin rod monitored by ORD. Reproduced by permission from Burke *et al.* (1973); copyright 1973 American Chemical Society.

the presence of flexibility. An examination of the data presented by Burke *et al.* (1973) shows that the difference between reduced viscosity and intrinsic viscosity is not large enough to account for the behavior shown in Fig. 4. Their argument, summarized below, depends on the temperature effect, and it is not compromised by the small difference in the two viscosities.

At temperatures below 35°C, the helical content of both the rod (Fig. 3) and LMM (not shown here) is essentially 100%. The reduced viscosity of LMM (Fig. 4) is also independent of temperature below 35°C, and the high

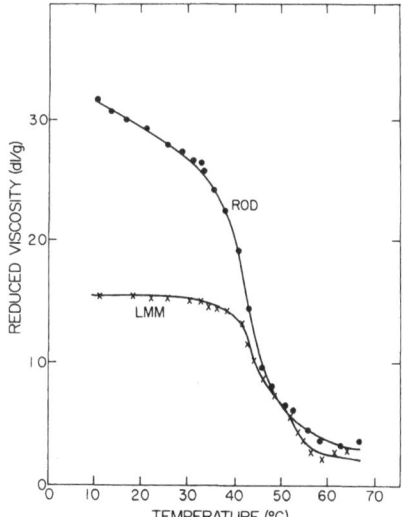

Figure 4. Temperature dependence of the reduced viscosity of LMM and the myosin rod. Reproduced by permission from Burke *et al.* (1973); copyright 1973 American Chemical Society.

value of the reduced viscosity is consistent with a rigid-rod model for LMM. (Globular proteins have intrinsic viscosities that rarely exceed 0.01 dl/g.) By contrast, the reduced viscosity of the rod shows a substantial dependence on temperature, even well below 35°C. Although there is no appreciable loss of helical content (Fig. 3), the reduced viscosity is quite temperature-sensitive. Clearly, some change in shape does occur at temperatures well below the first melting transition. Burke *et al.* (1973) argued that the most likely explanation of this behavior is that the rod has a local perturbation of its structure, probably in the trypsin-sensitive region between LMM and S-2, so that it bends. This hinged behavior would explain the ORD and viscosity results, and this model has received support from the hydrodynamic calculations on the intrinsic viscosity of rigid and hinged rods of García de la Torre and Bloomfield (1980).

One other viscosity study relevant to the question of rod flexibility is that of Tsong *et al.* (1979). They found that the reduced viscosity of S-2 also falls monotonically with temperature, even at temperatures well below T_m, so it, too, must have considerable flexibility. This result lends further support to the hinged-rod model put forward by Burke *et al.* (1973).

3.1.4. Thermal Denaturation

If the proteolytic sensitivity of the S-2–LMM junction indicates transitory instabilities in that part of the rod's coiled-coil structure—and this was an early reason for believing in rod flexibility—then those instabilities should manifest themselves in the sensitivity of the rod's physical properties to changes in temperature. Further, it should be possible to demonstrate that any such changes actually reflect processes at the putative hinge by comparing the behavior of the whole rod with that of LMM and S-2. Three thermal-denaturation studies have provided insights into the stability of the interior of the rod: the series on pH changes accompanying melting (Goodno and Swenson, 1975a,b; Goodno *et al.*, 1976), the investigations of Harrington and co-workers (Burke *et al.*, 1973; Sutoh *et al.*, 1978a; Tsong *et al.*, 1979) monitoring changes in ORD and intrinsic viscosity (see also Section 3.1.3), and the study of Samejima *et al.* (1976) using ORD, absorption, and fluorescence.

The study of protein-melting by monitoring pH changes as the solution is heated was developed by Bull and Breese (1973) and then applied to myosin and its subfragments (Goodno and Swenson, 1975a,b; Goodno *et al.*, 1976). Because protein denaturation causes changes in the exposure of ionizable residues, unfolding can be expected to be accompanied by the release or uptake of protons. In this method, an unbuffered protein solution with an initial pH near the isoelectric point is heated at a low but constant rate, typically less than 1°C/min, while the solution pH is continuously monitored. A plot of pH vs. temperature, or, better still, a plot of the derivative of that curve, will generally show the sudden release or uptake of protons, and the melting temperature, T_m, can be determined with a precision of ±1°C or better (Bull and Breese, 1973).

Figure 5 shows a melting curve for the myosin rod. It is seen that the rod melts with $T_m = 39.7°C$ and that melting is accompanied by the uptake of protons. The melting profiles of whole myosin and LMM are quite similar, with $T_m = 40 \pm 1°C$ in 0.5 M KCl when the initial pH is about 5.9 (Goodno and Swenson, 1975a). Higher melting temperatures are obtained at higher initial pH, as would be expected, and the melting temperature decreases in solutions containing higher concentrations of KCl. Divalent cations do not show a significant effect on T_m (Goodno and Swenson, 1975b).

In contrast, the melting of S-2 is characterized by the release of protons, and at acid pH, the melting temperature of S-2 is higher than that of myosin or any of the other subfragments. Also in contrast with the rest of the molecule, S-2 has a T_m that falls with increasing pH, dropping from 55°C at pH 5.5 to 43°C at pH 7.0 (Goodno *et al.*, 1976). This difference could be functionally important, because local changes in pH could modulate the relative stabilities of LMM, S-1, and S-2. The consequences of the relationship between pH and stability of the various regions within myosin are discussed by Goodno *et al.* (1976), who point out the possible relevance of this result for those models of contraction that suppose an active role for S-2 (A. F. Huxley and Simmons, 1971; Harrington, 1971). In particular, when the pH is increased above 7, S-2 becomes the least stable portion of the molecule and is expected to be relatively more flexible than LMM.

The flexibility of isolated S-2 can also be inferred from ORD measurements on the temperature dependence of the helical content of the molecule. A single melting transition is observed, with $T_m = 45°C$ at neutral pH (Sutoh *et al.*, 1978a). The absence of a second transition at a higher temperature indicates that S-2 has less thermal stability than the rod; Sutoh and co-workers estimate that at physiological conditions, about 20% of the S-2 helix is melted, so the molecule may be flexible under those circumstances.

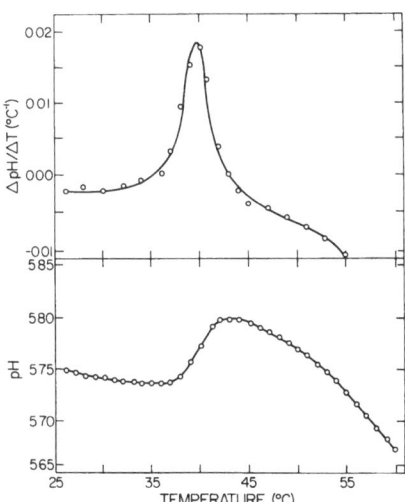

Figure 5. Thermal denaturation of myosin rod monitored by pH changes in unbuffered solution. Reproduced by permission from Goodno and Swenson (1975a); copyright 1975 American Chemical Society.

3.1.5. Temperature Jump

The observation by Harrington *et al.* (1971) that S-2 shows only a single broad melting transition and that about 20% of the molecule is in the random-coil state under physiological conditions (Sutoh *et al.*, 1978a) led them to a careful study of the kinetics of melting using UV absorbance in temperature-jump experiments (Tsong *et al.*, 1979). They reported that when solutions of S-2, initially equilibrated at temperatures between 35 and 55°C, were subjected to a jump of 5°C, two relaxation times were observed, one in the submillisecond (τ_f) and the other in the millisecond range (τ_S). Both relaxations showed a dependence on temperature, and the relaxation times passed through maxima at 43°C (τ_s) and 46°C (τ_f). Similar behavior was observed for the low-temperature transition ($T_m = 45$°C) of the myosin rod.

Taking together their intrinsic-viscosity data (described briefly in Section 3.1.3) and the data on the two rapid steps in the mechanical transient studies of A. F. Huxley and Simmons (1971), Tsong *et al.* (1979) concluded that the mechanical properties of muscle reflect shifts in equilibrium between the helical and random-coil states of S-2. When a muscle in isometric contraction is allowed to shorten abruptly, the instantaneous drop in tension arises, in this model, from the instantaneous decrease in the elasticity of the partially melted S-2, and the subsequent rapid tension recovery is due to the formation of additional coil as S-2 adjusts to the shorter length. This behavior is consistent with the well-known melting–crystallization properties of polymeric systems subject to tension (Flory, 1956). These results have been incorporated into Harrington's model for muscle contraction, which is summarized in Fig. 2, taken from Tsong *et al.* (1979).

3.2. Evidence That Bending Is Limited

As far as we are aware, only one set of experiments—our own—has produced negative evidence in the efforts to determine whether the rod is flexible. But several studies have given results indicating that bending is less free at the LMM–S-2 junction than at the hinge between S-1 and the rod.

3.2.1. Fluorescence Depolarization

In our study (Harvey and Cheung, 1977) of the fluorescence-depolarization decay from myosin rods labeled with 5-dimethylaminonaphthalene-l-sulfonyl chloride (DNS-Cl), we observed an unexpected phenomenon: the anisotropy was observed to increase with time (Fig. 6). This behavior is in striking contrast to the decay usually observed, where Brownian rotations work to randomize the orientations of the fluorophores, producing the depolarization that gives the technique its name (Yguerabide, 1972; Wahl, 1975).

We have studied a number of systems with fluorescence depolarization, including supramolecular aggregates (Harvey *et al.*, 1977), and in no other

Figure 6. Decay of fluorescence-polarization anisotropy for myosin rods labeled with DNS-Cl (○) and with *N*-iodoacetyl-amino-1-naphthylamine-1-sulfonic acid (1,5-IAEDANS) (●). Semilog plot of the lamp pulse. The observed rotational correlation time for the DNS-rods curve is −780 nsec. Reproduced by permission from Harvey and Cheung (1977); copyright 1977 American Chemical Society.

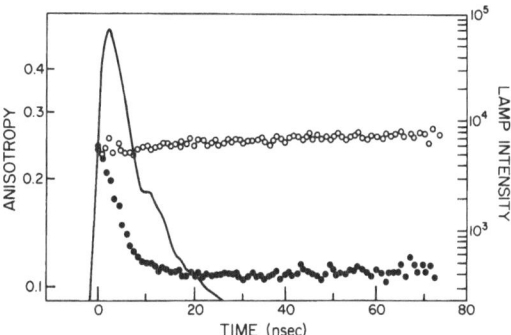

case have we ever observed rising anisotropy curves. The efforts we made to eliminate possible artifactual causes for this behavior are detailed in our original paper (Harvey and Cheung, 1977), and we were finally forced to conclude that the rising curves and their negative rotational correlation times are real, not artifacts.

We had previously reported an initially rising anisotropy curve in our computer simulation of fluorescence depolarization in rigid molecules (Harvey and Cheung, 1972). As we noted in that paper, Perrin (1934, 1936) pointed out that under certain circumstances, Brownian motion can produce increases in polarization. This happens when diffusion causes an initial increase in the alignment of the emission dipoles; the negative correlation time indicates that the correlation of the dipole alignments increases rather than decreases.

Anisotropy curves can be calculated from the equation of Belford *et al.* (1972) using the diffusion coefficients calculated from Perrin's equations. (Modeling the myosin rod by a very elongated ellipsoid rather than a cylinder introduces only very small errors.) The models for the rod are shown in Fig. 7, and Fig. 8 gives the decay curves for the various dye orientations. Note that curve 1, like our data, shows an initially rising anisotropy. Our observed negative rotational correlation time is consistent with that which would be observed for a rigid rod with the fluorophore oriented so that the transition dipoles are on opposite sides of the rod. Since a freely hinged rod should

Figure 7. Geometry of the model rod for the calculation of the theoretical anisotropy-decay curves shown in Fig. 8. (θ_a, θ_e) Angles between the rod axis and the absorption and emission dipoles, respectively. An angle of 30° between the dipoles has been used throughout. At the right are the dipole orientations for the curves shown in Fig. 8. Reproduced by permission from Harvey and Cheung (1977); copyright 1977 American Chemical Society.

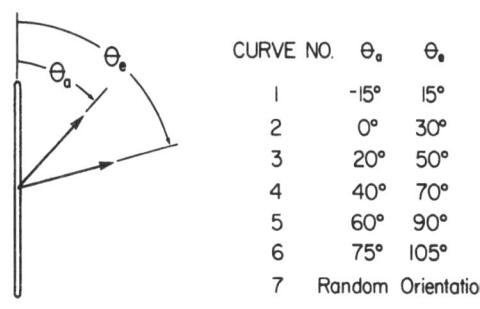

CURVE NO.	θ_a	θ_e
1	−15°	15°
2	0°	30°
3	20°	50°
4	40°	70°
5	60°	90°
6	75°	105°
7	Random Orientation	

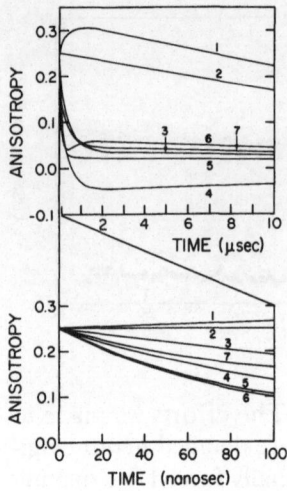

Figure 8. Theoretical anisotropy-decay curves for the various rod models shown in Fig. 7, calculated from the equation of Belford *et al.* (1972). The upper curves show the decay over 10 μsec, a time scale appropriate to the end-over-end tumbling of the rod, while the lower curves show the first 100 nsec of the decay, a time scale appropriate to the fluorescence lifetime. Reproduced by permission from Harvey and Cheung (1977); copyright 1977 American Chemical Society.

produce falling decay curves, we were led to the conclusion that the myosin rod is rigid or nearly so under our solvent conditions (0.6 M KCl, pH 6.0–8.0).

We have subsequently established an algorithm for the computer simulation of the Brownian motion of hinged particles, generating fluorescence depolarization curves for them (Harvey and Cheung, 1980). Those simulations have shown that if the rod could bend through an angle as small as ±10° the anisotropy decay would give positive rotational correlation times. Our initial conclusion that fluorescence depolarization does not reveal evidence of a hinge in the rod is substantiated by the simulation.

Further evidence that we should be able to detect appreciable bending comes from experiments performed at pH 4 and below, where we did observe falling anisotropies. Since Burke *et al.* (1973) had shown that the rod is quite flexible at low pH, it is significant that we were able to observe that bending.

3.2.2. Chemical Cross-Linking

The extent of radial movement of the cross-bridges during contraction and rigor has been investigated by Sutoh and Harrington (1977), by Sutoh *et al.* (1978b), and by Chiao and Harrington (1979). These studies were partly motivated by the controversy over whether the characteristic change in the ratio of intensities ($I_{1,0}/I_{1,1}$) of the X-ray equatorial reflections in diffraction studies was primarily due to azimuthal (Lymn, 1975) or radial (Haselgrove *et al.*, 1976) movement of the heads.

The principle of the cross-linking experiments is quite simple. A bifunctional reagent with specific chemical reactivity [for instance, dimethyl dithiobis(propionimidate), which reacts specifically with lysine side chains] is used to make a covalent cross-link between the myosin head (or S-2 in some experiments) and the thick filament. The extent of cross-linking can be monitored by enzymatic cleavage of S-1 from the aggregated thick filaments; if partially cross-linked samples are taken at regular intervals, the rate of cross-linking can be determined. The rate and extent of cross-linking should both be high when the heads are close to the thick filament and low if they can swing away from it because of flexibility in S-2.

The first study (Sutoh and Harrington, 1977) showed that cross-linking is very effective in a variety of solvents in the pH range 6.8–7.4, so the myosin heads are close to the thick filament. Even in the rigor myofibril, the heads are cross-linked more rapidly to the thick filament than to the thin filament, and the rate of cross-linking is indistinguishable from that in synthetic thick filaments when no actin is present. The studies were later extended to pH 8.0, where the rate of cross-linking was severely diminished, even though the rate of the reaction, aminidation of lysine side chains, has a positive dependence on pH. These results indicate that the heads are close to the thick filament at neutral pH, but move away from it at alkaline pH (Sutoh *et al.*, 1978b). This is also consistent with the observation that the rotational mobility of the heads in filaments is increased at alkaline pH (Mendelson and Cheung, 1976; Thomas *et al.*, 1975a,b). On the other hand, there was sufficient cross-linking between S-2 and the thick filament, even at pH 8.0, to conclude that a significant portion of S-2 lies close to the filament surface even when the head moves out at alkaline pH (Sutoh *et al.*, 1978b). This correlates with our observation that rod flexibility is too limited to be observed by fluorescence depolarization over the pH range 6.0–8.0 (Harvey and Cheung, 1977). All the results cited above were found to hold true when the studies were extended to glycerinated rabbit psoas muscle (Chiao and Harrington, 1979).

These results are important, because they show that the myosin head lies near the thick filament at neutral pH, and even when it moves away at alkaline pH, a major fraction of S-2 is still held close to the thick filament. If the rod bends at a joint between LMM and S-2, that bending is not detected in these experiments.

3.2.3. Viscoelasticity

The measurement of intrinsic viscosity (Section 3.1.3) is made on a solution that flows at constant velocity, so a constant shear stress is applied. More information on molecular conformation and flexibility can be obtained, however, by examining the solution's behavior as a function of frequency for a sinusoidal shear stress. To compare experimental viscoelastic properties with predictions based on specific models, it is convenient to measure the storage and loss-shear moduli. (The intrinsic viscosity is simply the zero frequency limit of the loss-shear modulus.)

Rosser *et al.* (1978) have measured these moduli over the frequency range 150–8000 Hz for solutions of myosin at pH 7.3. They reported that the data are well matched by a model that possesses only slight flexibility. The rotational relaxation time agrees with that for a model based on a modification of the hydrodynamic method of Nakajima and Wada (1977), using rods built of arrays of Stokes beads, and the Young's modulus for the rod, calculated from the ratio of the two longest relaxation times, is 9×10^9 dynes/cm^2.

This is only 25% below that determined for paramyosin (Rosser *et al.*, 1977), a double-helical coiled-coil molecule. Rosser *et al.* (1978) concluded that the rod has only limited flexibility.

3.2.4. Electron Microscopy

Although the electron-microscopy studies of Elliott and Offer (1978) and Takahashi (1978) do provide evidence for a hinge in the rod, as described in Section 3.1.2, there are substantial differences in the results of those two studies, differences that may indicate that rod flexibility is limited. The principal differences are summarized in Table 1.

To begin with, we note that while 89% of the molecules observed by negative staining showed sharp bends in the rod, only 25% of those in the shadow-casting study were bent when they were obtained from solutions containing ethylene glycol, and even fewer were bent when glycerol was used as the cryoprotectant. Elliott and Offer remarked that this result is consistent with the presence of a restoring force that opposes bending. Clearly, the preparation procedure introduces artifacts into one of these studies, and perhaps into both of them. Regarding the possibility of bending under the stresses of the evaporation and shadow-casting procedures, Elliott and Offer reported that during the drying of their solutions, molecular motion sometimes occurred, causing parallel alignment of the molecules, and that this happened more often with ethylene glycol than with glycerol. As they pointed out, "If one end of a molecule were attached to the mica initially, such movements would be expected often to cause bending."

It is, of course, possible that the sample-preparation procedure actually straightens bends in the rod. But a second difference between the two studies strongly suggests that fixing the samples is more likely to introduce artifactual bending: the location of the hinge. While Takahaski placed the hinge in the middle of the rod, Elliott and Offer's location is much closer to the head (Table 1). It seems likely that there are two regions of structural instability and that the one in the middle of the rod is prone to bending when the molecules are fixed for negative staining, while the other is bent by the preparative procedures for shadow-casting.

Table 1. Electron-Microscopy Studies on the Myosin Rod

Method	Fraction of molecules with sharp bends in tail (f)	Length of rod (nm) (L)	Distance from tail end of rod to hinge (nm) (D)	D/L
Shadow-casting[a]	≤0.25[b]	156	113	0.72
Negative staining[c]	0.89	140	68	0.49

[a] Data from Elliott and Offer (1978).
[b] "About a quarter . . . dried from ethylene glycol solutions . . .; with glycerol the number was fewer."
[c] Data from Takahashi (1978).

3.3. Discussion on Rod Flexibility

3.3.1. Extent of Flexibility

There is little doubt that the rod does bend under some circumstances. All the studies reviewed in Section 3.1 indicate rod flexibility, most likely in the region of trypsin sensitivity between LMM and S-2. On the other hand, the negative results of the fluorescence depolarization studies, the fact that S-1 and S-2 are close enough to the thick filament to be cross-linked to it under most circumstances, the stiffness of the rod revealed by the viscoelasticity measurements, and the large fraction of unbent rods found in the shadow-casting electron-microscopy studies all argue that the rod resists bending.

Several points should be considered in evaluating the evidence on rod flexibility. First, susceptibility to proteolysis, susceptibility to thermal denaturation, and the ability to bend under stress are not direct evidence of free bending at physiological conditions: lability is not the same as flexibility. Second, solution studies may overstate flexibility, because the aggregation into filaments may stabilize the rod against bending: the high degree of chemical cross-linking between the heads and the thick filament is strong evidence of this. And finally, all the experiments that give direct evidence of flexibility involve the imposition of conditions that destabilize the rod. Let us examine this last point carefully.

The direct evidence for flexibility comes from three studies: electric birefringence (Section 3.1.1), electron microscopy (3.1.2), and intrinsic viscosity (3.1.3). The thermal denaturation (3.1.4) and temperature-jump (3.1.5) experiments show lability, but are not measurements of bending itself. The electric-birefringence measurements (Highsmith *et al.*, 1977) were made at pH 9.3, and alkaline pH has been shown to increase the mobility of the myosin heads by a variety of methods, including fluorescence depolarization (Mendelson and Cheung, 1976), STEPR (Thomas *et al.*, 1975a,b), and cross-linking (Sutoh *et al.*, 1978b). There are obvious stresses imposed on the rod during the fixing necessary for the electron-microscopy studies, as discussed in Section 3.2.4. And intrinsic viscosity measurements are necessarily made in flowing solutions where shear stresses could be sufficient to bend the rod more than would occur when the molecule is in a quiescent solution.

In short, while the rod can be bent when subjected to moderate stresses, there is as yet no evidence for rod flexibility at physiological conditions or when aggregated into filaments. The simplest interpretation of these results is that the rod can bend, but that bending is opposed by a moderately strong restoring force.

3.3.2. Modes of Flexibility

Most of this discussion has treated rod bending, since that is the motion that is seen in the electron micrographs and is probably the motion responsible for the electric-birefringence and intrinsic-viscosity results. Our conclusion that lateral bending is opposed by a restoring force fits well with the sliding filament–rotating head model of H. E. Huxley (1969, 1971), since solvent conditions, and thus physiological state, clearly modulate rod stiffness. Increasing rod flexibility at one stage of the contractile cycle could facilitate the swinging of the myosin head toward the thin filament, and increased rod stiffness at another stage would pull the heads away from the thin filament.

But another possible intramolecular motion within the rod must be considered: stretching. A. F. Huxley and Simmons (1971) proposed that elastic stretching within S-2 is responsible for the transmission of force as S-1 rolls on the thin filament. The model of Harrington (1971) postulates that a helix–coil transition within S-2 is the actual force-generating step. These models (compared in Fig. 2) both assume that the important mode of flexibility is longitudinal rather than lateral. And the evidence from ORD and from the temperature-jump studies (Section 3.1.5) strongly suggests that helix–coil transitions do occur on a time scale relevant to the contractile cycle. Unfortunately, it is experimentally very difficult to actually observe stretching motions of the kind proposed here, and while longitudinal flexibility of the rod may play an important role in contraction, it has not yet been demonstrated experimentally. Although dynamic-compliance studies do show an elastic element within the cross-bridges (A. F. Huxley and Simmons, 1971; Bressler and Clinch, 1975), it has not been determined whether that element is due to the rotation of S-1 or stretching within S-1 or S-2.

3.3.3. Location of Flexible Regions

Virtually all investigators agree that such flexibility as does exist within the rod is most likely located within the trypsin-sensitive region between LMM and S-2. It is instructive to compare the sites of bending revealed by electron microscopy with the sites of trypsin sensitivity. The results of several investigations are summarized in Fig. 9. As mentioned earlier, there are apparently

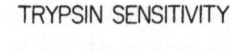

TRYPSIN SENSITIVITY

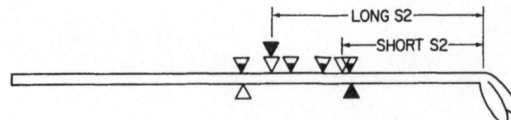

BENDS IN ELECTRON MICROGRAPHS

Figure 9. Location of sites of trypsin sensitivity (triangles above rod) and of bends observed in electron micrographs (triangles below rod). Since different authors report different total lengths for the rod, all sites have been normalized by reporting the distances to them from the tail end of the rod as a fraction of rod length. (LONG S2, SHORT S2) Terminology of Sutoh *et al.* (1978a). Sites of trypsin sensitivity: (∇) Sutoh *et al.* (1978a); ($\blacktriangledown$) Highsmith *et al.* (1977); ($\triangledown$) King (1976). Bends in electron micrographs: ($\triangle$) Takahashi (1978); ($\blacktriangle$) Elliott and Offer (1978).

two distinct regions of the rod that are prone to bending, and it is seen that they are close to the ends of the region of proteolytic sensitivity. In a note added in proof, Elliott and Offer (1978) pointed out that although Takahashi (1978) had located the hinge near the center of the rod, many of his micrographs revealed bends at a location near the one that Elliott and Offer had designated as the bending site. In fact, several of Takahashi's micrographs show two distinct bends or kinks in the rod, so both of the sites revealed by the electron micrographs are probably susceptible to bending.

4. Summary

A variety of investigations have provided evidence that the myosin heads are joined to the rod by a hinge or swivel that provides a great deal of rotational mobility for the heads, and the flexibility of that site appears to be universally accepted. The flexibility of the region of proteolytic sensitivity within the myosin rod, a region occupying nearly a quarter of the rod's length, is not yet unambiguously established. Thermal-denaturation and temperature-jump studies have demonstrated the lability of that region, supporting the idea that helix–coil transitions there may give the rod longitudinal flexibility, but no direct evidence for stretching within the rod has so far been produced. Electron microscopy has revealed bends at both ends of the proteolytically sensitive region, and such lateral flexibility has been demonstrated in solution studies at elevated pH and in solutions subjected to shear stress. No evidence has been found for bending at physiological conditions or when myosin is aggregated into filaments, however, and the results of several kinds of experiments argue that rod bending is limited. It is likely that the myosin rod bends laterally against an appreciable restoring force and that its flexibility is more like that of a spring than a free hinge.

ACKNOWLEDGMENTS. This work was supported in part by grants from the US Public Health Service (AM-17483) and the National Science Foundation (PCM-78-24803). We thank Dr. W. F. Harrington and Dr. C. A. Swenson for permission to reproduce their figures.

References

Belford, G. G., Belford, R. L., and Weber, G., 1972, Dynamics of fluorescence polarization in macromolecules, *Proc. Natl. Acad. Sci. U.S.A.* **69**:1392.

Borejdo, J., Putnam, S., and Morales, M. F., 1979, Fluctuations in polarized fluorescence: Evidence that muscle cross bridges rotate repetitively during contraction, *Proc. Natl. Acad. Sci. U.S.A.* **76**:6346.

Bressler, B. H., and Clinch, N. F., 1975, Cross bridges as the major source of compliance in contracting skeletal muscle, *Nature (London)* **256**:221.

Broersma, S., 1960, Rotational diffusion of a cylindrical particle, *J. Chem. Phys.* **32**:1626.

Bull, H. B., and Breese, K., 1973, Thermal transitions of proteins, *Arch. Biochem. Biophys.* **156**:604.

Burke, M., Himmelfarb, S., and Harrington, W. F., 1973, Studies on the "hinge" region of myosin, *Biochemistry* **12**:701.

Chiao, Y.-C. C., and Harrington, W. F., 1979, Cross-bridge movement in glycerinated rabbit psoas muscle fibers, *Biochemistry* **18**:959.

Dalton, L. R., 1976, Saturation transfer spectroscopy, *Adv. Magn. Reson.* **8**:149.

Eisenberg, E., and Hill, T. L., 1978, A cross-bridge model of muscle contraction, *Prog. Biophys. Mol. Biol.* **33**:55.

Eisenberg, E., Hill, T. L., and Chen, Y. D., 1980, Cross-bridge model of muscle contraction: Quantitative analysis, *Biophys. J.* **29**:195.

Elliott, A., and Offer, G., 1978, Shape and flexibility of the myosin molecule, *J. Mol. Biol.* **123**:505.

Flory, P. J., 1956, Role of crystallization in polymers and proteins, *Science* **124**:53.

García Bernal, J. M., and García de la Torre, J., 1980, Transport properties and hydrodynamic centers of rigid macromolecules with arbitrary shapes, *Biopolymers* **19**:751.

García de la Torre, J., and Bloomfield, V. A., 1978, Hydrodynamic properties of macromolecular complexes. IV. Intrinsic viscosity theory, with applications to once-broken rods and multisubunit proteins, *Biopolymers* **17**:1605.

García de la Torre, J., and Bloomfield, V. A., 1980, Conformation of myosin in dilute solution as estimated from hydrodynamic properties, *Biochemistry* **19**:5118.

Goodno, C. C., and Swenson, C. A., 1975a, Thermal transitions of myosin and its helical fragments. I. Shifts in proton equilibria accompanying unfolding, *Biochemistry* **14**:867.

Goodno, C. C., and Swenson, C. A., 1975b, Thermal transitions of myosin and its helical fragments. II. Solvent-induced variations in conformational stability, *Biochemistry* **14**:873.

Goodno, C. C., Harris, T. A., and Swenson, C. A., 1976, Thermal transitions of myosin and its helical fragments: Regions of structural instability in the myosin molecule, *Biochemistry* **15**:5157.

Harrington, W. F., 1971, A mechanochemical mechanism for muscle contraction, *Proc. Natl. Acad. Sci. U.S.A.* **68**:685.

Harvey, S. C., and Cheung, H. C., 1972, Computer simulation of fluorescence depolarization due to Brownian motion, *Proc. Natl. Acad. Sci. U.S.A.* **69**:3670.

Harvey, S. C., and Cheung, H. C., 1977, Fluorescence depolarization studies on the flexibility of the myosin rod, *Biochemistry* **16**:5181.

Harvey, S. C., and Cheung, H. C., 1980, Transport properties of particles with segmental flexibility. II. Decay of fluorescence polarization anisotropy from hinged macromolecules, *Biopolymers* **19**:913.

Harvey, S. C., Cheung, H. C., and Thames, K. E., 1977, Cooperativity in F-actin filaments on binding of myosin subfragments, demonstrated by fluorescence of 1,N^6-ethenoadenosine diphosphate, *Arch. Biochem. Biophys.* **179**:391.

Haselgrove, J. C., and Huxley, H. E., 1973, X-ray evidence for radial crossbridge movement and for the sliding filament model in actively contracting skeletal muscle, *J. Mol. Biol.* **77**:549.

Haselgrove, J. C., Stewart, M., and Huxley, H. E., 1976, Cross-bridge movement during muscle contraction, *Nature (London)* **261**:606.

Highsmith, S., 1978, The effects of divalent cations on the rotational mobility of myosin, heavy meromyosin and myosin subfragment-1 and on the binding of heavy meromyosin to actin, *Biochim. Biophys. Acta* **536**:156.

Highsmith, S., Kretzschmar, K. M., O'Konski, C. T., and Morales, M. F., 1977, Flexibility of myosin rod, light meromyosin, and myosin subfragment-2 in solution, *Proc. Natl. Acad. Sci. U.S.A.* **74**:4986.

Highsmith, S., Akasaka, K., Konrad, M., Goody, R., Holmes, K., Wade-Jardetzky, N., and Jardetzky, O., 1979, Internal motions in myosin, *Biochemistry* **18**:4238.

Huxley, A. F., and Simmons, R. M., 1971, Proposed mechanism of force generation in striated muscle, *Nature (London)* **233**:533.

Huxley, H. E., 1957, The double array of filaments in cross-striated muscle, *J. Biophys. Biochem. Cytol.* **3**:631.

Huxley, H. E., 1969, The mechanism of muscle contraction, *Science* **164**:1356.

Huxley, H. E., 1971, The structural basis of muscular contraction, *Proc. R. Soc. London Ser. B* **160**:442.

Huxley, H. E., and Brown, W., 1967, The low-angle X-ray diagram of vertebrate striated muscle and its behavior during contraction and rigor, *J. Mol. Biol.* **30**:383.

Hyde, J. S., 1978, Saturation transfer spectroscopy, *Methods Enzymol.* **49**:480.

King, M. V., 1976, Electron-microscopic mapping of the hinge region of myosin, *Experientia* **32**:975.

Kobayashi, S., and Totsuka, T., 1975, Electric birefringence of myosin subfragments, *Biochim. Biophys. Acta* **376**:375.

Lowey, S., Slayter, H. S., Weeds, A. G., and Baker, H., 1969, Substructure of the myosin molecule. I. Subfragments of myosin by enzymic degradation, *J. Mol. Biol.* **42**:1.

Lymn, R. W., 1975, Equatorial X-ray reflections and cross arm movement in skeletal muscle, *Nature (London)* **258**:770.

Mannherz, H. G., and Goody, R. S., 1976, Proteins of contractile systems, *Annu. Rev. Biochem.* **45**:427.

Margossian, S. S., and Lowey, S., 1973, Substructure of the myosin molecule. IV. Interactions of myosin and its subfragments with adenosine triphosphate and F-actin, *J. Mol. Biol.* **74**:313.

Marston, S. B., Treagear, R. T., Rodger, C. D., and Clarke, M. L., 1979, Coupling between the enzymatic site of myosin and the mechanical output of muscle, *J. Mol. Biol.* **128**:111.

Mendelson, R. A., and Cheung, P., 1976, Muscle crossbridges: Absence of direct effect of calcium on movement away from the thick filaments, *Science* **194**:190.

Mendelson, R. A., and Cheung, P. H.-C., 1978, Intrinsic segmental flexibility of the S-1 moiety of myosin using single-headed myosin, *Biochemistry* **17**:2139.

Mendelson, R. A., Mowery, P. C., Botts, J., and Cheung, H. C., 1972, The segmental flexibility of the S-1 moiety of myosin, *Biophys. J.* **12**:281a.

Mendelson, R. A., Morales, M. F., and Botts, J., 1973, Segmental flexibility of the S-1 moiety of myosin, *Biochemistry* **12**:2250.

Morales, M. F., and Botts, J., 1979, On the molecular basis for chemomechanical energy transduction in muscle, *Proc. Natl. Acad. Sci. U.S.A.* **76**:3857.

Murphy, R. A., 1979, Filament organization and contractile function in vertebrate smooth muscle, *Annu. Rev. Physiol.* **41**:737.

Nakajima, H., and Wada, Y., 1977, A general method for evaluation of diffusion constants, dilute-solution viscoelasticity, and the dielectric property of a rigid macromolecule with an arbitrary conformation. I, *Biopolymers* **16**:875.

Peller, L., 1975, Segmental flexibility in the myosin molecule: Evidence from binding studies of myosin fragments with actin, *J. Supramol. Struct.* **3**:169.

Perrin, F., 1934, Mouvement Brownien d'un ellipsoide. I. Dispersion diélectrique pour des molécules ellipsoidales, *J. Phys. Radium VII* **5**:497.

Perrin, F., 1936, Mouvement Brownien d'un ellipsoide. II. Rotation libre et dépolarisation des fluorescences: Translation et diffusion de molécules ellipsoidales, *J. Phys. Radium VII* **7**:1.

Rosser, R. W., Schrag, J. L., Ferry, J. D., and Greaser, M., 1977, Viscoelastic properties of very dilute paramyosin solutions, *Macromolecules* **10**:978.

Rosser, R. W., Nestler, F. H. M., Schrag, J. L., Ferry, J. D., and Greaser, M., 1978, Infinite-dilution viscoelastic properties of myosin, *Macromolecules* **11**:1239.

Samejima, K., Takahashi, K., and Yasui, T., 1976, Heat-induced denaturation of myosin total rod, *Agric. Biol. Chem.* **40**:2455.

Schoenberg, M., 1980a, Geometrical factors influencing muscle force development. I. The effect of filament spacing upon axial forces, *Biophys. J.* **30**:51.

Schoenberg, M., 1980b, Geometrical factors influencing muscle force development. II. Radial forces, *Biophys. J.* **30**:69.

Squire, J. M., 1975, Muscle filament structure and muscle contraction, *Annu. Rev. Biophys. Bioeng.* **4**:137.

Sutoh, K., and Harrington, W. F., 1977, Cross-linking of myosin thick filaments under activating and rigor conditions: A study of the radial disposition of the cross-bridges, *Biochemistry* **16**:2441.

Sutoh, K., Sutoh, K., Karr, T., and Harrington, W. F., 1978a, Isolation and physico-chemical properties of a high molecular weight subfragment-2 of myosin, *J. Mol. Biol.* **126**:1.

Sutoh, K., Chiao, Y.-C. C., and Harrington, W. F., 1978b, Effect of pH on the cross-bridge arrangement in synthetic myosin filaments, *Biochemistry* **17**:1234.

Takahashi, K., 1978, Topography of the myosin molecule as visualized by an improved negative staining method, *J. Biochem.* **83**:905.

Taylor, E., 1979, Mechanism of actomyosin ATPase and the problem of muscle contraction, *CRC Crit. Revs. Biochem.* **6**:103.

Thomas, D. D., 1978, Large scale rotational motions of proteins detected by electron paramagnetic resonance and fluorescence, *Biophys. J.* **24**:439.

Thomas, D. D., Seidel, J. C., Gergely, J., and Hyde, J. S., 1975a, The quantitative measurement of rotational motion of the subfragment-1 region of myosin by saturation transfer EPR spectroscopy, *J. Supramol. Struct.* **3**:376.

Thomas, D. D., Seidel, J. C., Hyde, J. S., and Gergely, J., 1975b, Motion of subfragment-1 in myosin and its supramolecular complexes: Saturation transfer electron paramagnetic resonance, *Proc. Natl. Acad. Sci. U.S.A.* **72**:1729.

Thomas, D. D., Dalton, L. R., and Hyde, J. S., 1976, Rotational diffusion studied by passage saturation transfer electron paramagnetic resonance, *J. Chem. Phys.* **65**:3006.

Tsong, T. Y., Karr, T., and Harrington, W. F., 1979, Rapid helix–coil transitions in the S-2 region of myosin, *Proc. Natl. Acad. Sci. U.S.A.* **76**:1109.

Wahl, P., 1975, Decay of fluorescence anisotropy, in: *Biochemical Fluorescence: Concepts* (R. F. Chen and H. Edelhoch, eds.), pp. 1–41, Marcel Dekker, New York.

Weber, A., and Murray, J. M., 1973, Molecular control mechanisms in muscle contraction, *Physiol. Rev.* **53**:612.

Yguerabide, J., 1972, Nanosecond fluorescence spectroscopy of macromolecules, *Methods Enzymol.* **26C**:498.

Yguerabide, J., Epstein, H. F., and Stryer, L., 1970, Segmental flexibility in an antibody molecule, *J. Mol. Biol.* **51**:573.

Index